AUDUBON ON LOUISIANA

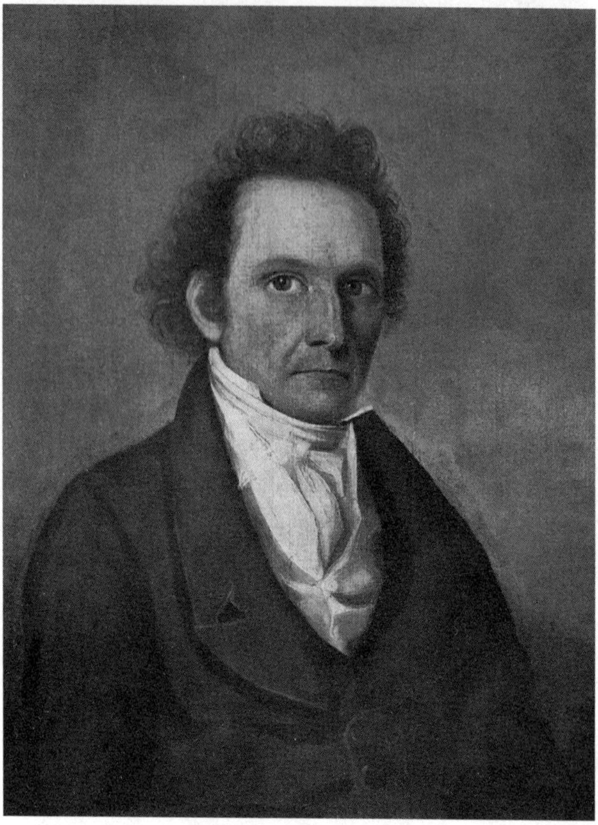

Self-Portrait, by John James Audubon, 1822,
in Beech Woods, Feliciana Parish, Louisiana.
Courtesy of the Frick Art Reference Library.

AUDUBON ON LOUISIANA

Selected Writings

OF

John James Audubon

EDITED
WITH AN INTRODUCTION
AND COMMENTARY
BY
Ben Forkner

Louisiana State University Press Baton Rouge

Published by Louisiana State University Press

Manufactured in the United States of America
First printing

DESIGNER: Michelle A. Neustrom
TYPEFACES: Whitman, text; Filosofia and TPTC CW Title Page Condensed, display
PRINTER AND BINDER: Sheridan Books, Inc.

The original ink maps and line drawings are by Randy Harelson.

LIBRARY OF CONGRESS CATALOGING-IN-PUBLICATION DATA

Names: Audubon, John James, 1785–1851, author. | Forkner, Ben, editor.
Title: Audubon on Louisiana : selected writings of John James Audubon / edited, with an introduction and commentary, by Ben Forkner.
Description: Baton Rouge : Louisiana State University Press, [2018] | Includes bibliographical references and index.
Identifiers: LCCN 2018006177 | ISBN 978-0-8071-6958-2 (cloth : alk. paper) | ISBN 978-0-8071-6998-8 (pdf) | ISBN 978-0-8071-6999-5 (epub)
Subjects: LCSH: Louisiana—Description and travel. | Natural history—Louisiana. | Birds—Louisiana. | Audubon, John James, 1785–1851.
Classification: LCC F374 .A93 2018 | DDC 917.6304—dc23
LC record available at https://lccn.loc.gov/2018006177

The paper in this book meets the guidelines for permanence and durability of the Committee on Production Guidelines for Book Longevity of the Council on Library Resources. ♾

For Nadine

Magnificent Louisiana!

—JOHN JAMES AUDUBON,
in a letter to Dr. Richard Harlan, November 18, 1829

CONTENTS

ACKNOWLEDGMENTS

A number of good friends have been more than generous with their time and skills in helping me with *Audubon on Louisiana*. Some of these friendships go back decades. Others are more recent, formed during the process of making the book. Of the latter, my deepest gratitude goes to my editor at LSU Press, Margaret Lovecraft. Margaret is a delightful human being, and a joy to work with. She is also, thankfully, a shrewd and sensitive reader. She helped steer me through some rough waters with a sure hand whenever I needed it, which was often. Her advice improved the book in every way, from top to bottom. When it was decided not to use my first choice for a title, Audubon's exclamation (*Magnificent Louisiana!*), Margaret suggested it be used instead as an epigraph for the entire book, a brilliant solution!

Two more new friends who have given their time and expertise to *Audubon on Louisiana* are Randy Harelson and Richard Gibbs, owners of the LeJeune House, one of the great plantation houses in Louisiana, and one that was certainly seen by Audubon. After years of work and attention to each minute detail, Randy and Richard have now restored the house to all its former glory. Randy of course designed the original ink maps and decorative drawings used in *Audubon on Louisiana*, brightening it up and making it easier to follow Audubon from place to place. It should be noted that Randy's maps and drawings are in part inspired by the playful sketches Audubon himself drew in the margins of his letters and journals. I want to thank Randy and Richard too for the hospitality they showered on me and my wife, Nadine, and for some relaxed convivial evenings on the back porch and in the kitchen. During one of these meetings I heard the suggestion that Audubon probably loved Louisiana all the more because it awakened memories of climate, vegetation, and Creole culture he had known as a small boy in the Caribbean. This is the kind of insight that flares up naturally when the shadows fall after a good supper in LeJeune House.

To return to the past, I would like to thank two old friends from Nantes, where Audubon explored the cobblestone streets and quays as a boy, and grew to adulthood. One of these friends, Patrick Couton, has been a force in my interest in Audubon for many years, going back to *Journaux et Récits*, the French compendium of Audubon's writings published by Atalante and the City of Nantes in 1992. A superb musician, Patrick is also an outstanding translator from English to French. Patrick supervised all of the translations of *Journaux et Récits*, and did some of them himself. At the time, he lived in a rambling upstairs apartment in the center of Nantes, a few steps away from the church St. Similien where Audubon was baptized Jean Jacques Fougère by the Abbé Hyacinthe Tardiveau on October 23, 1800. I have spent many pleasant hours in that apartment, talking to Patrick about Audubon, often over a choice bottle fresh from the cellar, and often in the company of Patrick's lovely wife, Chantal. I feel mighty fortunate with friends like these. The same goes for Andy Arleo, a voluntary citizen of Nantes via his native New Jersey. Andy is also a gifted musician, and linguist, just retired from a distinguished professorship at the University of Nantes. Andy read the first draft of my manuscript with the eye and the ear of his dual professions, and improved it with a number of observations and suggestions. Whenever my spirits began to list, inches from the floor, Andy was always there to lift them up with a beer in Nantes or a cheerful telephone call to Angers.

In Angers I could always rely on two other good friends, John Webb and Bernard Reeves, both fellow exiles, for more sessions of lifting spirits, in more ways than one. We meet once a month in a favorite café on the banks of the Loire, stared at by the same lone gray heron on the other side, as we make our way slowly through plates of charcuterie, and a bottle or two of Anjou's finest. John is a colleague at the University of Angers, born and bred in Baton Rouge, and a fountain of knowledge on all things touching Louisiana, including the birds and plants in Audubon's paintings. Bernard, a Cornishman no matter where he sets his sack, has made a name for himself as one of the preeminent interpreters in France. In our riverside conversations so far, no subject has been too arcane to curb his curiosity. He is our chief sniffer and snorter of the cork. I owe them both a great debt for taking my thoughts away from work and worries with good times and good company.

For assistance of a more practical nature, I want to thank Florence Casulli and Noah Solomon. Florence is a doctoral student at the University of Angers who helped me with the formatting and proofreading of various texts and files on the computer. Noah is a young new French-American neighbor from up the street who is an ace at transferring files and at guiding the mouse around the traps to reach the right place. He was always willing to take charge of the downloading, to answer my questions on the phone, and, when necessary, to walk over to the house to unknot a problem. Without Florence and Noah, it would have taken me twice the time to put the book together.

For providing vital documents and manuscripts, I am grateful to Jennifer Spence, curator of the Audubon State Park Museum in Henderson, Kentucky, and the helpful staffs at the American Philosophical Society in Philadelphia, the Houghton Library at Harvard University, and the Frick Art Reference Library in New York City.

Randy Harelson and I both want to thank cartographer and artist Mary Lee Eggart of Baton Rouge for her expert advice on the making of Randy's maps.

For taking me through the crucial final stages of preparing the manuscript, I am lucky to have had two of the best professional readers I have ever worked with: Catherine Kadair, senior editor at LSU Press, and Gary Van Euer, my outside copy editor. With patience, good will, and very sharp eyes, Catherine and Gary have been by my side to shoulder the burden from the beginning of the editing process. It does no harm that they both know Audubon well, and are passionate about their work. I want to thank them for all they have done to make *Audubon on Louisiana* as good as possible. This goes too for LSU Press director MaryKatherine Callaway and her entire staff, with a special salute to designer Michelle Neustrom, whose beautiful Great Blue Heron cover captures the spirit of Audubon to perfection!

Finally, my immediate family has been a constant blessing in the making of the book. My son Benjamin, his wife Telba, and their two daughters, Valentina and Natalia, recently moved to a shady craftsman bungalow on the Cane River in Natchitoches, Louisiana. Some of *Audubon on Louisiana* was written in what we call "The Blue Room," a bedroom and study fixed up for the old folks above the garage, an ideal place to write and to take a break when Valentina and Natalia called me out to play in the backyard. Benjamin helped me navigate the computer, and printed out the entire book to make it easier for me to reread. The long late afternoons sitting outside under the ceiling fans with Telba, Benjamin, and Nadine, shooting the breeze, while the girls danced around with their beads, hopping from one game to another, made it easy to return to work the next morning. Natchitoches, by the way, can take you back to Audubon's Louisiana in a blink of the eye, or a turn of the head. This was surely the case for me when I walked down to the dock early in the mornings with my first cup of coffee. I often caught the flash of a kingfisher darting by in the mist, and always herons sailing low over the water. When the turtles came out on the roots to sleep in the rising sun, I was all ready to go, knowing the day would be a good one.

I have dedicated *Audubon on Louisiana* to Nadine, my fetching, feisty, adorable wife. Need I go on? When I first met her in Nantes over forty years ago, Nadine was the irresistible girl from Brittany whose sparkle and skip and *joie de vivre* brightened up every room, and still do. Brittany is a land of lighthouses, ringing the coast and shining on the fearsome black rocks that Chaucer noted in *The Franklin's Tale*. Ever since those early days in Nantes, Nadine has been there for me, no matter the weather: the love of my life! To use the idiom of my own age and transatlantic culture: Thank you, baby!

READING *Audubon on Louisiana*

The introduction provides a condensed overview of Audubon's entire life, and a brief presentation of the selected texts. The headnotes for *Audubon on Louisiana* are substantial, in keeping with the principle that an anthology will be reread at random more often than straight through from beginning to end. For this reason, I have designed the headnotes to be complete in themselves, with enough information to orient and guide the reader through each section, making it possible to dip into the book at will. This allows me in the headnotes to revisit certain key moments in Audubon's career, as Audubon does himself, by expanding on situations that are only touched on in the introduction. The headnotes can thus be read separately. Since the sections are arranged chronologically, the headnotes can also be read in sequence, one after the other, as a fuller timeline of Audubon's life in Louisiana than the one sketched out in the introduction.

When I quote from the *Louisiana Journal* in the introduction and the headnotes, I standardize the English (spelling, capitalization, punctuation), knowing that the untouched original versions can be found in the body of the *Journal* itself. This is the case too when I quote from the letters in the "Last Voyage" section. Within the introduction and headnotes I use italics to indicate titles, and words I want to emphasize. Throughout the selections of Audubon's writings I follow the punctuation of the original texts, including Audubon's not always consistent choice of quotation marks and italics.

INTRODUCTION

Audubon was restless, tireless, and far-ranging in his travels, and shrines from Pennsylvania to Kentucky to Florida testify to his ambitions to explore every inch of the American forests and frontiers in his search for new birds. But as Audubon himself was always the first to proclaim, Louisiana stands far above every other region as the place that meant the most in his life and work. It was in Louisiana that he gave himself up totally to the project that finally became *Birds of America*. It was in Louisiana that he persuaded his wife Lucy to join him and make her home for almost ten years. It was in Louisiana that he perfected his art, and in Louisiana that he painted the most and, many would argue, the best of his birds. In Louisiana he became the Audubon we know today. And as I hope to demonstrate in this collection, much of his most memorable writing is devoted to the years he and his family spent in Louisiana.

Up until his death Audubon was fond of calling Louisiana his home, and claimed it as his favorite place. In fact, though he knew better, he often affirmed, even to his own family, that he was born in Louisiana. He sometimes named New Orleans as his birthplace, and apparently hinted that the actual spot was on a plantation in Mandeville. This was repeated in his granddaughter's edition of his journals in 1897. Audubon, I think, would have enjoyed knowing that the hint eventually fossilized into a biographical fact that survived him by many years. As late as 1901, in the very serious ornithological journal *The Auk*, an essay by O. Widmann titled *A Visit to Audubon's Birthplace* describes in glowing prose a pilgrimage made to Mandeville in order to honor Louisiana's most famous son.

The Mandeville myth was permanently laid to rest in the monumental biography of Audubon by Francis Hobart Herrick, *Audubon the Naturalist: A History of his Life and Times* (1917, revised 1938). Since then, all lovers of Audubon's art know that he was born on April 26, 1785, in Les Cayes, Santo Domingo, now Haiti. This is of course essential knowledge to the historian and the biographer, but has little bearing on the case at hand. In fact, Herrick's discovery simply adds force to the argument that for Audubon the appeal of Louisiana was far more vital a consideration than a mere accident of fate. If he had not been born in Louisiana, he should have been. Early on, Audubon had learned to live by his wits and his will, and if history had to be twisted a bit to serve the higher reality that Louisiana had made him who he was, and thus belonged to him as a sort of birthright, then so be it. After all, the country one chooses is surely as much a measure of a man's character as the country that is imposed on him at birth.

BEGINNINGS

Still, Audubon in his later years, looking back at a crucial time in his youth, must have considered the strange thought that Louisiana had been fixed in his sights well before he decided to set foot there and make it his own. When he left France for the United States during the summer of 1803, he was eighteen years old, beginning a new life in a new country. He had promised to learn English and to groom himself into a greater sense of direction and self-reliance in a place that knew something about the process. His father's large farm in southeastern Pennsylvania, Mill Grove, worked by a Quaker tenant family, overlooked Valley Forge. It was the perfect place to begin. As if to signal the transformation, a sort of rebirth, his father had paid to have a passport forged with his son's place of birth listed as Louisiana. His father, a revolutionary at heart, had lived to see the French Revolution begin to devolve into Napoleonic imperialism, and feared for his son's future in the strife to follow. The Louisiana Purchase, announced by Thomas Jefferson on July 4 that same summer, allowed father and son a safety net, and provided an easy path toward the American citizenship that Audubon was able to officialize in 1812, the day before July 4, always a key date in Audubon's affections. Louisiana was stamped again on his passport (Napoleon's wars of conquest were a death sentence for French and foe alike) when he returned from France in 1806 after a trip to Nantes to ask for his father's consent to marry Lucy Bakewell. Lucy was the English girl he had met on her father's estate, Fatland Ford, a short walk away from Mill Grove, and whose life and fortunes were to become as bound up with Louisiana as those of her husband.

After their marriage in 1808, the couple moved west to Kentucky, where Audubon (then twenty-three years old) had decided to open a store and go into trading with his partner, Ferdinand Rozier, a Frenchman from Nantes, the son of a friend of Audubon's father. Audubon and Rozier had first traveled to Kentucky the year before to prospect for possibilities. They both agreed that Louisville, then a rough-and-tumble but forward-looking frontier port on the busy Ohio River, would be a good place to set up shop. For Audubon, who had already ransacked the rich woods of southern Pennsylvania for birds to draw, Louisville was another gift from the gods. It was surrounded on both sides of the Ohio by unbroken canebrakes, natural salt licks, and thick hardwood forests hosting a plenitude of wildlife. The trees rang day and night with the songs of unknown birds. Audubon was in his element. He roamed the woods hunting new specimens. His portfolio of new drawings grew while Rozier minded the store. He met Alexander Wilson, heading south, who floated into Louisville in March 1810 on his rowboat the *Ornithologist*, carrying the first two volumes of his pioneering *American Ornithology*. Audubon looked at them and was convinced he could do better.

Audubon and Lucy quickly made friends with the families who had preceded them by a few years. A few years in Kentucky were all it took to become an old-timer. One of these new friends was William Croghan, an Irish immigrant who had married Lucy Clark, sister of William Clark (of the Lewis and Clark Expedition) and George Rogers Clark, hero of the Revolution and founder of Louisville. Croghan had fought under Washington and claimed to have met Audubon's father, who had sailed with the French fleet backing up the Americans at the battle of Yorktown. The Croghans presided over Locust Grove, a stately redbrick mansion on a hillside overlooking a large plantation close by the Ohio River. As the social center of the fledgling Louisville community, the doors of Locust Grove were always open, especially to a colorful *Louisiana-born* Frenchman and his charming English wife. Lucy could sing and play the pianoforte. Audubon's vivid drawings already made him a special case. The birds were drawn close up, ready to fly off the page, heads cocked, eyes steady and sparkling, each feather tense and distinct. He showed them to anyone who wanted to look. But Audubon also played the violin and the flute, spun yarns with the best, and could be counted on to liven up any gathering, high or low, in the living room, or around the campfire. The Audubons never had to wait for invitations in their early years on the frontier.

Audubon in Louisville still carried his passport stamped with Louisiana as his birthplace, but in a sense Louisiana had arrived in Louisville before Audubon. Just a year before Audubon and Rozier had opened their shop, William Clark and Meriwether Lewis had been celebrated in grand style at Locust Grove on their return trip from the Louisiana Territory. Louisiana became the talk of the town, still humming with the news when Audubon met Clark himself at Locust Grove and listened to his stories of the former French colony. By the time Audubon left Louisville with Lucy to head west on the Ohio, a fresh vision of Louisiana was in his sights again. Actually, despite the attractions of upper Kentucky and the generous welcome of the founding families, the Audubons had not been able to settle down in Louisville. Their house was a hotel, the Indian Queen, where their first son, Victor, was born. There was no room for Lucy's books and wedding china, and a hotel is not a home. They had made life-long friends, the Croghans and Clarks, and several members of a French settlement of shipbuilders down the river at nearby Shippingport, one of whom, Nicholas Berthoud, later married Lucy's younger sister Eliza. But the shop had not measured up to its promise. Most biographers suggest that Rozier made Audubon leave Louisville because competition had become too stiff. But Audubon always had a mind of his own, and he could not be pushed. He would not have moved unless there were other forces driving him westward.

JOURNEY UP THE MISSISSIPPI

Lucy had misgivings but accepted, and in the summer of 1810 Audubon and Rozier (with their young helper, Nat Pope) opened a general store in Henderson, a small hamlet on the Ohio River a little over a hundred miles downriver. Within only a few months in Henderson, Rozier had persuaded Audubon to head west into the Louisiana Territory, and to try their prospects in the "old French town" (in Audubon's words) of Ste. Geneviève on the Mississippi River. The appeal for Rozier was the French language and what remained of the colonial culture, together with the commercial promise of a port on the major artery into the booming South. Audubon agreed, at least at the outset of the venture. The pull of seeing Louisiana for himself was too strong to resist. They left Henderson on a flatboat loaded with supplies and merchandise, including barrels of Kentucky whiskey that they sold in Ste. Geneviève for two dollars a gallon (they had paid twenty-five cents in Henderson). Ste. Geneviève won over Rozier, who decided to stay. Audubon was not impressed, neither by the town nor the inhabitants. And Rozier's moody brooding had begun to jar against Audubon's exuberance. The partnership was dissolved, and Audubon left by himself (with his dog at his heels) and returned to Lucy in Henderson.

Almost twenty years later, Audubon wrote an account of the trip, *Journey up the Mississippi,* which he published in 1828 for a Liverpool editor of a Victorian anthology called *A Winter's Wreath*. At the time, Audubon was in England promoting himself and trying to bring *Birds of America* to press. *Journey up the Mississippi* is a fascinating document for many reasons, but I dwell on it here because it describes Audubon's first physical encounter with Louisiana. This is not the southern Louisiana that will revive Audubon's designs and desires in 1820, but it is Louisiana nonetheless. Audubon refers to it as "Upper Louisiana," using the old term to describe half of the vast French colony that originally stretched from Canada to the Gulf of Mexico and included the entire Mississippi River Valley. It is surely of significance that Audubon writes about his trip after having found himself in *Lower Louisiana*, and that years later, in 1843, he returns to explore old Louisiana in its outermost reaches in the last expedition of his life, up the Missouri River to the Yellowstone.

Throughout his account, Audubon insists on his passing into a different world from the one he has left behind in Louisville and Henderson. As always, in Audubon's mind the lure of the receding frontier, here verging on wilderness, was a powerful attraction. The Louisiana Purchase was too recent to have changed the nature of the land or the sparsely populated settlements spiced with the memory of French colonialism. Audubon, as he always did, mingled easily with the Indians. He spoke French with the Shawnees, and admired the independence of the elk-hunting Osage. He talked to a French count moving farther west, and in Cape Girardeau met a former district governor in Louisiana under Spanish rule, Monsieur Lorimier, a type of dying Creole dispossessed by the Purchase who had gone into retirement as a caricature of himself. Audubon describes Lorimier in a remarkable portrait worthy of George Washington Cable:

> Imagine a man not exceeding four feet six inches in height, and thin in proportion, looking as if he had just been shot out of a popgun. . . . His hair was plastered close to his head with a quantity of pomatum; and behind he wore a long queue rolled up in a dirty ribbon which hung down below his waist. The upper part of his dress was European, and had evidently once been made of the richest materials; and though now woefully patched and dilapidated, you

> might still observe here and there shreds of gold and silver lace adhering to the worn apparel. His waistcoat . . . had immense flaps or pockets that covered more than one-half of his lower garments. These latter were of a description totally at variance with the upper part of his costume. They were of dressed buckskin, fitting tight to his attenuated limbs and ornamented with large iron buckles at the knees which served to attach and support a pair of Indian hunting gaiters that had, like the rest of his dress, seen long and hard service. To complete his costume, he wore on his feet a pair of moccasins, or Indian shoes, that were really of most beautiful workmanship. These articles of dress, together with his small stature and singular features, rendered his appearance at a little distance the most ludicrous caricature that can be imagined; but on approaching nearer and conversing with him, his manners were found to be courteous and polished.

When Audubon wrote *Journey up the Mississippi*, probably just after Christmas 1827, his main motivation was self-promotional; he wanted to prove his status as a serious naturalist whose knowledge of the American wilderness had been learned and earned early, by virtue of his own efforts in the deep woods, unlike certain "closet naturalists." Most of Audubon's reporting in *Journey* serves his purpose well. There is enough reckless vigor and realistic detail to satisfy (or shame) anyone, including the haughty scholar cooped up in a university library. The Lorimier episode was added on as a comic interlude, but it is a good example of the hidden layers of Audubon's writing once the reader begins to probe.

Audubon's gentle mockery of Lorimier, a northern Creole who had moved south from Canada, derives from several sources. For one thing, it conforms to a fairly consistent prejudice in his journals against French Canadians. In the 1843 *Missouri River Journal*, he finds himself on a steamboat with a hundred noisy half-drunk fur trappers heading north on a hunting expedition. They share the same faults as the inhabitants of Ste. Geneviève (just upriver from Lorimier and Cape Girardeau): vulgar, unwashed, and wholly content with the routine of their base existence. For Audubon, lack of ambition was a failure of spirit he could not abide in himself or anyone else. In the *Missouri River Journal* he used the term *Creole* to describe them, which is accurate, but was more often applied to the French born in the Caribbean or in Lower Louisiana. This was of course Audubon's case, and he was given to speaking with some pride of his southern blood and constitution. No doubt there was a dose of snobbery involved, since the southern Creole was more likely to have had his cultural tastes and skills nurtured by the leisure of life on the plantation. Audubon knew this, but would always rather believe individual character and will are more important than the wheels of fortune. When Audubon finally judges Lorimier's manners as "courteous and polished," he is praising him for preserving his dignity despite everything, and by doing so, rising above his fallen status, his stature, and his species.

All through his writings, from *Journey up the Mississippi* on through the later essays and bird biographies, Audubon for the most part is careful to distinguish between fact and fiction. By the time he wrote the Lorimier sketch and placed it as an obvious sideshow attraction in the middle of a serious autobiographical narrative, Audubon's knowledge of English literature had grown by leaps and bounds. This is a feature of his work that all biographies, even the most recent, have overlooked. When he left Louisiana for England in 1826 Audubon had spent almost twenty years reading the classics of English literature, often with Lucy at his side. Lucy introduced him to the practice of reading out loud to each other, evening after evening. Audubon grew to worship the habit and could turn in his mind to favorite novels and poems whenever the mood or occasion demanded. Alone on his travels, he borrowed books whenever possible. On his two-month voyage to Liverpool from New Orleans in 1826, he read every book on shipboard, some of them, like Thomson's *The Seasons*, twice. In the letters and journal for that one year alone there are references (and citations from memory) to dozens of writers: Shakespeare, Milton, Dryden, Swift, Smollett, Johnson, Sterne, Goldsmith, Byron, Young, Scott, Edgeworth, Moore, and others.

Without any strain, Audubon could look back to the English satirists (especially his old favorites, Goldsmith and Sterne) as models for testing his wit in his treatment of Lorimier. Given his own love of literature, he fully expected his fellow readers to recognize the debt and to enjoy the exercise. But there was a more down-to-earth humor he could rely on too, closer to his life in the woods, and to his natural voice: the southern tradition of the frontier tall tale. It was the tall tale that made Lorimier absurdly short. Not long after Audubon published *Journey up the Mississippi*, that same southern backwoods humor, a staple of penny magazines and newspapers, was lifted into higher esteem by a literary movement founded by Augustus Baldwin Longstreet and his book, *Georgia Scenes*, published in 1835. *Georgia Scenes* was soon followed by collections from other southern states, notably Louisiana. Audubon was in good literary company.

Though Audubon published often and widely, he was well aware of his limitations, mainly because he was prepared to measure his efforts against the best. He privately despaired of ever calling himself an "author." But we have to remember that Audubon honed his skills by constant practice. He wrote whenever he could, at least two hours a day, or more. There are overwritten passages and awkward shifts of style throughout his work. But as *Journey up the Mississippi* demonstrates, from very early on, he could write well enough, even when he changed keys, and tunes, like the mockingbird in full voice. And of course the advantage he possessed over more grammatical or sophisticated writers was the force of his personality, and the wealth of his experience. The spirit and stride of the story always override the occasional stumble. For these reasons *Journey up the Mississippi* is the perfect primer for reading what he wrote to be published, and the best place to begin the selections in *Audubon on Louisiana.*

Along with *Journey up the Mississippi, Audubon on Louisiana* includes one essay, or *episode,* that derives from this same period. *The Prairie* describes a harrowing escape walking back to Henderson alone across the bottomlands and prairies of the territories on the eastern side of the Mississippi. Audubon's account is a virtuoso piece of storytelling, and a thrilling adventure. He may have let his imagination rework his memory, but what does it matter if it helps enhance the dramatic effect. *The Prairie* stands out as one of the key documents in Audubon's early development, and will later provide the heart of the matter for Robert Penn Warren's dark heroic masterpiece on Audubon in the poem *Audubon: A Vision* (1969).

LOUISIANA JOURNAL

Audubon remained in Kentucky for the next ten years, but Louisiana continued to stir his imagination with new ambitions and possibilities. He was pleased when his brother-in-law Thomas Bakewell proposed in 1811 that Audubon join him as a partner in a firm of factors he intended to set up in New Orleans. Audubon had business cards printed up and with typical flamboyance made a show of nailing them to tavern doors. The outbreak of the War of 1812 put an end to the plan. Still, Audubon had been touched by Lucy's joy at the prospect of moving south to her husband's legendary *birthplace.* As it turned out, Audubon's first visit to New Orleans, and southern Louisiana, was a voyage of angry desperation in 1819 to recover a debt on a repossessed steamboat. It was a failure, one more shock in a battering series. After years of relative prosperity in their home in Henderson, a combination of bad investments and bad luck (and a national economic crisis) forced Audubon and his family to sell almost all they possessed. Before the year was over Audubon had left Lucy and their two sons, Victor and John, behind and had moved to Louisville to find work. In Louisville he was jailed for debt and made to declare bankruptcy, and finished by sketching chalk portraits for notables as a survival measure. After she joined Audubon in Louisville with the two boys, Lucy gave birth to a girl, Rose, who died seven months later, the second daughter to die in infancy.

The following year Audubon took a temporary job in Cincinnati as a taxidermist for the new Western Museum. Lucy and the boys soon followed, but by this time Audubon had reached a breaking point. Before he broke, however, he convinced himself, and Lucy, that he dedicate himself wholly, for once and for all, to the task of completing his bird drawings and presenting them all in a single definitive volume. To this end he would go south to Louisiana and begin his search for new birds. Despite his failures in business, he had never lost faith in his art nor in the value of his early ambitions. After all, his drawings had been seen and at least silently admired by Alexander Wilson, now universally praised thanks to the success of the *American Ornithology*. Wilson's great work was proudly displayed in the library of the Western Museum. It was a gamble, but Audubon did have one card left, or two. He had Lucy along with his portfolio of birds. Though Lucy could not offer her blessings, she accepted his decision, and this was all the encouragement Audubon needed. Thus, on October 12, 1820, after promising Lucy he would shave once a week, he boarded a flatboat headed south on the Ohio and Mississippi Rivers, with New Orleans and all of southern Louisiana his ultimate destination. He paid for his passage by hiring himself out as hunter to provide game for the crew's meals. To help with his drawings, he took along a young art student, Joseph Mason, who was to receive no wages, at least not for now. Lucy stayed behind in Cincinnati to work as a teacher and to take care of the two boys. This reaching out to Louisiana, the storied land of his origins, was to change everything for Audubon. For the rest of his life he looked back on the voyage as the crucial turning point of his career. Knowing the stakes he soon began a new journal, written for Lucy and his sons, as proof of his commitment and as a testament to the vital spirits of the last chance.

The journal is an intense, compelling document of Audubon's self-revival, and one of the most remarkable diaries in the annals of American natural history.

It was first transcribed from the manuscript (discovered long after Audubon's death) by Howard Corning, and published in 1929 by the Club of Odd Volumes in Cambridge, Massachusetts. For Audubon, the original journal was written down as a serial letter to his family and to himself, and not intended for publication. Corning gave it the title of *Journal of John James Audubon, Made During His Trip to New Orleans in 1820–21*, and when I reprinted Corning's transcription, whole and untouched, in my general collection of Audubon's writings *Selected Journals and Other Writings* (Penguin, 1996), I called it the *Mississippi River Journal*. My title seems to have struck a chord, and has been adopted by Audubon biographers since. A more accurate title, however, given Audubon's designs, destination, and explorations in the countryside well beyond the river, would have been the *Louisiana Journal*. That is the title I have renamed it for this collection (what I have given I should be able to take away).

The *Louisiana Journal* is the heart of *Audubon on Louisiana*. For one thing, it demonstrates how the actual experience of Louisiana evolved into a triumphant narrative of transformation. Though he was tormented at times by solitude, depression, and fears of failure, by the end of the journal Audubon's resolution was stronger than ever. His faith in his skill and capacity to succeed had now taken hold of him like a religion, and his career as an independent artist with a sovereign mission was now launched for good. He had daily increased his already formidable knowledge of animals and birds in their natural habitat, storing up a fresh fund of firsthand observations he could draw on the rest of his life. He found himself painting possessed by new powers, as if Louisiana had revealed to him who he was and what he could do. As I pointed out in *Selected Journals and Other Writings*, on one plantation alone, Oakley, near the small river community of Bayou Sara, where he was a tutor during the summer of 1821, he produced one quick masterpiece after another: the pine creeping warbler, the mockingbird, the Mississippi kite, the yellow-throated vireo, the red-cockaded woodpecker, the American redstart, the summer redbird, the prairie warbler, the Tennessee warbler. Twelve years after my *Selected Journals* was published, I was pleased to see Danny Heitman make the same assessment in his engaging monograph for Louisiana State University Press, *A Summer of Birds* (2008). Rightly so, Heitman relies almost exclusively on the segment of Audubon's journal devoted to the short time he spent at Oakley.

One of the pleasures of reading the *Louisiana Journal* derives from Audubon's curiosity and sharpness of eye as he discovers a new world. For most Americans, of course, Louisiana was an exotic land only recently opened up, and the clash of cultures was a dramatic spectacle of enormous appeal. Audubon arrived exactly at the time the later local colorists (George Washington Cable, Lafcadio Hearn, Kate Chopin, Grace King) looked back to for their popular stories of Louisiana in the making, and years before other travel writers like Frederick Olmsted, Edward King, and even Mark Twain, sent back their impressions to the northern newspapers and magazines. Here is Audubon (in a telegraphic style perfectly suited to the rapid shifting of his perceptions) approaching Baton Rouge in January 1821, six months before his summer in Oakley. He had just arrived in Bayou Sara on a keelboat tugged along by the steamboat *Columbus:*

> At 12 o'clock today the *Columbus* came to at Bayou Sarah—a small village at the mouth of the inlet—many flatboats, the steamboat and 2 brigs waiting for cotton—the steamboat *Alabama* put off as we came to, and about half an hour after, the *Columbus* left us to ourselves to try to reach Baton Rouge before her, to procure the freight there—promising to wait 3 hours.
>
> The lands are flattening fast—the orange trees are now and then seen near the rich planter's habitation—and the verdure along all the shore is very luxuriant and agreeable—the thermometer at 68 at 12 o'clock in the shade, the day beautifully fair—expected to see some alligators—many Irish geese in the eddys—mallards but few geese—at half past 6 o'clock p.m. we came opposite Baton Rouge but the steamboat had left and of course we proceeded on our way floating—this last place is a thrifty village on the New Orleans State—from some distance above levees have made their appearances—I saw a Negro man fishing by dipping a scoop net every moment in the water immediately at a point where the current ran swift forming an eddy below, he had taken several tolerably large cat fishes.

The *Louisiana Journal* covers a little over an entire year, ending the last day of December 1821, just two weeks after his wife and sons had finally arrived to join him in New Orleans, and fourteen months after he had left them behind in Cincinnati. For the next five years Audubon and Lucy made their home in Louisiana. Audubon was often absent, once for fourteen months when he went east to Philadelphia to show his portfolio and search for sponsors, but Lucy did not budge. After she was asked to reside and set up a small school at the Percy plantation Beech Woods, she quickly made a reputation teaching plantation children. She changed plantations once, in 1827, moving her school from Beech Woods to William Garrett Johnson's plantation, Beech Grove. Both were located in the same neighborly community of the fertile Felicianas, and she and her son John (Victor left to work for his Uncle Nicholas Berthoud in late 1823) soon became accepted (and esteemed) as local fixtures. Audubon himself could rarely stay anywhere long enough to be a fixture. When he left for England in 1826 (and remained there three years), Lucy stayed behind, teaching, saving what funds she could, and keeping the hearth fires burning for the time when they could once more reunite. Louisiana was a new start and the defining moment in Audubon's life, but the years there were far from being carefree. He continued to be wracked by doubts and despair, despite the thickening of the portfolio. Lucy, as ever, was the stable center of the family, a steady light that Audubon needed and relied on more than ever. This brings me to Carolyn DeLatte's groundbreaking study, *Lucy Audubon: A Biography,* first published in 1982 by Louisiana State University Press and reprinted in an updated edition in 2008.

DeLatte makes it clear that once Audubon left for England, Lucy and Louisiana had become fused in his affections. When Audubon returned at the end of 1829, Lucy had spent almost ten years in Louisiana with for the most part only herself (and the occasional packet of letters from her husband) to count on. By the end of her stay she had naturalized herself into a native of Louisiana far more than Audubon himself. It is not for nothing that half of DeLatte's substantial biography concentrates on Lucy's ten years in Louisiana even though she lived to be eighty-seven. I could not disagree more with Christoph Irmscher's assessment in his preface to the updated edition of *Lucy Audubon* that DeLatte's insisting on the Louisiana years was a fault, a "curious omission" of what was to follow. Actually what DeLatte realized was that those ten years were critical to Audubon's ultimate success and deserved to be explained in the kind of rich detail that no other biographer had explored, much less mastered.

That Lucy and Louisiana are inseparable in Audubon's mind is obvious throughout the so-called *European Journals,* the diaries he kept after he left for England, and in his subsequent letters and writings. In England he measures the distance he has traveled, not from America, but from Louisiana (and Lucy), and he daydreams about walking with Lucy through "the fragrant woods of Louisiana." Lucy's steadfastness and strength of character made it possible for Audubon to search for these same qualities in himself, and confirm his self-confidence when faced with the landed gentry and university scholars who had made their way cushioned by the safe comforts of family estates and predictable careers. When Audubon dreams about his rambles in the Felicianas, thinking only about the good times, such as swimming with Lucy at night, and riding horses with her through the woods, he has only Lucy to thank. Like many lovers of solitude, and freedom of movement, he could only enjoy their full benefits if there were someone to welcome him home. Some of the most moving passages in Audubon's writings are his accounts of homecoming. As he recorded in his journal of 1826, whenever he doubted himself in England, a single letter from Lucy postmarked Bayou Sara could give him "new life."

Audubon's love of Louisiana became a favorite theme of his reminiscences within the family circle. His granddaughter, Maria R. Audubon, in her introduction to her 1897 edition of *Audubon and His Journals*, wrote (in her somewhat flowery style) that he often talked about his "beloved Louisiana whose praises he never wearied of singing, whose magnolia woods were more to him than palaces, whose swamps were storehouses of treasures." At the end of his life, the nostalgia for what Louisiana had come to represent had become an obsessive regret for a natural world he felt was rapidly disappearing, a sort of lost paradise that he vainly tried to recover in his last trip in 1843, a few years before his death, up the Missouri into the wilderness of what was left of the original old Louisiana of his youth.

LOUISIANA EPISODES

While Audubon and Lucy were living in Louisiana, there was little incentive for him to keep up a journal with his usual constancy. Very few documents from his own hand are known from this period. Only a couple of entries from the journal have survived, recorded in his granddaughter's introduction to *Audubon and His Journals*. What we do have, together with the memories recorded in his later journals, are a handful of episodes, or essays, that he published soon after the great double elephant folio of *Birds of America* (1827–1838) had begun to see the light. They were all included in the first three volumes of his five-volume *Ornithological Biography* (1831–1839). The main purpose of the *Biography* was to collect the "bird biographies," individual texts written by Audubon to provide information on the various species of birds contained in *Birds of America*. He inserted the episodes (there were a total of sixty) at regular intervals among the biographies. Audubon was always careful to distinguish the brightly painted landscapes and stories in the episodes (referred to collectively in the *Biography* as "delineations of American scenery and character") from the more serious bird biographies aimed at establishing his credibility in the scientific community.

Before he began the episodes and the *Ornithological Biography*, Audubon wrote several *letters* or long essays for a journal published by the prestigious natural history association, the Wernerian Society, in Edinburgh. The essays were first presented as lectures he read himself, or handed over to be read, to the members of the society. Audubon had traveled to Edinburgh (a few months after his arrival in England), as always, to promote his portfolio. The reception, within a week, exceeded all his hopes. He convinced the eminent naturalists at the University and the Royal Institution (including the founder of the Wernerian Society, Robert Jameson) that he knew his subject, and that he possessed genuine gifts as an artist; he found an engraver (W. H. Lizars) for his birds; and he finally glimpsed the grandeur of all his efforts when Lizars handed him the first in a series of superb hand-colored proofs. Fittingly enough, one of the first essays he wrote for the Wernerian Society was a lively account of Louisiana alligators in their natural setting in Bayou Sara and beyond, *Observations on the Natural History of the Alligator.* It was first given as a talk in January 1827, and was printed in the *Edinburgh New Philosophical Journal* later that same year. After all, just before Audubon left for England he had bought in New Orleans for one dollar a baby alligator (as a Louisiana talisman) to take with him on the ship. And he carried an open letter from the governor of Louisiana stamped with the state seal. For Audubon, this was all the passport he needed. To make sure no one missed the affiliation, when he was made a member of the Wernerian Society the same night he gave the alligator talk, he made sure he was announced as "John James Audubon Esquire of Louisiana."

In *Audubon on Louisiana* I include the alligator essay whole and unedited, along with the selection of Louisiana episodes from the *Ornithological Biography*. The alligator essay contains some of his freshest writing on Louisiana's nature and wildlife, throbbing with the vitalism that characterizes his diaries. Audubon is never a detached, indifferent observer, and always enters into the behavior of animal life with a boldness of spirit that is hard to resist. Even late in life, in the *Missouri River Journal*, we find him writing as a fifty-eight-year-old (toothless) explorer on his last trip, deep in Indian and buffalo territory, capable of lying down on the ground trying to stimulate the curiosity of a wary antelope: "We determined to stop and try to bring him to us: I lay on my back and threw my legs up, kicking first one and then the other foot, and sure enough the antelope walked towards us, slowly and carefully." In *Observations on the Natural History of the Alligator*, the same instincts of playfulness are part of the essay's charm, certainly to the Edinburgh scholars who first heard it as a speech. Even though Audubon throughout does his best to impress his audience with the terrible power and size of the mighty creatures, he cannot refrain from digressing about a game of football he once invented to watch them at play themselves:

> I have frequently been very much amused when fishing in a bayou, where alligators were numerous, by throwing a blown bladder on the water towards the nearest to me. The alligator makes for it at once, flaps it towards its mouth, or attempts seizing it at once, but all in vain. The light bladder slides off; in a minute many alligators are trying to seize this, and their evolutions are quite interesting. They then put me in mind of a crowd of boys running after a football.

Along with the episode *The Prairie*, bound up with his trip to Ste. Geneviève in 1810, *Audubon on Louisiana* collects nine other episodes from the *Ornithological Biography*, all but two offshoots and later reflections on his migration down the Mississippi to New Orleans in 1820. In reading these episodes it is best to allow ourselves some latitude, and longitude, when looking back at the geography of Audubon's Louisiana from the strict boundaries of Louisiana today. The Louisiana he traveled through, lived in, and wrote about, was the old Louisiana of the Mississippi River Valley, and stretched well beyond the limits modern history has since determined. French was still spoken up and down the river, and even a bustling trading and plantation town like Natchez had been French too long to shed its origins the day it became American. It was in Natchez that Audubon met the local historian, Charles Carré, who "spoke French with the greatest purity" and who became Audubon's mentor and guide. Then there is the crisscrossing movement Audubon describes in his portraits of the people he meets on his way down the river, almost always from north to south, and from east to west, deeper into Louisiana, and like the Mississippi itself, then and now, no respecter of invisible borders. In the episode *A Flood*, Audubon, looking west, is struck by the great spectacle of floodwaters from the vantage point of the Natchez bluffs: "Opposite the city of Natchez, which stands on a bluff bank of considerable elevation, the extent of inundated land is immense, the greater portion of the tract lying between the Mississippi and the Red River, which is more than thirty miles in breadth, being under water. The mail-bag has often been carried through the immersed forests, in a canoe, for even a greater distance, in order to be forwarded to Natchitochez."

In the episode *The Squatters of the Mississippi*, Audubon looks west again, this time to celebrate the frontier grit of the settlers, or "squatters," who have migrated into Louisiana to make a new life for themselves, much as Audubon himself had done. In fact, the idea of the episode is to set the record straight, and to counter the degrading stereotypes these new settlers had to fight as they worked their way into, if not prosperity, at least self-dependence. It is hard not to hear the voice of Audubon's own struggles in his praise of their industry and determination after they have crossed the Mississippi and selected their place of habitation:

> The exhalations arising from the swamps and morasses around them have a powerful effect on these new settlers, but all are intent on preparing for the winter. A small patch of ground is cleared by the axe and the fire, a temporary cabin is erected, to each of the cattle is attached a jingling bell before it is let loose into the neighboring cane-brake, and the horses remain about the house, where they find sufficient food at that season. The first trading-boat that stops at their landing, enables them to provide themselves with some flour, fish-hooks, and ammunition, as well as other commodities. The looms are mounted, the spinning-wheels soon furnish some yarn, and in a few weeks the family throw off their ragged clothes, and array themselves in suits adapted to the climate. The father and sons meanwhile have sown turnips and other vegetables; and from some Kentucky flatboat, a supply of live poultry has been procured.

One of the attractions of the Louisiana episodes is the all-embracing prism of Audubon's outlook and personality, singular when compared to most common mortals, but plural in his ability to adapt to all kinds of situations, and to mix easily with people up and down the social scale. He knew the squatters from his years on the frontier, but we should never forget that he had been educated during his boyhood in France by private tutors in Nantes where he had learned to dance, fence, and play the violin. He possessed a well-rounded European culture that never left him, and that is sadly neglected in the most recent Audubon biographies. In fact it was his music that ingratiated him to the Indians, and both music and dance that brought him some extra money when he gave dancing lessons to the planters, playing his own violin to provide the music. Thus when he found himself in the cosmopolitan world of New Orleans, he was not exactly a stranger to the sights and sounds of city life. He did not have a high opinion of New Orleans, largely because he had to struggle sorely to find work and make a name for himself, but he knew a citified dandy when he saw one, strutting along in dressed-to-kill New Orleans style, complete with magnolia, alligator, silk umbrella, and painted bunting:

> His head was covered by a straw hat, the brim of which might cope with those worn by the fair sex in 1830; his neck was exposed to the weather; the broad frill of a shirt, then fashionable, flapped about his breast, whilst an extraordinary collar, carefully arranged, fell over the top of his coat. The latter was of a light green color, harmonizing well with a pair of flowing yellow nankeen trousers, and a pink waistcoat, from the bosom of which, amidst a large bunch of the splendid flowers of the magnolia, protruded part of a young alligator. . . . The gentleman held in one hand a cage full of richly-plumed nonpareils. . . . He walked as if conscious of his own importance—that is, with a good deal of pomposity, singing, "My love is but a lassie yet," and that with such thorough imitation of the Scotch emphasis that had not his physiognomy brought to my mind a denial of his being from "within a mile of Edinburgh," I should have put him down in my journal for a true Scot. But no: his tournure, nay, the very shape of his visage, pronounced him an American from the farthest parts of our eastern Atlantic shores.

Audubon does not give the name of this odd bird singing with a brogue, but we know from the title of the episode, *The Original Painter,* that he was John Wesley Jarvis, a well-known eccentric (and portrait artist) who had moved south from New York. But recent transplant or not, the flair with which Jarvis overacts his role on the public stage proves beyond a doubt he has been made in New Orleans.

I have mentioned that New Orleans (and all of Louisiana) was a strange exotic world to Americans north and east, especially in the years following the Purchase.

The exoticism filled the pages of the local colorists after the Civil War, when Louisiana became the most written-about region in the entire United States. In many ways Audubon's writings, though not as polished, anticipate their work. He got there first, at the right time, and was alert to what made Louisiana different. Unlike the local colorists he did not need to look back several decades to the culture-shaking changes of the 1820s; all he had to do was look around and write down what he saw under his nose. Another New Orleans piece, *The Fair Incognito,* is a good example. He wrote it for his journal, but later enclosed it, complete and unexpurgated, in a letter to Lucy. When Audubon first arrived in New Orleans, his lack of money and status prevented him from easy access into the old Creole families. Their power and wealth were rapidly fading, but they still set the tone in the upper reaches of society. He complains in his journal (the *Louisiana Journal*) that he often walked by their social rituals and amusements, but as an outsider looking in, not as a familiar compatriot whose language and culture should have made him a guest of honor from the beginning. On one of his walks he gives a glimpse of what he was missing: "The levee early was crowded by people of all sorts as well as colors, the market very abundant, the church bell ringing, the billiard balls knocking, the guns heard all around." That same evening, after bringing back a fish crow to draw (he was still sleeping on the keelboat that had brought him from Natchez), he walks out again: "I brought it on board and began to work immediately—at dark took a walk to Mr Gordon, from there on to Mr Laville where we saw some white ladies and good looking ones—returning on board the quatroon ball attracted my view but as it cost one dollar entrance I merely listened a short time to the noise and came home as we are pleased to call it."

With the adventure he described in *The Fair Incognito,* the closed Creole community opened up, at least once, giving Audubon a glimpse into the secret life of the Creoles that few visitors could ever approach. He was accosted on the street by a young woman, "a female of fine form," whose face was covered by a veil. She asked Audubon in French if he was the bird artist she had heard about, and when he answered yes, she invited him to come to her rooms in the Faubourg Marigny (on the rue de l'Amour) in a half hour. When Audubon arrived, she took off her veil, introduced herself as Madame André, and made her proposal. She wanted Audubon to paint her in the nude. For the next ten days, Audubon did as she wished, painting every day for an hour or two. When the painting was done he received for his troubles "the finest gun in New Orleans" inscribed with her gratitude in French.

The question of why Audubon wrote the story out in such detail, and then sent it to Lucy in a letter, is not easily answered. Perhaps he wanted to impress her with his power of resistance. After all, he was a healthy high-spirited man, by himself, and obviously not immune to temptation. No one has ever found out who the mystery woman was. It has been suggested that she was a mistress of Bernard Marigny, the rich Creole who created the Faubourg Marigny as a French Louisiana enclave against the rising tide of the Anglo-Americans. It was Marigny who renamed Rampart Street the *rue de l'Amour,* and Marigny who established his plantation near Mandeville on Lake Pontchartrain, the same Mandeville which Audubon, we remember, may have hinted was his birthplace. I wonder too about Audubon renting the house (square in the middle of the Faubourg Marigny) on Dauphine Street for his family when Lucy and the boys arrived in New Orleans. Surely the strong presence of the Creole language and culture was a factor, but there may have been some memories that Audubon did not want to quickly forget. At any rate, *The Fair Incognito* is the kind of highly flavored narrative that the later local colorists excelled in, though unlike Audubon they would have been looking back fifty years before their time. Otherwise the literary comparisons are of little note. We do not read Audubon for the sublimity of his style. We read him for his contagious appetite for life and his life story. When he does write to be published, as he did in the episodes, the verbal echoes of some of his favorite writers (Sir Walter Scott, for one) can sometimes overwhelm his own voice. These echoes simply have to be taken for what they are, tributes of an enthusiastic apprentice learning his craft. But if he could never rival the local colorists their literary skills, on the other hand he did not have to worry about the restraints of the popular press or the need to uphold a public image. There are no naked women in George Washington Cable.

The last two episodes in *Audubon on Louisiana* are *A Long Calm at Sea* and *Still Becalmed. A Long Calm at Sea* records Audubon's first days on board the ship *Delos* while waiting for enough wind to set sail for England; he had left New Orleans in May of 1826. This was the beginning of what turned out to be the three-year effort to find an engraver of his birds, begin the printing of *Birds of America,* and secure enough subscribers to continue. The *Delos* had been towed by steamboat to the old French fort La Balize, perched on a small island in a lobe of the mouth of the Mississippi familiarly and fittingly called, Audubon *oblige,* the Birdfoot Delta. Audubon was leaving Lucy and Louisiana behind, counting the hours, and amused himself by admiring the dolphins and other fish as the ship drifts further south into the Gulf of Mexico. *Still Becalmed* continues the same story, with Louisiana still close enough to make its presence felt:

> One day several small birds, after alighting on the spars, betook themselves to the deck. One of them, a female Rice Bunting, drew our attention more particularly, for, a few moments after her arrival, there came down, as if in her wake, a beautiful Peregrine Falcon. The plunderer hovered about for a while, then stationed himself on the end of one of the yard-arms, and suddenly pouncing on the little gleaner of the meadows, clutched her and carried her off in exultation. But, reader, mark the date, and judge besides of my astonishment when I saw the Falcon feeding on the Finch while on the wing, precisely with the same ease and composure as the Mississippi Kite might show while devouring high in air a Red-throated lizard, swept from one of the magnificent trees of the Louisiana woods.

I have chosen these two episodes, even if they follow Audubon slowly moving out of Louisiana and into the Gulf of Mexico. La Balize is certainly in Louisiana; in fact it was one of the oldest settlements in the original French colony. Audubon is in Louisiana too as the ship floats with the current down the river through the marshlands of the fanlike Mississippi River Delta. Then there looms the open sea of the Gulf of Mexico, washing the Louisiana shoreline for hundreds of miles. In Audubon's time the barrier islands and waters of the Gulf close to the coast were steeped in recent colonial history, and today saltwater and seafood remain as inseparable from our idea of Louisiana as birds and bayous.

LAST VOYAGE TO LOUISIANA

After he left Louisiana for England in May 1826, Audubon resumed his regular journal writing for Lucy and his sons. Most of his entries (from 1826 to 1829) trace the long hard road (all the "vicissitudes" as he put it) as he travels up and down England, from Liverpool to Edinburgh to London, making sure the complex process of setting the *Birds of America* in motion was definitively engaged. Once he had replaced the Edinburgh engraver Lizars with the London engraver Havell, and the new engravings had begun, and once he had secured enough subscriptions (including one from King George IV) to proceed with his plans, he set sail for home and arrived in New York in May 1829. After a summer searching for new birds and painting them in New Jersey and Pennsylvania, he finally returns to Louisiana, and to Lucy, in November of the same year:

> It was dark, sultry, and I was quite alone. I was aware yellow fever was still raging at St. Francisville, but walked thither to procure a horse. Being only a mile distant, I soon reached it, and entered the open door of a house I knew to be an inn; all was dark and silent. I called and knocked in vain, it was the abode of Death alone! The air was putrid; I went to another house, another, and another; everywhere the same state of things existed; doors and windows were all open, but the living had fled. Finally I reached the home of Mr Nubling, whom I knew. He welcomed me, and lent me his horse, and I went off at a gallop. It was so dark that I soon lost my way, but I cared not, I was about to rejoin my wife, I was in the woods, the woods of Louisiana, my heart was bursting with joy!

Audubon and Lucy remained in Louisiana for less than two months before leaving together, first of all back to Kentucky, and then on to England to supervise the ongoing work on *Birds of America*, and to continue the writing of the *Ornithological Biography*. Audubon would return to Louisiana one more time, by ship in the spring of 1837, threading the barrier islands on a government cutter with his son John and his friend Ed Harris along the coast of the Gulf for one final look for new birds, even though by this time the *Birds of America* was close to complete. This was a far cry from his uncertain days on the deck of the *Delos* in 1826, hugging his portfolio, dreaming of Lucy, and waiting for wind. He had made other expeditions since he and Lucy had left Louisiana, notably to Florida and to Labrador. They were all highly successful, effectively adding scores of unfamiliar birds to his list. This trip was different. He had little need of more birds. Though he tried to convince himself that he was still working on the book, thus his desire to explore the west coast of Florida along with the coast of Louisiana, his main motivation was a matter of going home, at least for one last look.

Much of his writings about Louisiana in his later years are short reflections scattered in the journals. They often take the form of a nostalgic phrase or two, struck off on the spur of the moment, almost always addressed directly to Lucy.

This is also the case in his letters, many of them included in Howard Corning's groundbreaking edition, *Letters of John James Audubon, 1826–1840,* published in 1930. Audubon was a compulsive, prodigious letter writer, often writing two or three letters a day in addition to the regular evening entry in his journal. They are the best place to discover Audubon's natural voice (uninhibited, humorous, down-to-earth), and they are the best place to examine the inner history of his working life. Audubon could be urgent when the occasion demanded, but the main quality the letters reveal is an unhurried delight in using the written page as a form of familiar conversation. He is generous with details of what he has seen and done during the day, and he loves to embroider an anecdote, especially when writing to Lucy, or to a good friend like the Rev. John Bachman, a minister (and naturalist) who lived in Charleston, South Carolina.

The packet of letters Audubon wrote to Lucy and Bachman during the expedition to the Gulf of Mexico is a good place to discover his taste for talking on paper. Unlike the brief flashes of old memories in much of his writing of this period, these last voyage letters, including the one to his Scottish friend and helpmate on the *Ornithological Biography,* William MacGillivray, are rich in full and flowing Louisiana observations. He had not been back to Louisiana since 1829. And by this time he was fairly sure he would never return. Looking back, the symbolic implications of the trip are hard to miss. Audubon seems less interested in finding new birds than in making a pilgrimage. As a matter of fact, for the entire two months he did not see a single bird he had not already painted. He did carry back the usual barrels of birdskins preserved in rum, along with the odds and ends he loved to collect, shells, eggs, and this time fourteen live turtles. But these were there to justify the time and effort, to himself and to anyone else who asked. His mind had been fixed on the fate of Louisiana in his life as soon as he reached New Orleans from Mobile. Louisiana had launched the beginning of *Birds of America* in 1820. In 1837 it was only fitting that Audubon commemorate the ending in grand style, floating on a boat in the warm waters of the Gulf and watching the Louisiana skies fill with millions of birds he now knew by name to a feather. Before setting off for the coast from New Orleans, in one of his letters to Bachman, he expressed his amused delight that the local press had done its best to celebrate the visit. For Louisiana, the announcement was news; for Audubon it was a personal triumph:

> I am glad, and proud too; that I have at last been acknowledged by the public prints as a Native Citizen of Louisiana—and had it been supposed

when first we arrived here that our stay would have been half as long as it has been—I really think my countrymen would have honored us with a public dinner! Try to find out the paragraphs in the New Orleans Courier (French and English) of the beginning of this month.

LOUISIANA BIRDS (*Ornithological Biography*)

The bird biographies Audubon wrote for the *Ornithological Biography* are a major (and little known) source of Louisiana lore, but with a more public voice than the one found in the letters and journals (written, we have to remember, for a private audience). In 1830 Audubon decided to publish the *Ornithological Biography* as a series of companion volumes to the *Birds of America* folio, still in the process of being issued to subscribers in *numbers* of five plates apiece. He was assisted by Lucy, and by William MacGillivray, a young Scottish naturalist and university teacher he had found in Edinburgh. MacGillivray helped to identify and, as Audubon expressed it, filter out the "asperities" in the text, provided some of the measurements, and did the anatomical drawings and descriptions at the end of each biography. Working intensely, often ten to twelve hours a day, they finished the first volume (published in April 1831) within four months. The completed *Biography* would eventually comprise five dense volumes, over three thousand pages long.

Very little escaped Audubon during the years he spent tracking his birds up and down the Mississippi Valley, and elsewhere. The natural world of the solitary walker of the woods had been his university. Audubon knew his academic training in the official sciences was limited, but he prided himself on writing directly from his own observations and experiences: "I know that I am not a scholar, but meantime I am aware that no man living knows better than I do the habits of our birds; no man living has studied them as much as I have done, and with the assistance of my old journals and memorandum-books which were written on the spot, I can at least put down plain truths."

Audubon may not have received the benefits of the world's universities and institutes, but he knew what he had achieved, almost entirely through his own efforts. And anyone who has read the entire *Ornithological Biography* would have to agree. Though not as spectacular, it remains as unique and overpowering as the paintings in *Birds of America*. It is Audubon's *Great Book of Nature*, one of the most remarkable compilations of nature writing ever published. I can think of no other work by one man that can match it in its range and authenticity, and certainly no other work as well supplied with firsthand observations over years of constant searching.

Any reader drawn to the joys of natural history should be immediately won over by Audubon's exact knowledge of the birds he discovered in Louisiana and the physical world they lived in. They were the reason he had come there in the first place. Then there is the added charm of certain vignettes when he allows himself the freedom of unspooling his impressions without thinking too hard about his reputation as a sober-minded ornithologist. Once he begins reaching back, much as he strives to maintain an objective perspective, all manner of sensations and memories from his early years in Louisiana blithely overflow into the text. Thankfully for the reader, the demands of science fail to suppress the old feelings. There are scores of passages to choose from. Audubon cites Louisiana far and above all other locations, north and south. Even when he writes about the birds that are migrating through, as transients, he lingers on their Louisiana passage, with no guilt of disproportion or favoritism, before following them outward to their ultimate destinations and breeding grounds. Just to give an idea of the bounty of Louisiana writing the *Ornithological Biography* offers, here are two short excerpts, both taken from a few random pages of the very first volume:

The Baltimore Oriole

> The Baltimore Oriole arrives from the south, perhaps from Mexico, or perhaps from a more distant region, and enters Louisiana as soon as spring commences there. It approaches the planter's house, and searches amongst the surrounding trees for a suitable place in which to settle for the season. It prefers, I believe, the trees that grow on the sides of a gentle declivity. The choice of a twig being made, the male Oriole becomes extremely conspicuous. He flies to the ground, searches for the longest and driest filaments of the moss, which in that State is known by the name of Spanish Beard, and whenever he finds one fit for his purpose, ascends to the favourite spot where the nest is to be, uttering all the while a continued chirrup, which seems to imply that he knows no fear, but on the contrary fancies himself the acknowledged king of the woods. This sort of chirruping becomes louder, and is emitted in an angry tone, whenever an enemy approaches, or the bird is accidentally surprised, the sight of a cat or a

dog being always likely to produce it. No sooner does he reach the branches, than with bill and claws, aided by an astonishing sagacity, he fastens one end of the moss to a twig, with as much art as a sailor might do, and takes up the other end, which he secures also, but to another twig a few inches off, leaving the thread floating in the air like a swing, the curve of which is perhaps seven or eight inches from the twigs.

The Louisiana Water Thrush

Much and justly as the song of the Nightingale is admired, I am inclined, after having often listened to it, to pronounce it in no degree superior to that of the Louisiana Water Thrush. The notes of the latter are as powerful and mellow, and at times as varied.

This bird is a resident of the low lands of the States of Louisiana and Mississippi, and is to be found at all seasons in the deepest and most swampy of our cane brakes, from which its melodies are heard to a considerable distance, its voice being as loud as that of the Wood Thrush. The bird may be observed perched on a low bough scarcely higher than the tops of the canes, in an erect attitude, swelling its throat, and repeating several times in succession sounds so approaching the whole two octaves of a good piano-forte, as almost to induce the hearer to imagine that the keys of that instrument are used on the occasion. The bird begins on the upper key, and progressively passes from one to another, until it reaches the base note, this last frequently being lost when there is the least agitation in the air. Its song is heard even in the winter, when the weather is calm and warm.

I have taken the liberty of naming this first songster of our groves after the country which has afforded me my greatest pleasures. . . .

As the above passages suggest, it would be possible to open the *Ornithological Biography* almost anywhere to find a paragraph or a page worthy of a Louisiana anthology. I have chosen instead to concentrate on nine full-length biographies; all are devoted to Louisiana birds Audubon painted there during the early 1820s, three near Oakley. If they lead the reader into a longer plunge into the whole five volumes of the *Biography*, Louisiana and beyond, so much the better. It was highly praised by minds that mattered when it first appeared. Reading it as a young man,

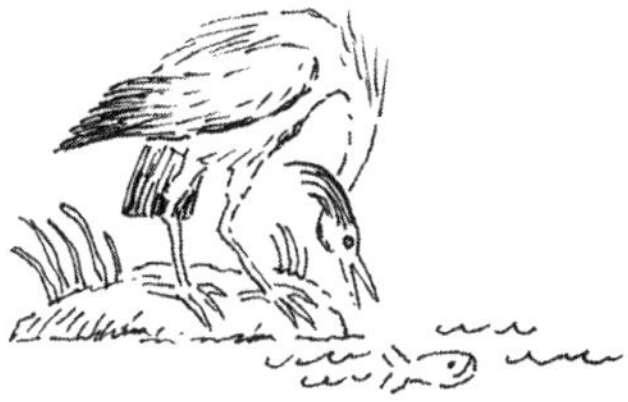

The circumstances behind the writing of the *Journey* cannot be known with exactitude. It first appeared in print in 1828 for the 1829 number of the Liverpool annual anthology, *The Winter's Wreath*, a miscellaneous collection of poetry and essays designed primarily as a showcase for Liverpool's literary community. The numbers were always printed the year before they were published. Writers outside Liverpool were often invited to contribute, including such major figures as William Wordsworth and Thomas B. Macaulay. Five numbers of the *Wreath* were issued in successive years, beginning in 1827 (announced as the number for 1828) and ending in 1831 (announced as the number for 1832).

Audubon was probably invited to send his essay by William Roscoe, a well-known historian, biographer, and botanist, and one of the leading lights of Liverpool's literary and scientific circles. Roscoe had founded the Liverpool Botanic Garden in 1802. When Audubon arrived in Liverpool on a rainy July 21, 1826, over two months after setting sail from New Orleans, it was only a few days before his letters of introduction led him to the Rathbone brothers, Richard and William, owners of a vast cotton trading and shipping concern. The Rathbones took to Audubon and his birds from the start. Audubon was treated as a prized guest, and soon a good friend, by the entire family, including the widowed matriarch, Hannah Rathbone, whom Audubon fondly called "The Queen Bee," and her daughter, also named Hannah. Mrs. Rathbone presided over Greenbank, the Rathbone country house and grounds a few miles outside of Liverpool.

Through the Rathbones, Audubon met Roscoe, who arranged to have a public showing of his drawings at the Liverpool Royal Institute, Audubon's first exhibit in England, and the first step towards the launching of *The Birds of America* in Edinburgh six months later. William Roscoe had a vested interest in *The Winter's Wreath*. He wrote for it himself and was chief mentor for its young editor, William Chorley, also a close friend of the Rathbones. We can only speculate about the date when Audubon's essay was originally written. It was published for the 1829 volume of the *Wreath*, but the volume itself had been gathered (and printed) the year before, according to the date of Chorley's preface. Of course the date of publication is no proof of the original conception. Roscoe's daughter Jane wrote a sonnet in praise of Audubon in September 1826, less than two months after he had stepped ashore in Liverpool, but it was published for *The Winter's Wreath* six years later.

Audubon's essay could have been written as early as the winter of 1826–1827 when he began working on his accounts of the turkey buzzard and the Louisiana alligator, soon to appear in the *Edinburgh New Philosophical Journal*. More likely it was written just after Christmas 1827 when he returned to Liverpool after his resounding success in Edinburgh, and talked to Roscoe, who was recovering from a stroke. At any rate, it is of a piece with the turkey buzzard and the alligator essays, and the subsequent essays (on the rattlesnake and on his method of drawing) written, and read as papers, in Edinburgh at the same time the first plates of *The Birds of America* were being engraved and printed. Like *Journey up the Mississippi*, these essays took the form of letters, and they all aimed at demonstrating Audubon's credentials as a serious naturalist whose knowledge was based on years of firsthand observations in the wild.

Journey up the Mississippi reaches back to an earlier period in Audubon's life to make similar claims, though it insists more on his restless energies as an explorer and woodsman than on his budding gifts as a naturalist. During the summer of 1810, Audubon had moved with Lucy to Henderson, Kentucky, a small settlement on the Ohio, and opened a general store with his French partner Ferdinand Rozier. Later that winter, both Audubon and Rozier decided to stock a flatboat with merchandise, and scout out the possibilities of Ste. Geneviève, an old colonial French town on the Mississippi poised to profit from the opening up of the vast new Louisiana Territory.

Audubon's first weeks and months in Liverpool had encouraged him to embrace his frontier experience as a powerful feature of his art and public image. Arriving on the Liverpool docks after two months at sea, his face was tanned to a copper red and his hair hung down to his shoulders. Once the brilliancy of his foreign birds had produced the desired effect, his appearance and personality did the rest. He was doomed to make the most of his reputation as an American exotic.

For the most part Audubon accepted the challenge to gratify the gallery, and performed the role with a certain relish. But he was not acting, as he well knew.

His French accent was not contrived, though it could be useful in presenting himself as a genuine Louisiana Creole. He might weary of imitating bird songs, the barking of alligators, and the yell of an Indian war party, but he had not learned these on stage, just as he had not drawn his birds from stuffed specimens in a museum. In fact, the more he evoked the depths of the dark woods, the more he missed them. His private journals are filled with nostalgia for the world he has left behind. Following a talk one day with Mr. Rathbone about American trees, Audubon reflects in his journal that he had not yet seen a tree in England larger than a Louisiana sapling. After a long day of an English drizzle he thinks back on the torrential rains of Louisiana, missing the percussive fury of thunderstorms back home. Even the destructive floods of the Mississippi are preferable to the damp dullness of the English countryside.

The writing of his journal and the sporadic, but lengthy, letters to and from Lucy in Bayou Sara helped stoke his homesick memories and meditations. He advises Lucy to go horseback riding every day under the wild magnolias, and dreams about eating roasted ibis around a fire in the swamps after a day of hunting with his friends. But the impulse that led him to the writing of *Journey up the Mississippi* was mainly the desire to authenticate in print his rare personal knowledge of the living world his paintings depicted.

The months ahead proved that his instincts were right. Soon after his series of essays for the *Edinburgh New Philosophical Journal* was published in 1827 he received news from Philadelphia that his credibility as a naturalist had been attacked by friends and supporters of Alexander Wilson. These supporters were led by George Ord, a well-heeled local Philadelphia merchant (and self-taught scholar) who had set himself up as Wilson's patron. He had completed Wilson's nine-volume *American Ornithology* after Wilson's death in 1813. Ord had already snubbed Audubon when Audubon had arrived in Philadelphia in 1824 to show his drawings and muster support. Ord was determined to resist any challenge to Wilson's crown as America's foremost ornithologist, especially when the challenge came from an upstart French Creole.

When Audubon heard about Ord's aspersions from Thomas Sully, an artist himself and one of the few friends Audubon had made in Philadelphia, he immediately answered Sully in a long angry letter. He may have made some careless mistakes in the Edinburgh essays, but he would not allow his basic honesty and expertise to be doubted. He protested that for twenty years he had been "in the regular habit of writing down every day all the incidents of which I have been an eye-witness, on the spot, and without confiding to memory, as many travellers have done and still do." In Edinburgh Audubon could respond with the clear advantage of having been praised by some of the most notable scientific minds in England. He was no longer the isolated outsider, forced to thrust himself into the waiting rooms of the big houses and forbidding institutions. But Ord's jealous attack was a warning, and stung enough to goad Audubon into his master plan of converting his journals (there were thousands of worn and weather-stained pages) into the five-volume work of bird descriptions, the *Ornithological Biography,* that he began writing in 1830.

Journey up the Mississippi is best read as the beginning of the line of defense and self-justification the *Ornithological Biography* fulfilled. In Audubon's view, however, Ord and his cronies were annoyances to be slapped aside when necessary, like horseflies, but negligible details in the long history of art and nature, and the study of American birds. Audubon knew that his main competition came from the accomplishments of other naturalists, namely, as far as America was concerned, Alexander Wilson himself, and earlier trailblazers such as William Bartram. Audubon had met Wilson twice, and had studied the nine volumes of his *American Ornithology* with the care of a disciple intent on surpassing his master. When he had money, back in Henderson, he had bought the first six volumes and had inscribed them: "John James Audubon Louisiana 1816." The writing down of *Louisiana* is significant; it had become as automatic a reflex of self-identification as his own name. Audubon much later bought (or was given) volumes seven and eight of Wilson's work (but not, apparently, volume nine). For many years what he could not buy he borrowed, seeking them out as soon as they appeared. What he did not copy down in his journals, he committed to memory.

The same was true with William Bartram. Bartram's account of his four-year expedition in the 1770s through the southern colonies, *Travels through North and South Carolina, Georgia, East and West Florida,* had appeared in 1791 to tremendous acclaim. Audubon would have known of Bartram's reputation long before he read his book. Bartram's father, John, was a colonial legend, a plant collector and seed seller who had created a pioneering horticultural garden in Philadelphia, and who was roundly hailed as the founding father of American botany. William himself was born and raised near Philadelphia not far from the Audubon plantation at Mill Grove. Audubon would have heard about the Bartrams soon after his arrival from France in 1803. Twenty years later, thanks to his *Travels,* William Bartram had become a household name, at least in the households of the intellectually curious. His hyper-lyrical descriptions of the flora and fauna of the southern wilderness had inspired Coleridge and Wordsworth, who echo his very words in some of their major poetry.

Bartram died in 1823 in Philadelphia, just three years before Audubon arrived in England. For British readers, Bartram was the one figure whose writings Audubon's early essays were bound to evoke. Audubon knew he could not rival Bartram's feverish prose, but he could hoist himself up into an affiliation with both Bartram and Wilson by carrying on their work, and by extending the scope of their travels and revelations. This was especially the case as far as Louisiana was concerned. Both Bartram and Wilson had explored the edges of Louisiana, but as Audubon implies all through his essays and bird biographies, they had only touched on Louisiana's immensities of natural life. The proof was all those resplendent Louisiana birds Audubon had added to Wilson. When Audubon wrote *Journey up the Mississippi* he was writing with years of tracking through lower Louisiana packed in his memory. With the *Journey,* he could also prove that already as a young man he had penetrated into the barely charted regions of greater Louisiana in the same spirit of discovery displayed by Bartram and Wilson. Ord could rant and rave all he liked. Audubon's place was with his masters.

SOURCES

The quotation from the Sully letter is taken from Herrick's *Audubon the Naturalist: A History of His Life and Times* (1917, revised 1938). The text below is taken directly from the original edition of *The Winter's Wreath* for 1829 (see headnote above), thus the British spelling and the minor eccentricities of punctuation.

LOUISIANA BOUND

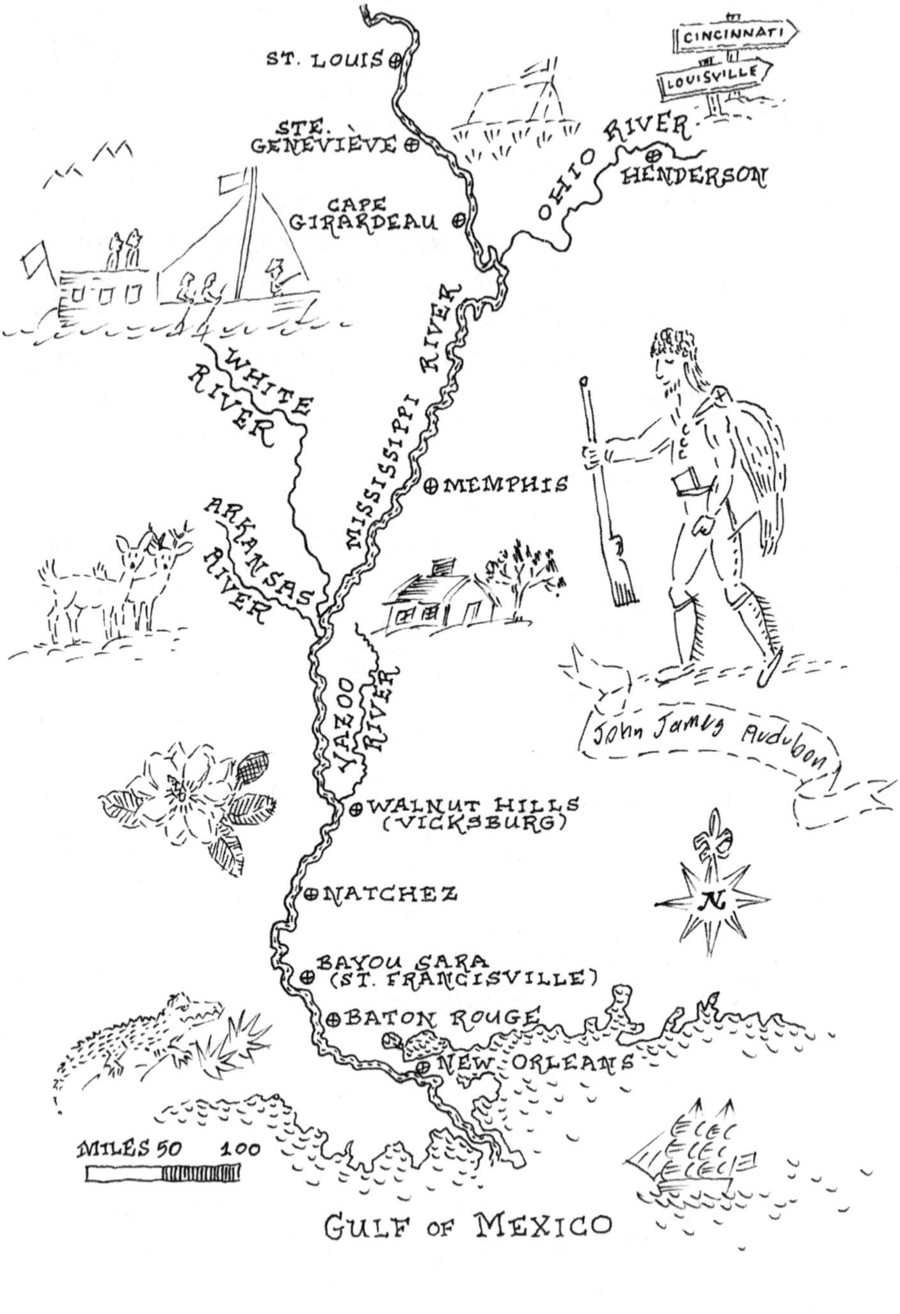

Journey up the Mississippi

Our late conversation upon the subject of Christmas holidays has induced me to write, for your amusement, my dear friend, an account of a Christmas which I spent some years ago in the land that I call my own—America,—my country.

About the end of December, some eighteen years ago, I left my family at a village near Henderson, in the lower part of Kentucky; being bound on an expedition to the upper parts of the Mississippi. I started with my friend F— in a vessel there termed a keel-boat;—an open boat, with a covered stern which forms the cabin, over which projects the slender trunk of some tree (about sixty feet long) as a steering oar; the boat being impelled by four oars worked in the bow, at the rate of about five miles an hour, going with the current. The banks of the Ohio were already very dreary;—indeed nothing green remained, except the hanging canes that here and there bordered its shores, and the few dingy grape leaves, which hardly invited the eye to glance towards them. We started in a heavy snow storm,—and our first night was indeed dismal; but as day began to appear, the storm ceased; and we found ourselves opposite the mouth of Cumberland River, which flows from the state of Tennessee, passing Nashville, and is a tolerably navigable stream for many hundred miles. Here the Ohio spreads to a considerable width, and forms in summer a truly magnificent river, and is even at this season broad and beautifully transparent, though so shallow that it is often fordable from the Illinois shore to Cumberland Island. Vast trees overhang both banks, and their immense masses of foliage are reflected in the clear mirror.

Ere long we passed the mouth of the Tennessee River and Fort Massacre, and could easily perceive that the severe and sudden frost, which had just set in, had closed all the small lakes and lagoons in the neighbourhood, as thousands of wild water-fowl were flying and settling themselves on the borders of the Ohio. Suffering our boat to drift with the stream, whenever large flocks approach us, we shot a great number of them.

About the third day of our journey we entered the mouth of Cash Creek, a very small stream, but at all times a sufficient deep and good harbour. Here I met a French Count, a celebrated traveller, bound like ourselves to St. Genevieve, Upper Louisiana (now the State of Missouri.) We soon learned that the Mississippi was covered by thick ice, and that it was therefore impossible to ascend it. Cash Creek is about six miles above the confluence of the Ohio and Mississippi. The stream flows from some hills to the northward of its mouth, which are covered with red and black oak, shumac and locust trees; and were formerly said to contain valuable minerals, of which they have been since proved to be totally destitute.—The point of land, between the Creek and the junction of the two rivers, is all alluvial and extremely rich soil, covered with heavy black walnut, ash and pecan trees, closely tangled canes and nettles, that are in summer at least six feet high. It is overflowed by both rivers during their freshes.

The Creek, now filled by the overplus of the Ohio abounded with fish of various sorts, and innumerable ducks, driven by winter to the south from the Polar Regions. Though the trees were entirely stripped of their verdure, I could not help raising my eyes towards their tops, and admiring their grandeur. The large sycamores with their white bark formed a lively contrast with the canes beneath them; and the thousands of parroquets, that came to roost in their hollow trunks at night, were to me objects of interest and curiosity. About fifty families of Shawanee Indians had moreover chosen this spot for an encampment, to reap the benefit of a good harvest of pecan nuts; and to hunt the innumerable deer, bears and racoons, which the same cause had congregated here. These were not the first natives, (for I cannot like many Europeans call them savages) that I had seen; I understood their habits and a few words of their language, and as many of them spoke French passably, I easily joined both their "talks" and their avocations.

An apparent sympathy connects those fond of the same pursuit, with a discernment almost intuitive, whatever be their nation;—all those hunters who loved fishing and pursuits of enterprize, ere long crowded round me; and as soon as they learned my anxiety for curiosities of natural history, they discovered the most gratifying anxiety to procure them for me. Even the squaws set small traps for the smaller animals;—and when, in return, I presented them with a knife, a pair of scissors, &c. they expressed their gratitude as gracefully as the most educated female would have done. My Friend F— neither hunter nor naturalist, sat in the boat all day, brooding, in gloomy silence, over the loss of time, &c. entailed by our detention. The Count kept a valuable journal, since published—hunted a great deal, and was as careless of the weather as myself; but his companion and father-in-law, like my partner, sat in his boat, pining with chagrin and ennui:—their case, however, was hopeless; here we were, and were forced to remain, until liberated by a thaw.

On the second morning after our arrival, I heard a movement in the Indian camp, and having hastily risen and dressed myself, I discovered that a canoe containing half a dozen squaws and as many hunters, was about to leave the Illinois for the Tennessee side of the river. I learned also that their object was to proceed to a large lake opposite, to which immense flocks of swans resorted every morning. These flocks are so numerous and strong, that it is, however incredible it may at first seem, a well-known fact, that they keep the lakes which they frequent open, merely by swimming upon them night and day.—Having obtained permission to join the party, I seated myself in the canoe, well supplied with ammunition and a bottle of whiskey,—in a few moments the paddles were at work, and we swiftly crossed to the opposite shore. I was not much astonished, during our passage, to see all the labour of paddling performed by the squaws; for this feature of Indian manners was not new to me; but I was surprised to see that upon entering the canoe, the hunters laid down, and positively slept during the whole passage. On landing, the squaws, after securing the boat, proceeded to search for nuts, while the *gentlemen* hunters made the best of their way, through the "*thick and thin,*" to the lake.

Those who have never seen anything of what I call "thick and thin," may perhaps think I allude to something like the furze which covers some of the moors of Scotland,—but they must imagine the shores of the Ohio, at its junction with the great muddy river called the Mississippi, to be fairly overgrown with a kind of thick-set cotton trees, that rise as closely from the muddy soil of the bank as can well be conceived—they are not to be beaten down: you must slide yourself between them,—and in summer you have a pretty task to keep off the musquitos that abound amongst them. After these thickets there are small nasty lagoons, which you must either swim across, jump over, or leap into and be drowned, according to your taste or capability;—but when the task of reaching the lake is accomplished—what a feast for a sportsman! There they lie, by hundreds, of a white or rich cream colour—either dipping their black bills in the water; or leaning backwards, and gently resting with one leg expanded, floating along and basking in the sunshine. The moment that these beautiful birds saw our videttes, they started up in immediate apprehension:—but the plan of our Indians drove the poor swans nearer to their fate, the farther they retreated from either shore. Men were placed behind the trees, who knew how to take a dead aim, and every shot told. Being divided, three on one side and four on the other, the moment that one party had driven the swans towards the other,—the former hid themselves; and when the birds flew from the fire of the latter, they alighted within good distance of those who had first alarmed them. What would those English *sportsmen* who, after walking a whole day, and exploding a pound of powder, march home in great glee, holding a partridge by the legs, with a smile on their lips and a very empty stomach,—say to this day's devastation amongst the swans? I saw these beautiful birds floating on the water, their backs downwards, their heads under the surface, and their legs in the air, struggling in the last agonies of life, to the number of at least fifty—their beautiful skins all intended for the ladies of Europe.

The sport was now over;—the sun was nearly even with the tops of the trees—a conch was sounded, and after a while the squaws appeared, dragging the canoe, and moving about in quest of the dead game. It was at last all transported to the river's edge, and we were landed upon the Illinois bank again before dark. The fires were lighted—each man eat his mess of pecan nuts and bear's fat, and then stretched himself out, with feet close to the small heap of coal intended for the night. The females then began their work;—it was their duty to skin the birds. I observed them for some time, and then retired to rest, very well satisfied with the sports of this day—the 25th of December.

On the following morning I found that a squaw had given birth to beautiful twins during the night—she was at work, tanning deer-skins, She had cut two vines at the roots of opposite trees, which, having their upper branches twined in the tops of the trees, made a kind of swing; and framed a cradle of bark, in which the infants were swung to and fro by a gentle push of her hand,—from time to time, she gave them the breast, and to all appearance seemed as unconcerned as if nothing had taken place. What a difference between this Indian mother and a lady of fashion!

An Indian camp upon a hunting expedition is not, I assure you, a place of idleness; —and although the men do little more than hunt, they pursue this task with a degree of eagerness bordering upon enthusiasm. One of their party, a tall and robust man, assured us one morning that he would have some good sport that day, as he had found the *gite* of a bear of some size, and wished to combat him singly;—we all started with him, to see him fulfill his bold promise. When we had gone about half a mile from the camp, he said he discerned the bear's track, although I could positively perceive nothing; and he went on, rambling through the thick cane brake, until we reached a large decaying log of timber, of an immense size:—in this he said that the bear was concealed. I have rarely seen a finer object than this Indian, at the moment when he prepared to encounter his prey. His eyes sparkled with joy—the rusty blanket was thrown in an instant from his shoulders—his brawny arms seemed swelling with the blood that rushed through their prominent veins, and he drew his scalping knife with a fantastic gesture that plainly declared *la guerre à l'outrance*. He ordered me to mount a delicate sapling, which would, he said, be secure from the bear, who could easily ascend a larger tree with the activity of a squirrel; whilst the other two Indians stood at the entrance of the hollow log, which the hero entered with the most resolute determination. All was still for some minutes;—he then emerged, and said the bear was slain, and that I could safely descend. His companions entered the log, and having tied the animal to a long vine, which they had cut, our united strength drew him out. This exploit was in fact less dangerous than it appeared, for the bear, when attacked in a confined spot like the trunk of a hollow tree, makes no resistance, but retires further and further back, until he is killed. As we returned to the camp, one of our Indians broke the twigs in our way from time to time, and on our reaching the camp, two squaws were sent on the track of the broken twigs, who returned at night with the flesh and skin of the animal.

The nuts were soon nearly all gathered; and I began to perceive that the game must be getting scarce, as the hunters remained in the camp, during the greater part of the day. At last, one morning, they packed up their moveables, destroyed their abodes, and put off in their canoes down the Mississippi for the little Prairie, —bent on moving towards the Arkansas. Their example made us desirous of moving; and I set off with two of the crew to cross the bend of the river, and ascertain if the ice still remained too solid to allow us to proceed. The weather was milder: and on reaching the Mississippi, I found the ice so much sunk as to be scarcely discernible above the water; and I toiled along the muddy shore,—my fellows keeping about fifty yards behind me, until I reached Cape Girardeau. After calling for some time loudly for a boat, we saw a canoe put off from the opposite shore. When it reached us, a stout dark coloured man leaped on shore, who said his name was Lorimie, the son of the Spanish Governor of Louisiana. Being a good pilot, he undertook with six stout men of his own, in addition to our four hands, to bring our boat up, and the bargain was soon arranged. His canoe was hauled into the woods, some blaze was made on the surrounding trees; and he then took us by a direct route through the woods back to Cash-Creek, in about one third of the time I had occupied in coming, and ten times more comfortably. The night was spent in preparations; in making towing ropes of bullocks' hides, and cutting good oars; and at daylight we left Cash Creek, to embark on wider waters.

Going down the stream, to the mouth of the Ohio, was fine sport, and my friend F— thought himself near the end of the journey;—but alas!—when we turned the point, and began to ascend the Mississippi, we had to stem a current of three miles an hour, and to encounter ice which, although sunk, much impeded our progress. The patroon, as the director of the boat's crew is termed, got on shore; and it became the duty of every man to *haul the Cordelle*—viz. to tow the boat by a rope fastened to a pole in the bow, leaving only one man in her to steer. This was slow and heavy work; and we only advanced seven miles, during the whole day, up the famous Mississippi. On the approach of night, our crew camped on the bank; and having made a tremendous fire, we all eat and drank like men that had worked hard, and went to sleep in a few moments. We started the next morning two hours before daybreak, and made about a mile an hour against the current; our sail lying useless, as the wind was contrary. This night we camped out as before;—and another; and after that, a following day finding us at the same work, with very little progress, and the frost becoming quite, severe again, our patroon put us into winter quarters in the great bend of Tawapatee Bottom.

What a place for winter quarters! Not a white man's cabin within twenty miles on the other side of the river, and on our own, none within at least fifty! A regular camp was raised—trees cut down, and a cabin erected, in less time than a native of Europe would think possible. In the search for objects of natural history, I rambled through the deep forests, and soon knew all the Indian passes and lakes in the neighbourhood. The natives, by some intuitive faculty, discover an encampment of this kind, almost as quickly as a flight of vultures find a dead deer; and I soon met some strolling in the woods on the look out. Their numbers gradually increased; and in about a week, several of these unfortunate rambling beings were around us. Some were Osages, but the greater part were Shawanees. The former were athletic, robust, well-formed men, of a nobler aspect than the others, from whom they kept apart. They hunted nothing but larger game—the few elks that remain in the country, and one or two buffaloes were all that they paid attention to. The latter were more reduced, or rather harder pressed upon by the whites; they condescended to kill opossums and even wild turkeys for their subsistence. Though I was often amongst the Osages, and very anxious to observe their manners, as they were a race new to me, yet as they spoke no French and very little English, I could hardly get acquainted with them, being ignorant of their language. They were delighted to see me draw; and when I made a tolerable portrait of one of them in red chalk, the others, to my astonishment, laughed to an excess. They bore the cold better than the Shawanees, and were more expert in the use of their bows and arrows.

Our time passed away;—after hunting all day with a young Kentuckian of our party, he would join me at night to chase the wolves that were prowling on the ice—crossing the river to and fro, howling, and sneaking about the very camp for the bones which we threw away. Meanwhile I studied the habits of the wild turkeys, bears, cougars, racoons, and many other animals;—and I drew, more or less every day, by the side of our great fire. I will try to give you some idea of a great fire at a camp of this sort in the woods of America. Just before evening the axe-men tumble down four or five trees—probably ash, about three feet in diameter, and sixty feet to the first branches, or as we call them, the limbs. These are again cut into logs, of about ten feet in length, and, with the assistance of strong sticks, are rolled together, into a heap several feet high. A fire is made at the top, with brush-wood and dry leaves, kindled by a flint and steel; and in the course of an hour, there is a flame that would roast you at the distance of five paces:—under the smoke of this the party go to sleep. It happened, on the only night that my friend F— slept on shore, that being very chill, he drew himself so close to the fire, that the side of his face, which lay uppermost, was fairly singed, and he lost one of his whiskers. We all laughed at this;—but it was no joke to him, and he shaved off the remaining whisker very ruefully the next morning.

We remained here six weeks:—we had plenty of company from our Indian friends, with whiskey and food in abundance; but our stock of bread began to give way, and we got tired of using the breasts of wild turkeys for bread, and bear oil instead of meat. The racoons and opossums, however tender, were at last disliked: and it was decided one morning that I and my Kentuckian friend should cross the bend, to procure some Indian corn-meal, and have it dragged down by men on skates, or otherwise. I was no novice in the woods; and my companion bound on his mocassins with great glee, and told me to come onwards,—and I followed his steps, until, meeting a herd of deer, we pursued them, tracking them with great ease through the snow. I shot one; and as we did not know what to do with it, we hung it on a tree; and, after marking the place, resumed our course. We walked on till nearly dark, but no river was seen. My friend urged me forward, and I still followed him; knowing very well that the business would end at last in supping on an opossum,—when we suddenly struck upon two tracks, which I took for those of Indians. He said that they would guide us to the river; and we followed them, until at an opening, I saw the wished-for Mississippi;—but many *shoe-tracks* were visible and I began to get alarmed. My friend still kept up his spirits, until at length we arrived at—our own encampment! The boatmen laughed, and the Indians joined in the chorus:—we eat a racoon supper, and were soon after refreshed by sleep. This was a raw expedition; yet nothing was more natural than that it should happen to those not perfectly acquainted with the woods. They start—form a circle, and return to the point which they left at first. I cannot account for this: but the same thing has often occurred to me in my early hunting excursions. Of which the following is an instance.

One day in the winter, I had been shooting ducks upon a lake at no great distance from home. It was rather late when I discontinued my sport, and turned my steps homewards, well laden with the birds, which were suspended from a belt around my waist. Soon after I had entered the woods which I had to traverse, it began to snow; and as I was therefore desirous to reach home as soon as possible, I unburdened myself of about half of my game, laying it upon a stump of a newly-felled tree. Thus lightened, I set off at a quick pace;—the snow continuing to fall, I was no longer able to see my way; but pushed on, as I thought, in a straight line; until, after about two hours of smart walking, I arrived at a felled tree, upon the stump of which lay about a dozen dead ducks. It did not for a moment occur to me that these could be the birds of which I had divested myself two hours before; and so, thinking that they had been deposited there by some other hunter, I set forth again; wondering, however, that I was so long in reaching home, which I knew to be not more than ten miles distant from the place where I had been shooting. After walking as hard as I could for about an hour longer, to my disagreeable surprise, I came upon the ducks again! This second meeting induced me to examine them narrowly; when I discovered that they were my own game; and that I had actually been walking all this time in a circle; although, on leaving them, I had taken different directions each time, I had returned, as it were by magic, to the self-same spot from which I had first departed! It was now night:—I saw that it would be useless to attempt reaching home before morning, and prepared myself for camping out beside my ducks. I lighted a fire, and passed the night very comfortably sheltered by the snow, which I scooped out so as to form a very good *gite.* The following morning, I started at daylight, and in about half an hour reached my home, which was not more than three miles distant from the place where I had passed the night.

My friend and I were not to be thus defeated; we moved off, as soon as day broke, without mentioning our intentions, taking our guns and my dog, in search of the opposite side of the bend. This time, luckily, we pushed straight across;—neither the innumerable flocks of turkeys nor the herds of deer stopped us, until we saw Cape Girardeau, about an hour before sunset. On reaching the river, we called in vain for a boat—the ice was running swiftly down the stream, and none dared put out. A small abandoned hut stood close to us, and we made it our home for the night; and our evening meal was principally composed of a pumpkin that had withstood the frost. With a gun and a little powder we soon kindled a fire and lighted some broken branches—we fed the flames with the boards of the abandoned house, and went to sleep very comfortably;—what a different life from the one I lead now! And yet that very evening I wrote the day's occurrences in my journal before going to sleep—just as I do now; and I well remember, that I gained more information that evening about the roosting of the Prairie hen, than I had ever done before.

Daylight returned, fair and frosty:—the trees, covered with snow and icicles, became so brilliant when the sun rose, that the wild turkeys quite dazzled, preferred walking under them to flying amongst their glittering branches. After hailing the opposite shore for some time, we perceived a canoe picking its way towards us through the floating ice;—it arrived, and we soon told the boatmen our wishes to procure some bread or flour. They returned, after having been absent nearly the whole day, bringing us a barrel of flour, several large loaves, and a bag of Indian corn-meal. The flour was rolled high on the bank; we thrust our gun-barrels through the loaves, and having hung the bag of Indian corn-meal on a tree, to preserve it from the wild hogs, we marched for our camp, which we reached about midnight. Four of our men were sent with axes, who formed a small sledge, on which they placed the precious cargo, and hauled it safely to the camp, over the snow.

The river having risen slowly and regularly, as the Mississippi always does, now began to subside; the ice, falling with the water, prepared fresh trouble for us; and in order to keep the boat afloat, it was thought prudent to unload the cargo. It took us two days, with the assistance of the Indian women, to pile our goods safely on the shore, and to protect them from the weather. For the security of the boat, we cut down some strong trees, with which we framed a kind of jetty, a little higher up the stream, to ward off the ice, which was rapidly accumulating. Being now fairly settled in our winter quarters, we spent our time very merrily; and so many deer, bears, and wild turkeys suffered in our hunting parties, that the trees around our camp looked like butchers' stalls, being hung round with fat venison, &c. Moreover we soon found that the lakes contained abundance of excellent fish; and many of us would walk over the ice with axes, and whenever a trout, pike, or cat-fish rose immediately beneath it, a severe blow on the ice killed the fish, which the hunter secured by opening a hole with the axe. We used also sometimes to make a large hole in the ice, several feet in diameter; the fish, in search of air, resorted to it from different quarters, and were shot as they appeared on the surface of the water. The squaws tanned the deer-skins, stretched those of the racoons and otters, and made baskets of canes; my friend played tolerably on the violin; I had a flute—and our music found pleased hearers, whilst our men danced to the tunes, and the squaws laughed heartily at our merriment. The Indian hunters formed the outer ring of our auditory—smoking their tomahawk pipes with a degree of composure, which no white man ever displayed at such merry-makings. After we had passed six weeks in this manner, the river began to fall very much,—the ice was heaped along both shores, so that a narrow channel alone appeared clear; and at last our patroon said, that this was the time to depart for Cape Girardeau. All was bustle—the cargo was once more put on board;—our camp was abandoned, and the Indians and we parted like brethren.

Our navigation was now of the most dangerous nature; the boat was propelled by pushing with long poles against the ice, or the bottom, whenever it could be touched;—and we moved extremely slowly. The ice on each side was higher than our heads;—and I frequently thought that if a sudden thaw had taken place, we should have been in a dangerous predicament indeed;—but good fortune assisted us, and at length we reached the famous Cape.

The little village of Cape Girardeau contained nothing remarkable or interesting except Mr. Lorimie, the father of our patroon, who was indeed an original, and the representative of a class of men now fast disappearing from the face of the earth. His portrait is so striking and well worth preserving, that I shall attempt to draw it.

Imagine a man not exceeding four feet six inches in height, and thin in proportion, looking as if he had just been shot out of a pop-gun. He had a spare, meagre countenance, in which his nose formed decidedly the most prominent feature. It was a true *nez à la Grand Fréderic*—a tremendous promontory, full three inches in length—hooked like a hawk's beak, and garnished with a pair of eyes resembling those of an eagle. His hair was plastered close to his head with a quantity of pomatum; and behind he wore a long *queue* rolled up in a dirty ribbon, which hung down below his waist. The upper part of his dress was European, and had evidently once been made of the richest materials; and though now wofully patched and dilapidated, you might still observe here and there shreds of gold and silver lace adhering to the worn apparel. His waistcoat, of a fashion as antique as that of his nose, had immense flaps or pockets that covered more than one half of his lower garments. These latter were of a description totally at variance with the upper part of his costume. They were of dressed buck-skin, fitting tight to his attenuated limbs, and ornamented with large iron buckles at the knees, which served to attach and support a pair of Indian hunting gaiters that had, like the rest of his dress, seen long and hard-service. To complete his costume, he wore on his feet a pair of mocassins, or Indian shoes, that were really of most beautiful workmanship. These articles of dress, together with his small stature and singular features, rendered his appearance, at a little distance, the most ludicrous caricature that can be imagined; but upon approaching nearer, and conversing with him, his manners were found to be courteous and polished. He had been, as I before mentioned, the governor of Louisiana, while it was in the possession of the Spaniards; when this country was purchased by the Government of the United States, he retired to this little village, where he was looked upon as a great general, and held in the highest esteem and consideration by all the inhabitants.

We decided not to remain here; and our patroon urging us to proceed to St. Genevieve, we moved once more between the ice, and arrived in a few days at the Grand Tower; an immense rock detached from the shore, around which the current rushes with great violence. Our *cordelles* were used to force a passage at this dangerous spot; and our men, clinging to the rock as well as they could, looked as if each movement would plunge them into the abyss—but we passed on without accident. All this night, we heard the continual howling of the wolves, amidst the heavy woods that covered the large hills on the Illinois shore, opposite to this rock. From what I know of their habits, I am convinced that they were hunting deer in the following manner. They hunt in packs, like dogs,—but with far more sagacity and contrivance. They divide themselves into separate bodies; some to rouse the game, and others to waylay them. The pack that is on the hunt starts one or more deer, following them with a note, like that of hounds in full cry,—and drives the game before it towards the wolves posted in ambush. These wolves, when the deer pass, start up fresh, and following their prey, soon overtake it;—and it is well known that a cry is uttered as a signal for assembling at the death of the game, somewhat like the death-note of the hunter's bugle.

We arrived safely at St. Genevieve, and concluded the object of our adventure very satisfactorily. St. Genevieve was then an old French town—small and dirty; and I far preferred the time I spent in Tawapatee Bottom to my sojourn here. Having arranged my affairs, I waited only for a thaw to return home. The ice broke at last; and bidding my companions good-bye, I whistled to my dog—crossed the Mississippi, and in a few hours was on my road, on foot and alone, bent on reaching Shawanee Town as soon as possible. I had little foreseen the nature of the task before me,—as soon as I had left the lands, on reaching the Prairies, I found them covered with water, like large seas: however nothing could induce me to return, and my ardent desire to rejoin my wife and family made me careless of inconvenience or fatigue. Unfortunately I had no shoes;—and my mocassins constantly slipped, which made the wading very irksome. Nevertheless, on the first day I made forty-five miles, and swam the Muddy River. I saw only two cabins during the whole day; but I had great pleasure in observing the herds of deer that were crossing the Prairies, as well as myself, ancle-deep in water. Their graceful motions, and their tails spread to the breeze, were discernible for many miles. With the exception of these beautiful animals, and the thousands of buffalo-skulls that lay scattered about, just appearing above the water, which was about a foot deep—there was nothing remarkable at this season; but in Spring, about the month of May, the Prairies are indeed a garden. The grass, rich and succulent, shoots from the soil with incredible rapidity; and amongst its green carpeting, millions of variegated flowers raise their odoriferous heads. Butterflies of the richest colours hover about in the sunshine, and the humming-bird darts swiftly along, gathering honey, amongst clouds of bees. The deer are quietly reposing upon the luxuriant herbage, in picturesque groups, and the flocks of the squatters are seen scattered about in all directions. The weather is mild, the sky cloudless; and nothing can be conceived more delightful than travelling over these fertile regions at this season. Yet they are infested by one scourge—the Buffalo-gnats. These insects fly in dense bodies, compacted together like swarms of bees, as swift as the wind. They attack a deer or buffalo—alight upon it, and torture the animal to death in a few minutes. This may appear incredible, till we recollect that the swarms are so dense that above a hundred will often alight upon a square inch. I had myself an opportunity of witnessing their fatal power, when I crossed the same prairies in the May following the very Christmas of which I am writing. I was mounted upon a fine horse; and in consequence of the advice of experienced persons, I had his head and body wholly cloathed with light linen, to protect him from these gnats—leaving only the nostrils uncovered. Being unaware of the full extent of the danger, I was not, as it proved, sufficiently careful in joining the different cloths which covered my horse. I had ridden a considerable distance, when, on a sudden, he actually began to dance; he snorted, leaped, and almost flew from under me. This took place near the Big Muddy River, for which I instantly made, and plunged the horse into the stream to quiet him;—but upon reaching the shore, his motions were languid—his head drooped—and it was with difficulty that I reached a squatter's hut, where the poor animal died in a few hours. He had been bitten between the joinings of his body-cloths by a swarm of these remorseless insects, whose bite is invariably fatal, whenever they can settle upon the body of an animal in any number. They do not attack the human species;—and it is only during the heat of the day that they appear,—at which time the cattle in the prairies resort to the woods for security. The deer rush to the water to avoid them, and stand, during the mid-day heats, with only their noses appearing above the surface.

A light smoke arising from the trees which covered a beautiful mound, promised me a good dinner, and gave me an appetite;—and I made straight for it. The woman of the house which stood there received me kindly; and, whilst the boys were busied in examining my handsome double-barrelled gun, as I sat drying my clothes by the fire, the daughter ground coffee, fried venison, and prepared eggs; which, washed down by a good glass of brandy, formed a sumptuous repast. To those who, used to the ceremonies of cities, have no idea how soon an acquaintance is cemented in these wilds by the broad ties of hospitality—it would have been a matter of surprise to see how, though we were previously strangers, we became in an hour as familiar as if we had been friends for years.

I slept at this hospitable dwelling;—and the kind hostess was stirring at day-break to get me a good breakfast before I started;—of course for all this she would receive no recompence, so I gave each of the boys a horn of powder—a rare and valuable article to a squatter in those days. My way lay through woods, and many cross roads that intersected them embarrassed me much; but I marched on; and, according to my computation, I had left about forty-five miles behind me at night-fall. I found a party of Indians encamped by the edge of a cane brake; and having asked in French permission to pass the night with them, my request was granted. My bed was soon prepared, in which, after eating some supper, I was ere long fast sleep. On awakening the next morning, I found, to my surprise, that all the Indians were gone, with their guns—leaving only two dogs to guard the camp from the wolves. I was now not above forty miles from Shawanee; and my dog, who knew very well that he was near home, seemed as happy as myself. I did not meet a single person the whole day, and not a cabin was then to be found on that road. At four the same evening I passed the first salt-well; and half an hour brought me to the village. At the Inn, I was met by several of my friends, who had come to purchase salt; and here I slept,—forty-seven miles from home.—The next day, to my great joy, brought me to my family:—and thus ended this pleasant excursion. Now confess, my dear friend, were not these rare Christmas doings?

LOUISIANA JOURNAL

Without any money my talents are to be my support, and my enthusiasm my guide in my difficulties, the whole of which I am ready . . . to keep, and to surmount.

—*Louisiana Journal*, October 12, 1820

The *Louisiana Journal* is the most rending self-portrait Audubon committed to ink. Beginning from the time he left Pennsylvania for Kentucky with Lucy in 1808, he wrote dozens of diaries over the years. When he was forty-one years old he sat down and counted twenty-six, and he was far from finished. Most of them were eventually lost or destroyed. The few early diaries Lucy saved were published in some form or another by Lucy herself, in the texts she sent Robert Buchanan for his authorized biography (and anthology) in 1868, and by Audubon's granddaughter, Maria Rebecca Audubon, in what she intended to be the definitive collection of her grandfather's writings thirty years later. Under their watchful eyes, these diaries were heavily edited and to some extent rewritten, but at least they were published.

Aside from a few carefully handpicked selections in the Buchanan biography, the *Louisiana Journal* was too close to the bone to be exposed to the public, but too cherished a memory to be abandoned. After Audubon's death the carefully stored manuscript had surely become hallowed as a family testament, a raw record of Audubon's solitary struggle to reverse the downward spiral of his life by a combination of sheer will and artistic faith, driving him day after day for over a year through a backbreaking ordeal that would have brought most strong men to their knees.

When he decided, with Lucy's approval, to head down the Mississippi and devote himself to turning his portfolio of birds into the great book he had dreamed about for years, he had reached a standstill on a dead end in Cincinnati. After ten thriving years on the Kentucky frontier, a series of bad investments forced him to leave Henderson in 1819 for Louisville, where he was briefly jailed for debts and forced to declare bankruptcy. For years he would continue to be haunted by the threat of creditors seeking him out. In Louisville he was able to barely scrape by, painting portraits and giving drawing lessons. Lucy did have her brother Tom and her sister Eliza (who had married a wealthy businessman, Nicholas Berthoud), both of whom had settled nearby, but neither Lucy nor her husband was capable of living on charity.

Audubon soon moved to Cincinnati, where he found a job as a taxidermist in the Western Museum of Cincinnati College. Lucy and the two boys joined him, but hard times followed the family like a dark cloud. Audubon had to beg for his salary, and found himself once more trapped in a hopeless routine of teaching and painting portraits. With two young sons to support, Lucy and Audubon made the decision to risk everything on the grand design. Audubon would take his portfolio down the Mississippi, and spend what he first intended to be seven months adding new birds and expanding his range of ornithological knowledge. Lucy would stay behind with the boys, taking on private students to provide the minimum for food and household expenses.

The seven months stretched out to fourteen, and at the end of the journal Lucy and the boys had joined Audubon in Louisiana, where Lucy would live (and teach) for almost a decade. Thus when she held the pages of the *Louisiana Journal* in her hands, years later, without having to reread what she knew by heart, she could respond to every misspelled word, every botched and blotted line, every defiant flourish of Audubon's fine cursive handwriting; she knew the story to a syllable, and she could never forget what had been at stake, and all the battles and humiliations that she and her husband had endured and overcome.

For Audubon, thirty-five years old and endowed with remarkable natural vigor and resistance, hard physical work was the least of his troubles. To pay for his trip down the Ohio and the Mississippi into southern Louisiana, he had willingly hired himself out to a flatboat captain as a hunter to supply food for the crew. He knew what he had to do and did it well. Rowing a skiff from flatboat to riverbank, he tramped through the canebrakes, stalked and killed deer and other game, gutted them, skinned them, hauled them back for the cook. When game was scarce, he threw out trotlines for fish. There were always catfish, though it took time and effort to pull their skin off with a pair of pincers.

The hunting and the fishing every day did not faze him; neither did the shooting of birds and the hours drawing them on the deck despite the pitch and roll of the boat. None of these activities was of a nature to worry Lucy and restrain her from printing more than a few censored pages of the diary after Audubon's death.

She was used to life on the frontier. What did hold her back were surely the waves of self-doubt that Audubon could not suppress, even when he did his best to reassure both himself and his family. These were too private, and too painful, for her (or her granddaughter) to publish or to rewrite.

Audubon seldom admitted defeat, and could usually overcome his frustrations by stepping aside from his cares and forgetting everything but the immediate wonders of the woods. The *Louisiana Journal* is often brightened with the old youthful Audubon exuberance and self-delighted discovery. A cloud formation, a sunrise, or a hawk in flight could stop him in his tracks, even when he was too tired to think. After a wet night wrapped in a bearskin he was still able to wake up in awe at finding himself alive in three dimensions, on his way to the promised land on a boat. But he could not hide from Lucy, and did not want to hide, in a diary that is a serial letter disguised as a chronicle, that he was often on the brink of despair. He had waged everything on this voyage, and he knew when he waved farewell to Lucy and the boys that Louisiana was his last chance.

Before the birds and magnolia woods of Oakley worked their spell, the *Journal* veers uneasily from one mood to another, with Audubon never fully able to shake the fear of having drifted down the Mississippi towards a doom even darker than the poverty and disgrace he had left behind in Kentucky. The glowing vision of Louisiana as a bright new world of hope and glory might turn out to be a mere mirage, a crazy Frenchman's *feu follet* in a lonely swamp.

Audubon was not a man to give in to fate, but on this trip thoughts of suicide crossed his mind several times. Only after he had fought through the trials of New Orleans, a city he hated for testing his spirit to the breaking point, was his battered faith in Louisiana rewarded by the security of Oakley and the four months reaping the natural bounty of the Felicianas. When the mistress of Oakley, Lucretia Pirrie, forced Audubon to leave (he relates the full story in the *Journal*), he did have to return to New Orleans to resume his ill-paid lessons and portrait paintings. New trials and struggles had to be faced and overcome. One jealous painter accused Audubon of being a fraud who had bought the portfolio of birds and was passing them off as his own. Audubon challenged him to a duel and never saw him again. Still, the Oakley episode had breathed new life into his art and confidence. The Louisiana birds he had discovered and painted with increasing skill and authority became the heart of *Birds of America,* the standard by which he judged all his subsequent work.

Obviously the note of self-redemption and reconciliation that rings through the conclusion of the *Journal* was enough to give it special value for all the Audubons. He had held at bay his inner demons, rediscovered his powers and enthusiasm as an artist, and had found a new home for his family. But for Audubon and Lucy, rereading the *Journal* much later, after more years of separation and sacrifice, it signaled the original breakthrough behind the making of *Birds of America*. It would not have been lost on either of them that Louisiana, in one year, had made the future possible. The *Journal* contains the best argument for Audubon's insistence on announcing himself as a native of Louisiana long after he had left it behind for good. After that year he could replace the daydream of his origins with the reality of rebirth. Louisiana was no longer a cradle in the clouds, but the solid crucible that had changed his life, thus Lucy's holding on to the fading manuscript, complete and untouched, as long as she lived.

Since Audubon is writing the journal for himself, and his immediate family, and not for publication, he relies on a body of knowledge, and personal history, that today's reader has to divine between the lines and work out for himself. This is not all that difficult. The day-by-day recording generates its own steady narrative flow, and Audubon's private reflections, repeated and expanded, transform the dry lists of birds and routine chores into a compulsive, and often desperate, autobiography that is hard to resist. Audubon can be impetuous and passionate, especially when he has been spurned or misjudged, but spontaneous self-combustion is one of the charms of his uncensored voice. Once he plunges deeper into Louisiana, after stopping in Natchez, Audubon begins to take more notice of individuals whose social prestige and success might help him further his own plans. The sketches he makes of them, and of the human comedy in general, at times swell to full-scale portraits or episodes, and give us some of the most fascinating pages of the journal. In most cases, there is more than enough context and detail to keep the reader on track, no matter how minor or little known the individual may be in Louisiana history.

At the outset of the journal, however, it does help to piece together a few basic details about the organization of the initial river trip and the main figures involved. There were two flatboats leaving Cincinnati together, both carrying cargo to be floated down the Ohio and the Mississippi to New Orleans. By the 1820s steamboats were by far the quickest way up and down both rivers, but renting shipping space on a steamboat came with a high cost. The cargo on both the flatboats was owned by a Mister Shaw, who himself came along as a passenger. One of the flatboats was captained by a Mister Lovelace, the other, by Jacob Aumack. While on the Ohio the flatboats were lashed together to make it easier on the two crews. Once they reached the Mississippi they were unlashed, allowing the two captains to search out the strongest current, sometimes racing each other just to break the boredom. When Audubon hired himself out as a hunter on the Aumack boat, he brought along his assistant, the young Joseph Mason, who had been Audubon's best student in Cincinnati. Mason would stay with Audubon for almost two years, up to August 1822, painting most of the superb floral backgrounds for the Louisiana birds.

One other passenger onboard shared a professional interest in the world of nature, and like Audubon was taking careful notes for a book. In fact he made a name for himself long before Audubon. He was Samuel Cummings, who became Audubon's closest companion on the trip, and a good friend. (When the flatboats approached Henderson, Audubon asked Cummings and Mason to go ashore and fetch Dash, a hunting dog Audubon had left behind. Audubon had decided to keep a low profile in the region of his creditors.) Cummings was a civil engineer making a survey of the Ohio and Mississippi Rivers. The first edition of his guide was published in 1822 with the title *The Western Navigator.* In 1825 a new revised edition came out as *The Western Pilot* and was often reprinted.

Cummings probably brought along the latest edition of an earlier river guide, the *Navigator,* by Zadok Cramer. Before Cummings, the *Navigator* was widely consulted as the standard authority, and would have been required reading on the flatboat whether Cummings brought his own copy or not. Certainly, it was one of the few books definitely onboard. Audubon refers to it several times. When he identified the islands by numbers, they were the numbers Cramer had assigned. He also read Cramer for the stories associated with places on the two rivers, and for the historical sketches and essays at the end of the book. Cramer had trained his narrative talents by putting together a number of popular almanacs before he wrote the *Navigator.* Audubon would have found accounts of the Lewis and Clark expedition, the Dunbar-Hunter expedition into southern Louisiana, anecdotes about Daniel Boone, quotations from Thomas Jefferson and William Bartram, descriptions of trees and plants, even a fairly thorough history of early Louisiana. This was the kind of incidental detail that Audubon enjoyed and made sure to provide when he began writing his bird biographies ten years later.

The only book we know for certain he brought himself was the first volume of William Turton's six-volume English translation (1802–1806) of *The General System of Nature (Systema Naturae)* by Carl Linnaeus, familiarly known as Turton's Linné. The first volume contained the section on birds. Early in the eighteenth century Linnaeus, writing in Latin, had revolutionized the scientific classifications of plants and animals. Since one of Audubon's perennial concerns was to be accepted as both an artist and an ornithologist, he needed Linnaeus as a stamp of credibility. He had discovered the Turton translation by reading the first volumes of Alexander Wilson's nine-volume *American Ornithology* (1808–1814), and knew he could set himself up as a peer and a rival to Wilson (an ambition that had become an obsession) only by absorbing his learning and references. Wilson was the first ornithologist to use the Linnaean system in the United States.

By the time of Audubon's trip, Wilson's reputation as the foremost American ornithologist and painter of birds was well established. Audubon had first met him in 1810 in Louisville after Wilson had floated to a stop in a one-man skiff on the Ohio on his way to the Mississippi (like Audubon ten years later), scouting for subscribers to his book, and after selling the skiff, searching for new birds mainly on foot and on horseback all the way down to New Orleans. At the time, Audubon had not yet hatched his plan for his own book. Still, the meeting must have stirred his imagination, especially when he compared Wilson's drawings to his own and was immediately convinced his were superior. During the following years Audubon kept up with Wilson's progress and the serial publication of *American Ornithology*. When he had the means, in 1816, he purchased the first six volumes, oddly inscribed with his name, the date, and the single word: "Louisiana." Since he certainly did not buy them in Louisiana, there must have been another reason for the inscription, but it remains a mystery. Perhaps he was already prefiguring the role Louisiana would play in his triumph over Wilson as the supreme master of ornithological art in his time.

Whatever the reason, there is no doubt that Audubon studied the six volumes with great care. He annotated them and committed much of their contents to memory. He apparently did not yet own volume seven, but he had read a borrowed copy before he left Cincinnati. Of all the names mentioned in the *Louisiana Journal,* Wilson's is cited the most frequently, often with Audubon finding fault with the accuracy of his descriptions and drawings. In Natchez and in New Orleans Audubon writes about hunting around for subscribers to Wilson's work in order to consult the last two volumes, eight and nine, that he had not yet read. Wilson's biographers like to suggest that Audubon plundered his way through Wilson page by page. If he did so, he was doing what Wilson himself had done with his predecessors, and what every artist and scholar does who is determined to measure the value of his work against the best. Audubon certainly never hid nor undervalued his debts to Wilson, even when striving to surpass him.

Setting out from Cincinnati, Audubon knew he had at least one powerful advantage over Wilson. Wilson had spent barely two months in the lower South along the Mississippi, not nearly enough time to find and paint the daunting number of birds the area contained. Because Lucy agreed to stay behind and keep the family intact, Audubon was able to plan a much longer trip, at least seven months. There were unprobed private yearnings leading him down to Louisiana, but the practical reasons were just as compelling, and much easier to justify, to Lucy and to anyone else who would listen. The main idea was to make the most of the great Mississippi Flyway, the major migration route for birds in North America. This is why Audubon chose the ideal time to leave, the beginning of fall (unlike Wilson, who made the mistake of leaving too early in a freezing spring), and why from the beginning of his journal he eagerly notes the enormous flocks of geese, cranes, ducks, grackles, and purple finches, flying over the flatboats on their way south.

With Lucy's help, Audubon planned the trip carefully, with a farsighted resourcefulness that is often forgotten in assessments of his character. During the summer before he left, he wrote to several local notables, including Henry Clay in Kentucky and William Henry Harrison in Ohio (both congressmen from their respective states), soliciting letters of recommendation he could carry with him as introductions to government officials. Audubon made sure to copy their replies in his journal, in no particular order. Behind this flurry of letter writing was the hope he could latch on to, or even initiate, an authorized (and thus sponsored) expedition into the vast uncharted territory west of the Mississippi opened up by the Louisiana Purchase in 1803. There had already been a number of such search parties, going back to the groundbreaking Lewis and Clark Expedition of 1804 to 1806. When he arrived in Louisville in 1808 (where two years earlier the two explorers had been celebrated on their return), the twenty-five-year-old Audubon was thrilled to hear all about the adventure directly from the horse's mouth, the horse being William Clark himself who often visited his sister and brother-in-law, Lucy and William Croghan, at their plantation Locust Grove, where Audubon was a frequent guest. The Croghans' son, George, became one of Audubon's favorite hunting partners in Louisville.

Audubon also knew about the smaller expeditions Jefferson had sent into lower Louisiana, the Freeman-Curtis attempt to trace and explore the Red River in 1806, and the earlier Hunter-Dunbar foray (1804–1805) up the Ouachita River. Both expeditions were undertaken to gather vital information, and to make maps that would soon help the United States in its thorny Louisiana border dispute with Spain, a problem that was finally resolved when the Adams-Onis Treaty was signed by Secretary of State John Quincy Adams in 1819, and ratified by the United States Senate two years later.

A month before Audubon began writing his letters, early in May, he witnessed the spectacular arrival of a specially fitted steamboat in Cincinnati transporting another government expedition, this one led by Major Stephen Long, who was leading a group of soldiers and scientists up the Platte River to the Rocky Mountains and back down through the southwestern regions along the newly fixed Spanish border. The scientific contingent included two young men from illustrious families in Philadelphia, Thomas Say, an aspiring entomologist who was the great grandson of John Bartram, and Titian Peale, son of Charles Peale, the renowned portrait painter and founder of the Philadelphia Museum. Audubon remembered them as "staring" at his bird paintings on display in the Cincinnati College Western Museum. He does not say if they admired them or not. Whatever the case, it must have been vexing to Audubon to contemplate his own situation in the presence of two untested peers fortunate enough to enjoy a free ride into the wilderness to boost their careers.

One personality Audubon had in his sights was the governor of the newly formed Arkansas Territory, James Miller, who had established a temporary residence at the old Arkansas Post, a small settlement ten miles from the Mississippi up the Arkansas River. Two months after leaving Cincinnati, on December 10, Audubon and the two flatboats landed a few miles above the mouth of White River, a tributary of the Mississippi just north of the Arkansas. Captain Aumack, who wanted to see the Post for himself, decided to go alone on foot. Audubon, by this time weary of Aumack's arrogance, made other arrangements. Accompanied by Joseph and Anthony, one of the flatboat carpenters, and armed with his letters of recommendation, he set off with a skiff up the White River toward a cutoff linking the White to the Arkansas. Fighting the current, they were forced to pull themselves along by grapevines. When they arrived at the cutoff they secured the skiff and went the rest of the way by foot, following for several miles a narrow path that twisted around swamps and canebrakes. Guided by a Frenchman they met at an isolated farmhouse, they trudged another seven miles "through mud and water" and arrived at the Post well after nightfall. Audubon calculated they had walked thirty-two miles to reach their destination. Captain Aumack was waiting for them, but at least they had been spared his sour company for an entire day.

The following morning Audubon woke up glowing with revived hope. Two days before, he had complained darkly of missing Lucy, of eating another dreary potful of boiled sweet potatoes, and of laboring under the glare of Aumack. Now, with the prospect of meeting Miller, and the dawn buzzing alive with birdsong, the future all at once looked less dim: "The morning broke and with it, mirth all about us, the Cardinals, the Towe Buntings, the Meadow Larks and many species of Sparrows, cheering the approach of a benevolent sun shining day." Unfortunately, he soon learned that Miller had gone up the river, scouting out a new seat for the territorial government.

It might seem odd at first that Audubon's good mood was in no way shaken by the news, despite his disappointment. When we consider his fund of resilience and enthusiasm when the circumstances were right, however, he had every reason to feel reinvigorated by the experience. For one thing, little had been lost, and much had been gained. He had not met James Miller face to face, but he could leave the letters of recommendation behind and hope for the best. Now that the governor knew who he was, two days later Audubon wrote his own letter, spelling out his intentions and offering his services. Who knew what the future would bring. As for the present, the walk to the Arkansas Post had allowed Audubon to escape, for a couple of days at least, the routine of the flatboat. Instead of watching himself drift into the unknown, he could push forward with a definite purpose, finding his own path, and leading the two younger men along. For a while he was no longer confined in a commitment, providing food for the flatboats, and counting the steamboats churn by day and night making his own slow progress a mockery of movement.

Audubon's eye was always alert, especially when scanning the skies, but the Arkansas Post episode shows how a new situation could sharpen his curiosity, and is a good illustration of the offhand detail and fresh observation that sets his journal writing apart from the more formal diaries of other naturalists. He springs from scene to scene, learning what he can, and fixing in his mind, like so many specimens pinned to a board, the character of the inhabitants ("worn-out Indian traders") and the surrounding backcountry. He misses nothing, including the sight of a velocipede (the early bicycle, only recently invented in Europe) standing in the middle of a frontier outpost. How the velocipede had made it there is left to our imagination. On his way back to the flatboat, he meets a few Osage Indians, gives them a drink of whiskey, and trades some gunpowder for two venison hams. Again, no aspect of nature, human or otherwise, is lost on him. He notes with relish, and admiration, that the Indian woman who has waded out from the canoe to drink whiskey with the men calmly crushes with her fingers a louse she had plucked from under her arm.

Three months later Audubon had reached New Orleans only to find himself bogged down in a routine even more suffocating than life on a flatboat. Walking in circles, he spent his time searching for the odd job as a portrait painter and teacher in order to keep food on the table and to provide time for his bird drawings. The time was scarce, and so was the food, including a wedge of cheese filled with maggots. Always the naturalist, Audubon amused himself by measuring the distance the maggots could jump. Then, when the news boomed through New Orleans of the long-stalled Adams-Onís Treaty (fixing the border between the United States and the Spanish colonies) finally being signed in Washington, Audubon saw another opportunity to pounce upon.

He read the article announcing the ratification in a New Orleans newspaper on March 21, 1821. The article mentioned a new expedition to validate the border that was being planned to set off from Natchitoches sometime later in the year. Audubon immediately began another round of letter writing and meetings designed to allow him to join the expedition. This time he decided to appeal directly to ultimate authority and wrote a letter to President James Monroe. When the governor of Louisiana, Thomas Robertson, finally agreed to intervene on Audubon's behalf with an official letter to Monroe, Audubon was overjoyed. Even before the governor had made the offer, Audubon had roamed the streets of New Orleans imagining himself heading up the Red River. He ran into his old friend George Croghan, who encouraged him and lifted his spirits even more by inviting him to his nearby plantation. He searched out old Dr Hunter of the 1804 Hunter-Dunbar expedition for another recommendation. Hunter was still living, but barely alive. When Audubon called on him, Hunter was privately engaged in emptying his bladder, and took so long that Audubon gave up and left without meeting him.

Once more, nothing came of all these efforts, and Audubon had to wait ten years before his growing reputation finally secured him a ride on a government boat during the trip he made in 1831 to look for new birds along the Florida coast. What seemed at the time another harsh setback in New Orleans turned out to be a blessing in disguise, thrusting him headlong into a Louisiana he had only begun to explore and embrace. Had he ventured west before taking on the job at Oakley, the years he spent combing the countryside of the Felicianas, and making the dense green woods between Bayou Sara and Natchez his own backyard, would have been lost. The *Birds of America,* certainly in the form we know it, with its great festival of Louisiana birds, would never have existed.

In tracing Audubon's urgent appeals for government help throughout the *Louisiana Journal,* it is hard not to notice that his official argument to increase ornithological knowledge of the wilderness, for himself and his country, could not hide a deeper temptation to evade his problems by fleeing westward. Though he knew better and always resisted acting on the thought, he suggests at times that the world beyond the Mississippi could be a semipermanent refuge, an easy step away from all his social and domestic obligations. As the days and weeks grind on, and his doubts and fears weigh down, the carefree life he likes to imagine in the big woods emerges whenever he meets a new party of Indians. The same day the flatboats leave the Ohio River and turn south on the Mississippi, Audubon meets two Indians and confides in the journal his admiration: "I saw here two Indians in a canoe; they spoke some French, had bear traps, uncommonly clean kept, a few venison hams, a gun, and looked so independent, free and unconcerned with the world, that I gazed on them, admired their spirits, and wished for their condition." A few days later, having just arrived in the half-dead river town of New Madrid, he reflects on the filthiness of the white inhabitants and writes that "the Indian is more decent, better off, and a thousand times more happy."

Over twenty years later, during his voyage up the northern reaches of the Missouri River, Audubon was dismayed to discover another vision of the Indian, victims of ruthless westward expansion, pushed out of their native lands, decimated by sickness, and reduced to begging when not hunted down as outlaws. The Indians he met in lower Louisiana were still at home in their world, and the contrast with his own situation was obvious. Audubon had left behind everything in Kentucky and Ohio, including Lucy and the two boys. The Indian radiated completeness. Audubon in Louisiana, looking for new birds, was also searching for a new version of himself. He admired the Indians and envied their free spirits, but was under no illusion about sharing their existence. On the other hand, they stood as beacons of serenity beyond the pale. Audubon, now more an outsider than ever, could at least benefit from their example. Learn from the past, forget the future, burrow into the day's work, and let the lightning strike where it may.

He knew he could rely on his art, and his skill as a woodsman. He had proven this in Kentucky, even back in Pennsylvania, when he was first mesmerized by the West as an eighteen-year-old boy fresh from France. In Louisiana, on his own again, the hardships he endured were double-edged. They increased his gloom that fate was against him, but at the same time made him stronger and more self-dependent. When he arrived in Natchez after over two months on the flatboat, he noticed that when sitting down to eat at the hotel he "involuntarily" picked up his food with his fingers. Rather than being embarrassed by the gesture he explains in the journal that on the flatboat, "we seldom eat together and very often the hungry cooked; this I performed when in need by plucking and cleaning a duck, or a partridge, and throwing it on the hot embers." And he goes on to justify himself in a remarkable homage to frontier travel as the best way to forge character and winnow weakness: "such life is well intended to drill men gradually to hardships, to go to sleep with wet muddy clothing on a buffalo skin stretched on a board, to hunt through woods filled with fallen trees, entangled with vines, briars, canes, rushes." Audubon wants Lucy and his sons to appreciate his struggles when they eventually read the journal. But at the same time, he is affirming to himself a willingness to change that becomes one of the central themes of the voyage.

One reason the *Louisiana Journal* is such a vital document lies in the conviction with which Audubon writes down and begins to master his own transformation. He remains himself, of course, through all his modes of being, but at the end of the journal he has given up one way of life (including a home and household) and has made himself over as a perpetual nomad, gathering Lucy and the boys into a sort of family caravan, a tribe of four, yoked to the *Birds of America*, prepared to set up camp for a season or two, but always ready to move on when the time was right. When we trace Audubon's career over the years, beginning with the crisis of 1820, it is striking to consider that for the next twenty years, after his leaving for Louisiana, he is always more or less on the move, with no home of his own.

When Lucy and the boys join him in New Orleans after the fourteen-month separation, Audubon enters into the fullness of his new estate. Four months later he leaves for Natchez after having found Lucy a position as a tutor and companion to the wife of a New Orleans builder who had become a good friend. A rough

Louisiana pattern had been established: Lucy would try to provide the family a regular income as a teacher, and Audubon would concentrate on his birds, taking on the odd job to help. There were strains and setbacks to overcome, but there was no looking back. It took the Audubons another year to settle into a temporary modus vivendi when Lucy was hired to open a school on the Percy plantation, Beech Woods, in the Felicianas not far from St. Francisville and Bayou Sara. She remained at Beech Woods from 1823 to 1827. For three of these years, before leaving for England in 1826 to find an engraver, Audubon orbited around Lucy. This was the case even when he went to Philadelphia via Louisville and was gone for over a year. For most of his time in Louisiana he was hunting and drawing birds in the woods, where he did much of the best work in *Birds of America*.

This is another obvious truth that the *Louisiana Journal* brings home: without Lucy backing him up even at a distance, Audubon would not have achieved what he did. There never had been any leashes in their marriage, long or short. Separation and distance, relatively speaking, became conditions of his art that they both gradually accepted as a necessity. Soon after he arrived in England, in an essay he wrote for the *Edinburgh Journal of Science, Account of the Method of Drawing Birds*, Audubon argues that the gift of solitude freed his powers of concentration and gave his work an authenticity it would otherwise have been denied: "I am persuaded that alone in the woods, or at my work, I can make better use of the whole of myself than in any other situation." Writing this in 1828, Audubon is thinking of Louisiana and Lucy. When Lucy settled temporarily in Beech Woods, Audubon could enjoy the benefits of roving near and far without the fear of finding himself isolated or abandoned. Lucy never hesitated to write or voice her protests and worries, and some of the letters that they exchanged during Audubon's first year in Louisiana suggest that Lucy's patience had almost reached a breaking point. The bonds that held them together now, however, were cable-strong, stronger than fate itself. For one thing, their natures had commingled into one. They both possessed a stubborn will, verging on willfulness, and they both insisted on sharing an obsessive faith in Audubon's art, ambition, and each other.

Often beset by the feeling that time and opportunity were leaving him behind, especially during the long days and nights on the river, Audubon sometimes despaired that he was trapped in a dream of repetition, with the forbidding prospect of failure waiting at the end despite all his efforts. He did have two antidotes (of his own making) to fight his fears: his portfolio of birds—or actually two portfolios, one big and one small—and his carefully kept diary. Since the *Louisiana Journal* is the earliest one that survived, it is difficult to know the exact nature of the ones he had kept in Kentucky. He would have named and described the birds he was drawing, and the circumstances of sighting and hunting them down in the woods, but not as obsessively as he does in the *Louisiana Journal*. He wrote in it regularly, clung to it, and reread it when dogged by his duties on the flatboat and the blankness of his future. More than his letters to Lucy, it contained proof that at least his own knowledge of birds was increasing by leaps and bounds. There was no small comfort in the fact that his mind and art were forging ahead, despite the doubts that continued to cloud his vision of ultimate success.

Even when seized by spells of melancholy, Audubon was not given to dwelling on the past. After fourteen months, he could hand over to Lucy the portfolio and the diary as a measure of what he had accomplished, with the promise of more in the months ahead. Fully engaged with the work at hand, it is unlikely he spent much time ruminating over what he had written. This would happen later when he foraged through his notes and expanded them for the first volumes of the *Ornithological Biography*. For today's reader, however, after going through the diary, entry by raw entry, it is revealing to stand back and look at it as a whole. Doing so, it is fairly easy to distinguish five signposts or turning points that Audubon might have missed or misread in the thick of the event, but that definitely mark a timeline of progress in the making: the Mississippi, Natchez, New Orleans, Oakley, Lucy in New Orleans.

The first of these he did not miss. The associations with his personal history were too potent. On November 17, a little over a month after he had left Cincinnati, the two flatboats reached the mouth of the Ohio and floated into the mighty Mississippi. The significance of this spurred Audubon to take the skiff by himself and row to the confluence in advance of the flatboats. He had been there before, first, eleven years ago when he had turned north on his way to Ste. Geneviève with his partner Ferdinand Rozier on the trip he describes in his essay *Journey up the Mississippi*. At that time he was a young man still delighted with the promise of his new life in the western territories. More recently, in May of 1819, he had turned south on his way from Henderson to New Orleans in a failed attempt to recover a debt he was owed by a man (Samuel Bowen) who had purchased a steamboat from Audubon's brother-in-law Tom Bakewell and had paid with a promissory note. Bakewell had handed over the note to Audubon in one of their business dealings, but when the note came due Bowen bolted and slipped away in the steamboat to New Orleans. When Audubon finally caught up with Bowen, who had made the mistake of returning to Henderson without the steamboat, there was a confrontation, a fight, and a court case. Audubon won the fight, and the court case, but recovered neither the money nor the boat. The whole painful fiasco would grate on his memory for years. He remembered it now, nineteen months later, because at least then he had been a man of business, though in dire straits, and now he had nothing.

The contrast with his situation now, compounded by the loneliness gathering about him without Lucy and his sons, has become a cross almost too heavy to bear. These are the doomsday notes he jots down in his diary the night after he had reached the Mississippi once again, a grim calendar of failure and loss:

> Eleven years ago on the 2 of January I ascended that stream to St Genevieve, Ferdinand Rozier of Nantes my partner in a large keelboat loaded with sundries to a large amount our property.
>
> The 10th of May 1819 I passed this place in an open skiff bound to New Orleans with two of my slaves.
>
> Now I enter it poor, in fact destitute of all things. . . .

When Audubon mentions his slaves as a measure of his change of status, he is writing at a time, the 1820s, when slavery in Kentucky and the other southern states was an accepted fact of the society and the culture. When Audubon moved to Kentucky he embraced its frontier traditions, including hunting, fishing, horsemanship, and 4th of July celebrations, with the gusto of an exile yearning to belong. Audubon purchased slaves as domestic servants in Henderson in much the same aspiring spirit that he purchased a house and land and horses.

Slavery was the way of the world, at least Audubon's world, and had been a part of his existence since his birth in the French slave colony known at the time as Saint-Domingue (Haiti). He had been whisked away to Nantes as a little boy because of the slave revolt, and his father had bought Mill Grove with money he had made from a sugar plantation worked by slaves. This said, Audubon was uneasy with the legacy. Once he left Kentucky (after selling his slaves) and moved into Louisiana, he refused to buy another one, even when he had the means. Lucy did own a young slave named Cecelia (along with Cecelia's two sons) to help with housework in the Felicianas during the 1820s. This may have been the same Cecelia Lucy had left behind in Henderson and had since reclaimed. The Audubons decided to leave all three in New Orleans with their friends the Brands when they moved away from Louisiana for good in 1830. After having lived in Louisiana for years, during the boom years of plantation slavery, Audubon later wrote an essay *The Runaway,* lamenting the breaking up of slave families. First published in 1834 in the second volume of the *Ornithological Biography, The Runaway* is one of the rare slave narratives written by a white southerner (Audubon proudly asserted himself as a southerner long after he left Louisiana). He was no abolitionist, but it is worth noting that in 1842 when he finally decided to buy land and build a new house, the chosen spot was on the Hudson River in New York.

Of course Audubon, writing down his thoughts in the cold Mississippi twilight on November 17, 1820, did not think of himself as an actor in human history. Absorbed in his work, and his crisis, he had enough to reflect on with the momentous parting of the ways the turn south represented in his own dubious future. After losing sight of the "beautiful and transparent water of the Ohio," he gazes down at the muddy Mississippi, and writes, portentously: "here the Traveller enters a New World." A few days later, on November 28, a steady rain holds him on the flatboat all morning. In keeping with his mood that the Mississippi is drawing him darkly on into the unknown, he takes the time to draft a short autobiography. Of all his self-portraits in writing, this one sticks closest to the vital facts and gives the distinct impression of wanting to be read as a last will and testament.

The next milestone on Audubon's journey was Natchez. For the first time on the river, a series of fortuitous encounters during his five-day stopover gave him reason to believe his luck might be changing, if only he could seize the occasion, or the "half-chance" as he called it, using the frontier idiom. When the flatboats arrived at Natchez the day after Christmas, Audubon stayed onboard, reading two new letters from Lucy and writing one back in response. The next day he went ashore. Topographically and socially, Natchez was divided into two towns, the rough and rowdy landing by the river, Natchez Under-the-Hill, and the more respectable district on the Bluff, with its plantation houses, banks, and hotels. Audubon spent little time walking through lower Natchez up the hillside road, though he would not have missed the brazen display of the boatmen's favorite sports on his way up. Abandoned flatboats had been lined up next to the wharf and transformed into places of business for the grog merchants, the professional gamblers, and the prostitutes posing as washerwomen. The latter would hang their laundry out to dry as publicity, adding a splash of color to the scene. Audubon, more than anyone, knew the appeal of aphrodisiac plumage when he saw it, but he did not take the bait. Within a short time of arriving at the top of the bluff, he ran into his brother-in-law, Nicholas Berthoud, who had arrived by keelboat from Louisville on a business trip to New Orleans.

Relations between the two were often strained, but this time the wealthy Berthoud proved to be a generous kinsman. He offered Audubon a ride on his keelboat to New Orleans, allowing Audubon to give up his job on the flatboat. He also invited him to stay with him at the Natchez Hotel, the best hotel in town. The Natchez was owned by John Garnier, a Frenchman who had emigrated to Louisiana from western France, near Nantes, Audubon's boyhood home. Through Garnier, Audubon touched a web of native French affiliations possible only in lower Louisiana. Garnier was kin to Julien Poydras, also from Nantes. Poydras had come to Louisiana via Saint-Domingue and had become one of the wealthiest and most influential men in the region. He lived further south down the river in his Creole house at Pointe Coupée. When his younger brother had been guillotined during the French Revolution, Poydras arranged for his nephew Benjamin to sail from Nantes and resettle in St. Francisville, where he founded a boat company. Though Audubon did not know it at the time, even in Natchez he was never far from home.

Nothing delighted Audubon more in Natchez than the sound of French. Natchez was now the temporary capital of the brand new state of Mississippi, but its French origins in colonial Louisiana were everywhere to be seen and heard. For Audubon this was the Louisiana he had been seeking. He spoke French with Garnier and with an elder Frenchman, Charles Carré. Carré guided Audubon around Natchez, entertaining him with his knowledge of Louisiana history, local Indian lore, and the "purity" of his French. When Audubon was not speaking French he turned to literature. Within two days he had found and read a copy of his favorite La Fontaine from cover to cover. Through Garnier again, Audubon was introduced to John Wilkins, a local subscriber to Wilson's *Ornithology* who lent Audubon the eighth volume. More pragmatically, Audubon also resumed his practice of drawing portraits now that he had left his job on the flatboat. He still had to provide both for himself and young Joseph Mason. The portraits brought him five dollars each, "fine sauce" for their "empty stomachs," he noted in the journal.

The stay in Natchez raised Audubon's spirits and self-confidence to a pitch of optimism he had rarely felt since he left Cincinnati. The only sour note occurred after he and Mason left Natchez on the first day of the year, 1821, on Berthoud's keelboat. The keelboat was already well on its way downriver, being tugged along by the steamboat *Columbus*, when Audubon realized he had forgotten the smaller of his two portfolios. This was not a disaster, since most of his birds were in the large one. It did contain a portrait of Lucy, however, and the loss of his wife's image was not easy to overcome. It caused him to look back twenty-one years earlier, when he was a boy of fourteen worried about his future, and reflect on what he had become: "What I have seen, and felt since, would fill a large volume—the whole of which would end at *this day January 1st 1821. I am on board a keelboat going down to New Orleans the poorest man on it.*"

This was an odd thought to cross his mind given his situation barely a week earlier, still fretting under the lordly authority of Captain Aumack. No doubt Audubon could not help seeing himself in the light of his brother-in-law and benefactor, Nicholas Berthoud, whose success made Audubon's own failures more glaring. Still, when he turned to the present, and looked out at the spectacle from the comfort of the keelboat, his mood shifted, as it often did when he gave it a chance. After all, he and Joseph were now treated as honored guests. They could sit back and enjoy the ride. Their meals were prepared for them, and after the steamboat left them at Bayou Sara the eight slaves of Berthoud rowed the keelboat the rest of the way to New Orleans. The captain of the keelboat asked Audubon to make his portrait and paid for it "in gold." Audubon could spend all his time on his drawings, sizing up the birdlife and the southern scenery: "how sweet for me to find myself the 1st of January in a country where the woods are filled with warblers, thrushes, and at the same time see the rivers and lakes covered with all kinds of water birds." This was much the same stretch of riverside that William Bartram had followed forty years earlier for his *Travels*. Bartram had enjoyed the company and conversation of the Creole planters, had marveled at the whooping cranes, and had watched a bear swim across the Mississippi. It took Audubon time to adjust his ears to the rural Louisiana French he heard when he went ashore, but he was quick to praise the Creoles as "good-natured souls" and he admired their gardens. The sight of a patch of artichokes reminded him of "the happy days spent in

France." Though he could not realize it at the time, his discovery of the Louisiana landscape from Natchez to New Orleans was prophetic of the area of Louisiana he would soon settle in, work and paint his way through, and make his own in the *Birds of America.*

Six days later, on January 7, the keelboat arrived in New Orleans, and reality set in again, with a vengeance. The keelboat docked on the levee near the old market, providing Audubon and Joseph a cramped but free place to stay until he could find better accommodations. New Orleans had been Audubon's goal from the beginning. A booming port, with its population growing by leaps and bounds, and with fortunes being made overnight, it held out the best hope for the showing of his birds and the selling of his skills. The promise of that plan, however, was soon put to the test. The day after his arrival, on January 8, the sixth anniversary of the Battle of New Orleans, Audubon's pocket was picked while he rambled though the festivities. From then on through the next five-and-a-half months he tried in vain to bury the memory of the shock to his pride and purse. He immediately began the old scheme of painting portraits for the wealthy merchants while at the same time looking for new birds to draw.

The proximity of the market helped. Professional hunters would bring their cartloads and baskets of freshly killed birds and other game to be sold every day. The French Creole craved all wild flesh, robins included. Audubon found a host of birds he needed to draw waiting for him on the market stands in the mornings. Those he did not find he hunted himself, rising at two in the morning to walk to the nearby cypress swamps. Soon, the commissions for portraits began to pour in, giving him so much work he is "sorry that the sun sets." His bird drawings kept up with the pace of the portraits. In February he sent Lucy twenty of them, "the fruits of a long journey." After six weeks on the keelboat Audubon found a room to rent, on Barracks Street in the heart of the Vieux Carré. He shared the room with Joseph, and for a time with Samuel Cummings, still at work on his river survey.

During the next few months Audubon found himself casting about for his bearings. On one hand he was an outsider, without family or funds, condemned to struggle and scramble to make ends meet. On the other, his origins made him feel more at home than the average newcomer with the Creole culture and society. Half the population of New Orleans still spoke French. Then there were the old friends and acquaintances from Kentucky who had moved to Louisiana or who had come down by the latest steamboat on a business trip. Audubon seems to have run into one of them by accident every week.

Socially, Audubon could blend in up and down the scale and could more than hold his own whatever the level or drift of the conversation. His education and gentlemanly skills (fencing, music, dancing, billiards) played to his advantage when the occasion demanded. But these occasions, for an artist needing work and recognition, were often a waste of time. He had little patience with a dinner party where a trained monkey was brought in to amuse the jaded guests with its "gambols and pranks." The overbearing "French gaiety" fell flat with an Audubon who was trying to concentrate on his art and mission. Later on he was disgusted and ashamed when he was invited to attend a mock Mason initiation during which an old man was mishandled and humiliated.

Still, Audubon's familiarity with all ranks and classes serves him well as he moves through the alleys and hallways of the city. And it gives to his journal a zest (and a range of reference) lacking in the better-known firsthand accounts of New Orleans of the time. Compared to Audubon's rapid shiftings of focus, and ripeness of phrase, most of these early travel essays seem plodding and conventional, somewhat of a crime considering the crazy strangeness of the spectacle. Audubon's natural curiosity is one reason his writing darts about and sniffs where it will, and from time to time digs surprisingly deep. Another is his artist's fascination with the odd telling detail, whether he finds it in the eye of an owl or the nose of an old Creole.

Audubon more than once cites Johann Lavater as an influence. Johann Jasper Lavater was an eighteenth-century Swiss thinker and man of letters who revived in popular essays the ancient theory of physiognomy, the idea that a man's inner nature could be read in the features of his face. Audubon would have read Lavater in French translations often illustrated with the famous drawings of Charles Le Brun, the court painter of Louis XIV. Le Brun presented the heads of humans next to the heads of animals (including birds) showing distinct resemblances. Le Brun's visions of a man's head as goat-like, or resembling a parakeet, verge on caricature, but it is not hard to see why a naturalist like Audubon would be intrigued by the possibilities. His intuition that the process could be reversed inspired him to endow his birds with expressions that seem almost human, conveying emotions like fear, anger, even tenderness, depending on the setting or the situation. As Audubon points out in the *Ornithological Biography,* what binds us visually works in other ways as well. Birdsong is a kind of music, and all children want to whistle.

Anyone who has been awed by one of Audubon's most striking gifts, his ability to make his painted birds come to life in elemental acts of survival (feeding, fighting, mating, nesting), should not be surprised at the force of his descriptions in the journal. In the middle of March he jotted down a short entry on the mimicry of a mockingbird he had noticed. Done rapidly, it hums along gaily with the same gratitude and joy of discovery that kept him going when he took up his brushes to work at his watercolors for hours at a time:

> Near our house a mocking bird regularly resorts to the south angle of a chimney top and salutes us with sweetest notes from the rising of the moon until about midnight, and every morning from about 8 o'clock until 11, when it flies to the convent garden to feed—I have remarked that bird always in the same spot and same position, and have been particularly pleased at hearing him imitate the watchman's cry of *All's Well* that issues from the fort about three squares distant, and so well has he sometimes performed that I would have been mistaken if he had not repeated too often in the space of ten minutes.

Birds and men, men and birds, Audubon sees them playing much the same game. Any flickering of dramatic life seems to catch his attention. Usually a quick sketch in a few lines is enough for the journal. Two longer essays, however, *The Fair Incognito* and *The Original Painter*, show him eager to try his hand at fuller forms, closer in style and spirit to the literature he loved. Audubon read widely, preferring the English novels he and Lucy had read out loud together. In New Orleans he continued to search out works of fiction (along with the ninth volume of Wilson), and tried his best to buy Lucy the latest Walter Scott novel, *Kenilworth*, to no avail. The show of eccentricity in New Orleans called for a more demanding literary effort, and Audubon was not one to back down before a challenge.

The Fair Incognito is based on a portrait commission by a beautiful young Creole woman, a Madame André, whose identity remains a mystery. She accosted Audubon one day in late February and invited him to her apartments where she asked him to paint her in the nude. Audubon seems to have written about the event with the aim of turning it into a short story, narrated with more dialogue than was his habit, and composed to heighten the suspense. He never published it, but he did embed it in a long letter to Lucy later in May when he was trying to persuade her to come to New Orleans. This is a risky method of seduction, though it can probably work in the right situation. Audubon surely wanted Lucy to stop hesitating, take the next steamboat, and save him from himself. *The Original Painter* is a lighter piece, a genial portrait of John Wesley Jarvis, the flamboyant well-known artist about town, one of several Audubon met in New Orleans, all of whom treated him more as a rival to be kept at a distance than a peer in need of a colleague's help. Jarvis was no exception, but his personality was outrageous enough to attract Audubon anyway. It gave him the opportunity of reaching into one of his favorite books, Sterne's *Tristram Shandy*, to find the perfect pitch to match the comical high spirits of Jarvis.

By the time Audubon sent *The Fair Incognito* to Lucy, he had almost given up on New Orleans. He did not know where he would go next, only that he would not go back, head hanging, to Kentucky, warning Lucy that "migratory birds or beasts follow the mast." He had enriched his portfolio but not his pockets. He had made a few good friends who hired him for portraits and lessons and generally helped him stay afloat. One of the most generous was Roman Pamar, a local merchant to whom Audubon had offered hospitality years ago in Henderson. Pamar and his close-knit household became a second family whose door was always open. Audubon ate there often, and writes that the Pamars were responsible for "all that I can call the pleasures I felt while at New Orleans." Otherwise, the high society of New Orleans had barely left him with a bone to gnaw on after over four months of hard work. Then, out of the blue, a golden egg dropped in his lap.

One of Audubon's drawing students was fifteen-year-old Eliza Pirrie. Her family owned Oakley, a large cotton plantation not far from Bayou Sara, a few miles north of Baton Rouge. This unusual portion of Louisiana (the Felicianas) juxtaposed well-drained bottomland, rolling hills, and thick woods, and was highly prized by planters for its healthy climate. The Pirries spent their summers at Oakley to escape the humidity, heat, mosquitoes, and yellow fever that turned New Orleans into an airless nightmare for four months every year. Unlike New Orleans, Oakley basked in the breezes and vital shade of one of the most pleasant landscapes in all Louisiana. Eliza's mother offered Audubon sixty dollars a month (including room and board for both Audubon and Joseph) to come to Oakley for the summer. In exchange Audubon would continue his lessons, spending half of every day instructing Eliza in drawing, music, arithmetic, even the braiding of her hair. The rest of the time he could devote to his birds.

From the deck of a keelboat, almost six months earlier, Audubon had been impressed with a glimpse of the area around Bayou Sara on his way from Natchez to New Orleans. This time, he was able to slowly soak in the sights as he and Joseph walked the five miles or so inland to the plantation after stopping for a rest at St. Francisville. Missing Lucy and his sons, and worried about the long-term commitment he had made, Audubon was not overly enthusiastic as he set off. As usual the glories of nature perked him up and he was soon transfixed by what he saw: "The aspect of the country entirely new to us distracted my mind from those objects that are the occupation of my life—the rich magnolia covered with its odiferous blossoms, the holly, the beech, the tall yellow poplar, the hilly ground, even the red clay I looked at with amazement and such entire change in so short a time appeared often supernatural, and surrounded once more by thousands of warblers and thrushes, I enjoyed nature."

Arriving at Oakley, Audubon was won over by the kind reception of Mr. Pirrie, whose good nature Audubon divined by inspecting his "features by Lavater's directions." As it turned out, these presages of better times proved to be true. Oakley was the great breakthrough Audubon had been waiting for.

For the next four months he was able to pursue his work without the relentless pressure he had felt in New Orleans. Having comfortable lodgings and a fixed schedule freed him to spend most of his time where he loved to be above every other place, in the heart of the deep woods. With or without Joseph at his side, Audubon could now dance to his own tune, no longer forced to march from door to door looking for handouts. It helped that the countryside around Oakley was packed with a variety of birds and wildlife difficult to find anywhere else in America.

Audubon had found his Louisiana arcadia, a treasure in the trees, and entered into one of the most productive periods of his artistic life. Not surprisingly, the entries in the journal are not as regularly kept. He was often too busy to write, content to fill some of the pages with a simple list of all his recent discoveries. When he did expand on the day's events, his tone was usually that of pure delight. His account of the red cockaded woodpeckers on July 29 is a case in point, as is his long description of a walk to a cypress swamp (with its "thousand knees raising like so many loafs of sugar") on August 12 when he turns from bird to bird (red-shouldered hawk, prothonotary warbler, yellow-throated warbler, green blue warbler, Mississippi kite, mourning warbler, fly catcher) as if he were a spectator at a pageant. Birds are not the only creatures that catch his eye. And when Audubon's eye is caught nothing is missed: "We saw a singularly rich colored (?) spider that finding a horse fly just entangled in her net, moved to it and covered it in a moment with the silk of her bag. Shooting it out in a stream and at the same time rolling the fly until the whole likened to the appearance of a small oblong ball of white silk, the spider then returned to the center of its net—no doubt this is a way of preserving the flies when the spider is not hungry."

Oakley is the brightest section of the *Louisiana Journal,* and of his entire Louisiana experience, to be matched only by the year he spent with Lucy at the Percy plantation Beech Woods when he returned from his trip to Philadelphia and New York at the end of 1824.

The Oakley sojourn ended ingloriously when Audubon ran afoul of Eliza's mother, by all accounts a prickly soul who was easily offended. She persuaded herself that Audubon was seeking a closer relationship with his pupil than her age and situation permitted. For Audubon, the risk of an infatuation (or unhealthy influence) was an absurdity, a plot contrived by a jealous young doctor who was attending Eliza during an illness. Audubon nonetheless found himself summarily dismissed. By the end of October, after four months, he was back in New Orleans, not sorry to have left behind the insinuations to fester where they would. The "sweet woods" around Oakley were another matter; they had worked their spell and would do so again. They were in his blood for good. For the time being, however, his only option was New Orleans. He returned to his routine of eating with the Pamars, hustling for lessons and the odd portrait, now more than ever determined to persuade Lucy to join him. His mood as usual was prone to swing from hope to fear almost overnight. On October 26 he took stock of what he had accomplished during the past year. He had a mere forty-two dollars, but he had finished sixty-two drawings, almost all in Louisiana, over twenty alone in the Felicianas. He knew his progress at Oakley had been close to miraculous. Three weeks later, with "no news from my wife," after painting by candlelight until ten at night, he writes that he is "very low of spirits" and wishes himself "off this miserable stage." A month more brings him no relief: "So anxious am I during the whole of my present days to see my family that my head is scarce at right with my movements."

The waiting and the bouts of melancholy ceased (for the time being) when Lucy and the two boys stepped off the small steamboat *Rocket* at noon on December 18. That afternoon they all moved into the house on Dauphine Street that

Audubon had rented in advance. The next day there was a memorial service organized by the Creoles of New Orleans for Napoleon Bonaparte, who had died earlier in the year. Audubon did not attend; instead, he examined the bird drawings he had sent Lucy in February. He was delighted to discover his recent watercolors were far superior, so much so that on December 30 he made a vow to paint ninety-nine birds in so many days, a remarkably ambitious commitment, but no less so than the one that had been driving him forward for the past fourteen months. The first of the ninety-nine was painted the next day, the last day of the year, and the last day of the *Louisiana Journal.*

The following year, 1822, offered little change in the turbulence that stirred around Audubon's every move now that he had decided that *Birds of America* would be his life's work. Once Lucy had arrived in New Orleans and had seen the Oakley paintings, she was won over again to the great design, knowing full well what struggles lay ahead. It took another year before a new pattern of relative stability would emerge. For the first three months they all remained in New Orleans, Audubon doing his best, but hanging fire until he could return to the woods. In April he traveled to Natchez to take a job teaching while Lucy and the boys stayed behind in New Orleans. Audubon had become friends with a wealthy entrepreneur, William Brand, who had hired Audubon to give drawing lessons to his young wife, Anne. When Audubon left for Natchez, Brand asked Lucy (along with the boys) to move into the household as companion to Anne, who was pregnant. By May, the boys had moved to Natchez, and by September, after the death of Anne Brand's child, Lucy had followed. In Natchez Audubon had reunited with old friends like the French-speaking Charles Carré and had made new ones, none more valuable than the young Dr William Provan. Provan helped Audubon find work as a teacher, nursed him when he was stricken with yellow fever, and organized a lucrative raffle for a painting. He also introduced Lucy to his future mother-in-law, the widow Jane Percy, who owned the plantation Beech Woods near Bayou Sara. At the beginning of 1823 when Lucy moved to Beech Woods to open a school for the children of nearby plantations, Audubon was in a sense released to devote himself more fully than ever before to *Birds of America.* He had his problems with Jane Percy, as temperamental and difficult to please as Lucretia Pirrie at Oakley, but security and the long-term well-being of Lucy overrode all other considerations. Lucy remained at Beech Woods until 1827, when she moved her school, by then a local institution, to nearby Beech Grove plantation, where she continued to teach until reuniting with Audubon after his three-year trip to England and France. By that time, *Birds of America* was well on its way. The Louisiana circle was complete. Audubon had brought himself from the brink, paved the way for all that would follow, and written down most of the hard process step by step. There is no better example of self-recorded self-regeneration than the notebook he handed over to Lucy in New Orleans at the end of 1821. When she packed away the *Louisiana Journal* to keep it safe, she was holding on to a central testament of ornithological art, and the most original autobiography ever to emerge out of Louisiana, old or new.

NOTE ON THE TRANSCRIPTION

One of four surviving manuscripts of Audubon's journals, the *Louisiana Journal* was a gift of Colonel John E. Thayer in 1913 to the Museum of Comparative Zoölogy of Harvard University. In 1929 Howard Corning's painstaking transcription of the entire journal, followed by the slighter 1840–1843 journal in a separate volume, was published (under the title *Journal of John James Audubon: Made During His Trip to New Orleans in 1820–1821*) by the Club of Odd Volumes in Cambridge, Massachusetts.

There have been other, less complete transcriptions of the *Louisiana Journal*, notably in two early biographies, Francis Hobbart Herrick's *Audubon the Naturalist* (1917, revised 1938), and Stanley Clisby Arthur's *Audubon: An Intimate Life of the American Woodsman* (1937). Alice Ford, in *John James Audubon: A Biography* (1964, revised 1988) and in *Audubon, by Himself* (1969), includes a few scattered passages, rewritten in conventional English. All these transcriptions mirror Corning, almost to a word. That is much the case too for the most recent transcription, done by Christoph Irmscher for the anthology *Writings and Drawings* (1999) in the Library of America series. Irmscher calls the journal the *Mississippi River Journal*, freely borrowing the title from my Penguin Nature Classic, *Selected Journals and Other Writings* (1996).

In *Audubon on Louisiana* I have decided to use the Corning transcription, as I did in *Selected Journals*. I have simply changed the title *Mississippi River Journal* to read, more accurately, *Louisiana Journal*. The fact that later transcriptions follow Corning so closely confirms the value and authority of his groundbreaking work. Irmscher obviously does his version with care, and some of his alternative readings, such as *cane* for *cave* in a few places early in the text (or *fresh* for *froth* and *raised* for *sailed* in the first day's entry), are more than plausible when based on internal evidence. Still, as Irmscher admits, given the nature of the task and the age of the manuscript, he is forced to guess, or "conjecture," as Corning did too. Though he no doubt improves on Corning from time to time (a letter here, a comma there), Irmscher must inevitably risk making misreadings of his own. On the whole, however, the differences between the two transcriptions are minute, and do not affect the essential accuracy of either.

I prefer Corning for several reasons, not least of all because he respects the original location of notes and letters that occasionally seem out of place. Doing so, he makes the experience of reading the transcription closer to that of reading the original manuscript. With a little attention, the reader can easily fit these awkward passages into context. He can also easily construe the meaning of certain old-fashioned terms. The word *fresh*, for example, is usually an abbreviation for *freshet*, meaning high water, as Audubon himself explains in his episode *A Flood*. Note also in Corning that words enclosed in brackets were crossed out (and replaced with substitute words) by Audubon himself. Twice I have added brackets of my own to signal a probable misprint or misreading. As most Audubon scholars (including Irmscher) have noticed, Corning misspells one key proper name (*Braud* instead of *Brand*), and a couple of times he is misled by an unclear capital letter (writing, for example, *Some Buntings*, and once, *Iowa Buntings*, instead of *Towe Buntings*). Once too he miscopies *Shippingport*. And he errs on one obvious date. These are all easily perceived from internal evidence and have been corrected.

Louisiana Journal

THURSDAY—OHIO RIVER OCT—12*th* 1820.

I left Cincinnati [today at] this afternoon at half past 4 o'clock, on Board of Mr. Jacob Aumack's flat Boat—bound to New Orleans—the feeling of a Husband and a Father, were My Lot when I kissed My Beloved Wife & Children with an expectation of being absent for Seven Months—

I took with me Joseph Mason a Young Man of about 18 years of age of good familly and naturally an aimiable Youth, he is intended to be a Companion, & a Friend; and if God will grant us a safe return to our famillies our Wishes will be [most Likely] congenial to our present feelings Leaving Home with a Determined Mind to fulfill our Object =

Without any Money My Talents are to be My Support and My enthusiasm my Guide in My Dificulties, the whole of which I am ready to exert to [meet] keep, and to surmount.

The Watter is Low, although a Little froth, sailed the River a few days since, about $4^{1}/_{8}$ feet. We only floated 14 Miles by the Break of the 13th of Oct*re* the Day was fine, [and] I prayed for the health of My family—prepared Our Guns and went on shore [on the] in Kentucky [Side]—Cap^{e} Saml Cummings who left Cincinnati with an Intention of Noting the Channels of this River and the Mississipi accompanied us—We shot thirty Partridges—1 Wood Cock—27 Grey Squirrels—a Barn Owl—a Young Turkey Buzzard and an Autumnal Warbler as Mr. A. Willson as being pleased to denominate the Young of the Yellow Rump Warbler—this was a Young Male in beautifull plumage for the season and I Drew it—as I feel perfectly Convinced that Mr. Willson has made an Error in presenting this Bird as a New Specie I shall only recommend You to Examine attentively My Drawing of Each and his Description—its Stomach was filled with the remaines of Small Winged Insects and 3 Seeds of Some Berries, the names of which I could not determine—

Early in the morning the wind rose and we came to on the Ohio side by G^{l} Wm Harrisons' Plantation and remained untill nine o'clock *PM*—

I saw several flocks of Ducks in the morning before we had cleaned our Guns—hundreds of Meadow Lark Alanda Magna seen travelling Southwest—

[the Night became] the Wind Rose and brought us to shore, it raind and blowed Violantly untill the Next Day—

SATURDAY Oct 14th 1820

After an early Breakfast We took to the Woods I say *We* because Joseph Mason, Cape Cummings & myself I believe are allways [——] together—

I shot a *Fish Hawk Falco Aliaetos* at the mouth of the Big Miami River a handsom Male in good Plumage. he was wingd only and in attempting to Seize Joseph's hand, he ran one of his Claws through the Lower Mandille of his Bill and exibited a very Ludicrous object—these Birds walk with great dificulty and Like all of the Falco and Strix Young throw themselves on their backs to defend themselves—

We returnd to our Boat with a Wild Turkey 7 Partriges a Tall Tale Godwit and a *Hermit* Thrush which was too much torn to make a drawing of it this was the first time I had met with this Bird and felt particularly Mortified at the Situation—

We passd the Small Towns of [Madi] Laurenceburgh—in Indiana—Petersbugh in K^{y}.—, We Walked in the afternoon to *Bellevue* the former residence of a *Far Famed* Lady of our acquaintance Mrs Bruce; Saw Thomas Newell and old Cap^{e} Green—if My Eyes did not err I saw my suspicions of her conduct that Evening Justified. We Killed 4 Small Grebes at one Shot from a Flock of about 30 We approached them with ease to within about 40 yards, they were chassing each other and quite Merry when the Destructive fire through the Whole in consternation, the Many Wounded escaped by Diving the rest flew off—this is the second time I have seen this Kind, and they must be extremely rare [Bird] in this part of America—

About Three Miles above Bellevue in Kentucky we walked through a Fissure of Rocks really romantic the Passage in [that] form of a half Moon is about 6 feet Wide, the rocks are composed of Large Round pebles cimented with Coarse Sand about 100 feet High on one Side and Sixty on the other—I made a Sketch of it for [my] your future Pleasure—We walked this day about 40 miles Saw one Deer crossing the River

SUNDAY Oct. 15, 1820—

there was this morning as heavy a white frost as I Ever Saw, the Wind blew cold and heavy from the North Shott 2 *Tell Tale Godwits*—and chased a Deer in the River for some considerable time, but a Canoe with Two Indianna Men had the advantage of us and caught it as I rose to shoot it—the Wind being fair We floated tolerably well, killed 5 *Teal* One Blue Wingd Teal, [a young Malard] 2 Doves 3 Partriges and fortunately another Hermit Thrush Turdus Sohtarius—We met and went on Board the Steam Boat *Velocipede* Saw Col. Oldham, Mr. Bruce, Mr. Talcut & Lady [on Board] Passengers besides a Consider Number of Strangers, Opend a Letter directed to Your Mother by Your Uncle W[am] B— Mr [——] Aumack Killed a Young Malard Duck—the Contents of the [——] Gizard of one of the 4 Grebes I open was Nothing else than a solid mass of fine hair apparently belonging to some very Small quadrupeds—(feathers)

Saw a Chimney Swallow; the Number of Ducks Increasing—and an Appearance of a Cold Nights—Killed a Great Carolina Wren the Grebes were cooked and eat but extremely Fishy rancid and fat—

at 10 'oclock We were roused from sound sleep by the Boats having ran on Rocks—the hands had to go in the Watter to take them off, it was cold and Windy—

MONDAY Oct 16th 1820

the frost much as yesterday; Turkey, being heard close by we took a walk after them unsuccessfully, they answered to my calls but kept off—

I did not feel well; took some Medicine and Drew the *Hermit Thrush Turdus Solitarius* that I killed yesterday—this Bird can easily be known from the Turdus Auracapillus being about ⅓ Larger, and from the Tawny Thrush, by Looking at the inner part of its Wings which exhibit a handsome light Band of Buff—its stomach contained the remains of insects and the Seed of the Winter Grape—was very fat and [and] delicate eating—these Birds are scarce and not generally known their note a plaintive soft one—Seldom more than 2 are seen together—Some of the Country we saw [to] is extremely high, hilly and broken—saw a wood grouse—many ducks Several *Northern Divers* or Loons—Some Cormorants, Many Crows—Several flocks of Cow Buntings moving Southward—Killed two Partriges and One Turkey—

the Boats ran ashore sandbar at Seven o'clock, with great Exercising one was brought off it was the one I was living in the other stayed fast all night, the hands suffered much from the cold—

TUESDAY Oct 17th 1820

The weather disagreably Cold, but Clear, the other Boat still fast We went early on Shore,—in Kentucky—a Long Walk through the woods was fruitless, I saw 4 Ravens—Many Winter Hawks—some Red breasted Thrushes or Robins—the Wood full of Grey & Black Squirrels returned to the Boats the *other* having joined us with Two Turkeys & One Wood Grouse or Pheasant—

The Turkeys extremely plenty and Crossing the River hourly from the North Side, great Number destroyed [there] falling in the Stream from want of strength—the Partrige when Crossing also, and in fact, all the Game that Cannot propery be called Migratorius=

Saw a Great Number of Chimney Swallows going Southwest—this Bird travells much more advantageously than most all others being able to feed without halting—Killed a fine *American Buzard Falco Leverianus* he was feeding on a Grey Squirel on a stump tree, he fell to the ground and Raised again without loosing his praise—his stomack was filled with that Last Prey—a Large Turkey Cock was stolen from us by some travellers—killed today 17. partridges—One *wood grouse* 4 Turkeys I Killed 2 at one shot—One Hare—One Robin and the American Buzzard—We put some fish Lines out having Landed for the Night all hands being very much fatigued—

Could My Wishes be fulfilled I would have [the] you well fed on that game as the rich calls it richest—the thermometre down at 36—

WEDNESDAY, Oct. 18th, 1820

Jacob Aumack went hunting with us saw some fine Turkeys, Killed a Common Crow Corvus [——] Americanus which I Drew; many Robins in the woods and thousand of Snow Buntings Emberiza Nivalis—several Rose Breasted Gross Beaks—We killed 2 Pheasants, 15 Partridges—1 Teal, 1 T. T. Godwit—1 Small Grebe all of these I have seen precisely alike in all parts—and one Barred Owl that is undoutedly the most plentifull of his Genus—I felt poorly all day and Drawing in a Boat were a man cannot stand erect gave me a Violent headache—The Watter raising a little gave me some Hopes of reaching Louisville before Sunday—anxious to know my Fate—I am confortably situated and would be sorry to be obliged to part with them—the Weather Milde & Cloudy—Caught No Fish Last night—

THURSDAY Oct.[r]—19th 1820

Cap[e] Cummings M[r] Aumack and Joseph after a Long Walk return to Dinner with only 7 Partridges and 1 Pheasant—Mr Shaw Shot 1 Pheasant—I finished my Drawing of the *Common Crow* and after Dinner Went a Shore with the Company—saw Many Cedar Birds killed a Young Blackburnian Warbler—a Young Carolina Cuckow so much reduce by the Hard Weather that he could scarcely fly—killed 5 pheasants 14 Partridges 1 Squirel and 3 Turkeys, Shot *at once* by *Joseph* who was not a little proud when he heard 3 Chears given him from the Boats this was his first essay on Turkeys—

[in the] While absent the boats having put too to make Sweeps, a flock of Turkeys came among[t] them and in tring to kill some with Aumack's pistols one was bustted and the other wounded Joseph's Scull pretty severely—saw an astonishing number of Gray Squirels—the Country being extremely hilly opposite *Wells Point* the hunting Was Laborious and fatiguing.

The Stomack of the Cuckoo Contained 2 entire Grass Hopers one Large Green *Kid diddid* and the remainder of remains of Diferents colopterous Insects.

WEDNESDAY Novemb[r] 1st 1820

Weather drizly and windy Landed a few hundred yards below *Evansville* in Indianna on a/c of M[r] Aumack Who had some money to Collect—he brought on Board only a french Double Barrel Gun and a Gold Watch the man made for Sale—I Wrote to my Beloved *Wife & Mr H. W. Wheeler* Saw Large flocks of Snow Geese but only one in perfect Plumage.

Not so stupid as mentioned in Linne. Left Evansville at 2 o'clock P.M.—Cap[t] Cummings & Joseph parted in the Skiff for Henderson to Get *Dash* a slut I had Left in the Charge of M[r] Brigs—about 3 Miles down We Saw 3 of these Birds that I have considered as being the Brown Pelican they made a noise somewhat like a raven—they alighted on a Red Maple Tree—after many [fruitless] hard trials:—We landed below them and Went a Shore with great expectation of Procuring one—M[r] Aumack drew near them but Missed 2 that Were together and that I expected and hoped to see fall—the wind raising it was Concluded We Should remain—rather sorry that the Cap[e] & Joseph were absent—as we had expected to go Within 2 Miles of Henderson & meet them [them] on their return—

The people at Evansville very Sickly, could not see M[r] D[d] Negley as it was my wish, he being at his House 4 Miles up Pigeon Creek—

Extremely tired of my Indolent Way of Living not having procured any thing to draw since Louisville—

THURSDAY November 2nd, 1820

Cap[e] Cummings Joseph & Dash arrived at one o'clock this morning having a severe rowing match of it We started about 5 and floated down slowly within 2 miles of Henderson when We experienced quite a *Gale* and Put to on the Indianna Shore opposite Henderson—The Wind Blew so violently that I could only make a very rough Drawing of that Place—I can scarcely Conceive that I staid there 8 Years and Passed them Comfortably for it undoutedly is one the poorest Spots in the Western Country according to My present opinion =

Saw some Large Sea Gulls—[a]: Larus argentatus Some Geese, Ducks, &c. So Warm to night that Bats are flying near the Boats—extremely anxious to be doing something in the Drawing Way—

[a]: they are Pale blue above, Tail & Belly White—a few of the outer Primaries Black and about the size of a Raven—

FRIDAY Novemb[r] 3rd, 1820

We Left our harbour at day Break and passed Henderson about sun raise, I Looked on the Mill perhaps for the Last Time, and with thoughts that made my Blood almost Cold bid it an eternal farewell—

Here one of the hands left us a Poor Sickly Devil who had been acting as Cook, called Luke a Shoe Maker of Cincinnati—

The *Indian Summer* that extraordinary Phenomenon of North America, is now in all its Splendor, the Blood Red Raising Sun—and the Constant *Smoky* atmosphere, is undoutedly not easily to be accounted for—it has been often supposed that the Indians, firing the Prairies of the West were the Cause, but since We have Left C. the Eastwardly Winds have prevailed without diminishing in any degree the Smoke—it is extremely bad to Most Eyes and particularly so to Mine—

Cap[e] Cummings, M[r] Shaw & Joseph took a Long Walk [for] but saw nothing, killed 4 Squirels—one Butcher Bird, and a Swamp Blackbird—Large flocks of them seen travelling Southwest—I shot a Turkey Buzard, *Vulture Aura* about 120 yards off with a ball Saw When We Landed at the foot of Diamond Island a fine *Snow Goose* [but] a Young One—Saw several N. Divers, Some Geese, a few Sand Hill Cranes—& some Ducks—

SATURDAY Novemb[r] 4th, 1820

Landed Last evening opposite the Middle of Diamond Island lately the property of Walter Alves Deceased of Henderson—about 9 o'clock the Wind Rose & Blowed a tremendous Gale which continued all Night, fortunately for us We Were under the Lee of the Land—

This Morning the Wind had abated alitle, but We could not go, the River making immediatly below us a Turn Southwest[d]—5 of us took guns and Went to the Island, We Walked almost all over it—Saw great many Turkeys—and Many Dears—I killed a Large Buck that died in the Cane, and Lost it—We Brought Nothing on Board, if 2 of us only had been there, probably We could have Made a good hunt—I Shot a *Winter Wren* but cut it so much that it could not be drawn [it]—

Returned to our Boats about 5 o'clock the Wind still Blowing, but rounding to the Northwest and Weather very Cold.

Watter falling—

I Remarked this morning that the Turkey Buzzards that had roosted over a dead Hog Last Night, Took a Long flight eastwardly this Morning as if to excite their appetite and returned to Consume their filthy Meal about 2 o'clock the number considerably Increased—While sailing high several Hawks for the Sake of amusement Chased them and sailed Many of them almost to the ground—

My Slut Dash apparently good for nothing for the want of Employment—

Now & Then We see a Blue Crane—Saw Many Wood Groos on the Island

SUNDAY November 5th, 1820

The Weather fair this morning, the Thermoter down at 30—the sun rose beautifull and reflected through the Trees on the Placid Stream much like a Column of Lively fire—the frost was heavy on the decks and when the Sun Shun of it it Looked beautifull beyond expression—

We floated tolerably well the river being here contracted by Large Sand Barrs—

Gayly we were overtaken by a Skiff containing a Couple of Gentill Young Men, Bound to New Orleans they had Matrasses, Trunks, a Gun & Provisions—

Saw about the same time a fine Brown Eagle—Shot at it without effect—

Many Dears where Merely Gamboling on the Sand Barrs and [much] excited us [to]

We passed *Mount Vernon* a Small Village in Indianna about one mile above the upper end of Slim Island Mr Shaw & Cap^e^ Cummings went to the Island but returned without any thing—this part of the River rather Dificult—

We Landed about 3 Miles above the Mouth of *High Land Creek* in the *Mississipi* Bend—

Saw Many Geese, some Sand Hill Cranes—a few Loons some *Red Breasted Thrushes,* many Sparrows & parokets.

Killed only one Winter Hawk, & Shot at one Wood Cock.

as I promised You a Picture of the Caracters We have on Board of Both Boats I will attempt to Copy them, [Was my Talent in writting anything] Could my Pen Act as a Black Chalk by the help of my fingers you might rely on the Exibition of the figures—Yet I undertake it with pleasure, knowing how sweet this May be to you & Myself some Years hence, while sitting together by the fireside Looking at Your Dear Mother reading to us—

being on Board of Boats Much in the situation of Passengers I am of Course Bound to give the preference to those who are termed Capitains and Mr Aumack is the First that I will bring to your attention—

You have seen him and of Course I have not much to say the acquaintance of Man When unconnected by Interest is *plain* [and] easily *understood* & Seldom Deviates.—

he is a good Strong, Young Man, Generously Inclined rather Timorous on the River, Yet Brave and accustomed to hardships—he Commands the Boat where I am—

Mr Loveless is a good Natured Natured, rough fellow brought up to Work without pride, rather anxious to Make Money—Playfull & fond of Jokes & Women—

Mr Shaw the owner of Most of the Cargo puts me in mind of some Jews, who are all Intent on their Interest & Wellfare; of a keen Visage & Manners; a Bostonian—Weak of Constitution but strong of Stomack—Would Live Well if at any one else' Expense. =

The Crew is Composed as follows

Ned Kelly a Wag of 21. Stout Well Made, handsome if Clean, possessed of Much Low Wit, produces Mirth to the Whole even in his Braggardism—Sings, dances and fiels always happy—he is Baltimorian—

2 Men from Pennsylvania although not brothers, are possessed of a great sameness of Caracters—these are *Anthony P. Bodley* & *Henry Sesler*—they Work Work Well, talk but litle and are Carpenters by Trade.

The Last is Much Like the Last of every thing, the Worst Part—Joseph Seeg, Lazy, fond of Grog, says nothing because it cannot help himself, sleeps Sound, for he burns all his Cloths, while in the ashes

Cap^{e} Cummings Joseph & Myself form the Rear at Times and at Times the Van—You have seen the Life and there Likeness could not give you a better Impression than that you have formed—We agree Well, and are Likely to agree Still—

MONDAY, Novemb^{r} 6th 1820

The thermometer this morning was down at 28 and it felt very disagreable—Took to the Shore and walked 9 miles to the Mouth of the Wabash, but saw nothing to Shoot at—about one Mile below that the Wind springing ahead Brought us to on the Illinois Shore and 5 Guns went hunting, I shot 6 Dear!—

the people here have a dreadfull sickly aspect and their Deportment not the Most Inticing—

the Caves plenty—

Our Boats started about One Hour before sun down, & Cap^{e} Cummings having extended his hunt up the Wabash had a Long Walk for nothing—

We landed for the Night about 6 miles above Shawaney Town on the Kentucky Shore—

Weather appearance of rain, and blowing, Much Warmer—

Saw Some Robins, a few Blue Jays, a few Blue Birds, Geese, S. H. Cranes, Ducks—Buzzards & the [general] usual number of W. Peckers common at this season—

TUESDAY Novemb^r^ 7th 1820

Weather at 50 this morning, rainy & Desagreable, Landed at Shawaney Town where we staid six hours—I staid snugg on Board Mr Aumack having naild up the only accessible hole to our Boat.—

I Wrote to My Wife & Directed it to Mr Wheeler—We left Shawaney town at half 5 & went only to the Lower end of it, the Wind raising again with an appearance of a Very Boisterous Night—

Jacob Aumack killed a *Rusty Grakle* a beautifull Male and as these Birds are scarce I intend Drawing it tomorrow = it is but seldom that Many of these Birds are seen together, they Walk with great Statliness and Elegance, are swifter of flight than the Swamp Blackbirds.

I felt very anxious during all the time of our stay to be off from this Place.

This Evening Ned Kelly & his Companion Joe Seeg having Drank rather freely of Grog, they had a Litle Scrape at the Expense of Mr Seeg's Eyes & Nose—

The People of Shawney Complaining of Sickness, the place improved but Litle

WEDNESDAY Novemb^r^ 8th, 1820

the Weather Calm & beautifull this Morning [and] We started with a good prospect and our Landing within Two Miles of the famous *Rock in Cave* to night prove it—

I Drew this Morning *My Rusty Grakle, Gracula Ferruginea*, and made a handsome piece of it.

Cap^e^ Cummings hunted all day but saw nothing—since 3 Days we have been particularly unfortunate in our Hunts.

Near our Landing Place We Went on Shore to procure some Venaison Hams—Mr Shaw bought 4^prs^ for 2$ remarkably fine—the young man who sold them had Killed 3 Dear to day and had hung one Large Buck for his Dogs to feed on—

Killed to day One Grey Squirel & 3 Tell Tale Godwits these Birds Wade so deep that one would suppose they are swimming, fly a few Yards into Shallower Watter, holding their wings up untill perfectly satisfied of their being on the Bottom—then run about briskly & ketch small fishes with great dexterity—

Saw one Sea Gull a Large one—a few Gold Finches Many Cardinals—some Divers—but neither Ducks or Geese—The Weather has become Cloudy & raining by Intervalls is Now beautiful! with an appearance of Frost about Two hours before sun sett a *Barred Owl* teased by four Crows and Chased from the tree where he was Lit raised up in the manner of a Hawk in the air so high that We Lost Entire sight of him, he acted as if Lost—now & then making very short Circles and flapping his Wings quickly, then zig zag lines—this was quite a new Sight and I expect take place but seldom—I felt anxious to see his Descent to the Earth but Could not—

The Trees here have Lost all their foliage, the Cane & a few Green Briars is all that animate the Woods—the Shores are thickly sett with Cotton Wood Trees.

THURSDAY Novembr 9th 1820

The Wind blowed nearly all day a head, the Weather cold—Saw no Game although We Walked a great deal in the Woods the Country extremely poor here—

We Landed at Sun Set at the *Rockin Cave* having Come only about 2 Miles—

I began My Sketch of it immediately, regretting that We had not reached this place Last Night.

We Purchased a Skiff from a flat Boat—Ducks & Geese very numerous flying down the Stream—Shot 2 Ducks of the flocking Fowl Kind—& 3 Squirels—the Thermometer in the Watter at sun set was down to 27, appearance of a Very Cold Night—

a Man on the Shore told me that Last Winter he had Caught a Large Number of Malard Ducks with a Trap set with a figure of 4 & shaped Like a Partridge trap

The Tell Tales we eat to Day were very fat but very fishy—I eat the purple Grakle it tasted well—

FRIDAY Novembr 10th 1820

as soon as Day Light permitted me this morning—I Took Joseph on Shore and Lighted a good fire—took also My Drawing Book etc with a Skiff—the Morning pleasant and the Thermometer raised to 50° While I was taking My Sketch of the *Rockin Cave* Cape Cummings took a Good Walk through the Woods—at 9 My Drawing was compleat—this Cave is one of the Curiosities that attract the attention of allmost every Traveler on the Ohio and thousands of Names & Dates ornament the sides & Cealing—there is a small upper room dificult of access imediatly above & through the Cealing of the Ground floord one, Large enough to Contain 4 or 5 persons when sitted on their hams—this place is said to have been for Many Years the *rendez Vous* of a noted Robber of the name of Mason it is about 20 Miles below Shawaney town on the same side; had our Boats spent a Day there, I would have been pleased to take several Diferent Views of it—the Rocks are Blue Lime Stone containing in Many Parts Round Masses of a fine flinty appearance Darker than the main Body—

at Nine o'clock it became Cloudy & Cold, We left for our Boats, but before We reached them, it snowed & Hailed & Wetted us Compitly—Our Boats had parted to Cross & Run through the *Walker's Barr* and *Hurricane Island*—We Landed only about One Mile below the Latter, the rain Increasing and the Weather extremely Disagreable—Never have seen so much snow at this season in this Latitude—

Saw a fine Black Hawk Falco Niger—& Black Gull,—Shot Two Ducks—

SATURDAY Novemb[r] 11th, 1820

It rained hard during the whole of the Night and this day floated only about 7 Miles—

Saw a few Turkeys,—

A Flock of *Carrion Crows* made us go to the shore but they were so exceedingly shy that they would fly several hundred Yards off—while the Turkey Buzzards that [when] accompanied them would suffer us to Walk under the trees on which they alighted—the Carrion Crows are very scarce in this Part of the Country and keep Generally Lower to the South—their flight is heavy and their appearance while on the Wing awkward

We Landed for the Night a Golconda a small Town of the Illinois—Titles disputed of course the place not Improving—Court was sitting—

SUNDAY Novemb[r] 12th 1820

The Wind Blowed this Morning and We did Not Leave the Shore untill 9 o'clock—Wind fair—Weather raw & Cloudy—Mr. Aumack Killed a Duck [Ruddy Duck] out of a Flock of 5 that proved to be a Nondescript—and also a *Imber Diver*—the Wind rendered our Cabin smoky I could Not begin to Draw untill after Dinner—I had the pleasure of Seing Two of the same Ducks Swimming Deep, with their Tail *erect*, and Diving for food—having never seen these Birds before, it was highly satisfactorily to Me—Tomorrow I will give a thorough Description of it

The *Imber Diver* was Shot Dead and proved a beautifull specimen—of Cours I will give You a Drawing of it—for Some time before I procured one of them; they Were Called *Northern Divers*, the Moment I Saw this, the size and Coloring Made Me Sure of it being an *Imber Diver*—

Saw a Large flock of Turkeys fly across from an Island to the Main, killed None—

We are Landed about half a Mile above the *Cumberland Island* the weather not so cold as it feels the Thermometer at 38—had some spits of snow in the forenoon—

Vast flocks of Ducks & Geese flying Southwardly

MONDAY Novemb[r] 13th 1820

a Beautifull Morning enable[d] me to go on with my Drawing very early—a Light frost embelished the raising of the Sun

We Landed at the Middle of Cumberland Island to dispatch a Skiff to take Soundings—

finished My Duck by Dinner Time and Was Lucky enough to kill another of the same kind, precisely a Like but rather Less in Size, it is with apparent Dificulty or a Sluggish disposition that these Birds rise out of the Watter & yet will not dive at the flash of a Gun—while on the Wing are very Swift

this afternoon I Begun the Drawing of the Imber Diver had Two Long Chases after a Couple of others, that out Managed our Skill—they would Dive as if Going down the Stream and raised from One to Two hundred Yards above us—they frequently Dipp their Bill in the Watter, and I think have the power of judging in that Way if the place Contains Fish—One I shot at; dove & raised again Imediatly as if to see Where I was or What Was the Matter—

I Saw Several of the *Fin Tail Ducks*, all acting the Same Way i.e. Swimming Deep & the Tail Erect—No Doubt this appendage is very usefull to them when under the Watter—

Saw a Bear on a Sand Barr, had a great run after it—to no purpose—

Saw 2 of those Birds I take for Black Pelicans—Many Loose flocks of Blackbirds Ducks & Geese—M[r] *Aumack* saw an *Eagle* with a White head and *Brown Body & Tail*, this Corroborates with the Idea of Willson of its being the same Bird with the *Brown* Eagle

Landed about the Middle of Tenessee Island Weather Much Milder—

Joseph Made a *Faux Pas* this day—the Whole of our Folks not in the best humour—Killed 7 Partridges 1 G. Squirel & One Duck

TUESDAY Novemb[r] 14th 1820

Drawing this Morning as soon as the Light would permit me—Started early—

Went out in the Skiff to try to Shoot the *Largest* White *Crane*, with Black tips, but he walk[d] off from the Shore and I return[d] knowing that it would be vain to attempt to follow him on a Large naked Sand Barr—felt great anxiety to procure such for he appeared Beautifull—

Saw several Eagles, Brown & White headed—Although I drew nearly the whole of this day I did finished My *Imber Diver;*

Cap[e] Cummings killed 26 Starlings, Sturnus Pradetorius all young—eat them at our Supper, good & Delicate—1 golden Plover & Two Squirels—

Mr. Shaw killed an Owl that unfortunately he did not fetch; = We passed this Day *Fort Massacre* here the Ohio is Magnificent, the river about one & ¼ mile Wide affords a view of 14 or 15 Miles, and this afternoon being Calm with one of those *Whimsical* sunsetts that only belong to America rendered the Scene extremely interesting.

We landed about One Mile below what is called the Little Chain an partial obstruction to the Navigation of this Queen of Rivers.—

Saw several *Swans* flying very high—Geese are in Constant View but have so far outwited us, these Birds are Wilder on the Rivers than when in the Ponds or small Lake that in Many places run Parralel with the Ohio at a Small distance in Land—

Cap[e] C. Brought an Oppossum, Dash after having broke I thought all its bones left it—it was thrown over Board as if dead, yet the moment he toucht the Watter he swam for the Boats—so tenacious of Life are these animals that it tooked a heavy blow of the Axe to finish him—Tomorrow We Hope to Pass the Last Dificulty & Two Days More May take us to the Mississipi.

Total Length of the Fin Tailed Duck 15½ Inches—¾[——] to [——] of Tail Bill Dark Blue, Broad for the size of the Bird—& sharply hook at the point—Legs & feet L. Blue the palms Black, Tung fleshy—upper Part of the Head [Dark Brown,] Back Wings—& Tail Dark Brown Zig Zags with transversal Bars—of Light D°—Sides Dark Chestnut Eye rather small Neck Breast & Belly Light Brown with transversal Black drops—A Triangular White spot forms the under Tail Coverts—

Tail composed of 18 Feathers rounding each feather narrow Sharp & Terminating in Spoon Like Shape Points—this is White—the Head & Neck Short & Thick—Swims Deep with part of the Belly Silvery white

Breadth 22 Inches—Wings Brown not reaching the Tail by ½ Inch—No Wing Stripe—

When I saw these Birds the Weather was Boisterous since fair have not seen one—

Imber Diver Weighes 6 ^{lbs}

Total Length 2 8/12 feet—

to end of Tail 2 4½/ 12— " —

Weidth tip to tip 4 "

Length of the Gut 5 8/12 "

Contents of gut & Gizzard Small Fish, Bones & Scales and Large Gravel—Body extremely fat

rancid—Belly & Vest White but not silvery as in the Grebes—

WEDNESDAY 15th 1820 November—

At Work again this morning as early as could be, a beautiful day—finished My Drawing to my liking having Sketch^{d} our Boats in the View of this Magnificient Part of the Ohio = Saw more than a Dozen of Eagles and one I had a good view of had a White Tail & a Brown head, Again [Yet] I remark^{d} that the *Brown Eagles In Ohio* Esp. were at Least ¼ Larger than the White headed ones

Saw a Large flock of Large White Gulls with Black Wing Tips very shy while on the Wing but not at all so when swimming fired two guns Without affect

Passed the famous Chain of Rocks, Much diversion to see Mr Aumack's Movements—

having seen Steam Boats allmost every day fast on Sand Barrs I have taken no particular notice of any of their Names & Positions—have Passed this day 3 flat Boats belonging to Wm Noble of Cincinnati that left that Place early in August—3 out of 6 are lost

Landed about 2 Miles above New America on the Illinois Shore—Broken Land, fine Timber of Oak & Poplar = Killed an Oppossum—

Our People Much MELOWED; Saw Winter & Shore Larks Many Geese & Ducks 2 Swans—

THURSDAY November 16th, 1820

We floated only about Two Miles & Landed at *America*—to sell some *Articles;* people very sickly, a miserable place altogether—took a long Walk this Morning—[and] this afternoon Cap[e] C. Joseph & I took a Skiff and spent the afternoon hunting, but killed nothing—Saw Two Black Hawks =

At our Return at Night found M[r] Aumack in Bad humor, and after We had retired to our Cabin for the Night, Received a *Humorous Lesson* that I shall Never forget—

My Dear Children if Ever you read these trifling remarks pay your attention to what follows—

Never be under what is Called obligations to Men not Aware of the Value or the Meaness of their *Conduct*

Never take a passage in any State or Vessel without a well understood agreement between you & the owners or Clerks & of all things Never go for Nothing if you Wish to save Mental Troubles & Body Viscisitude

Well aware that I shall never forget this Night as Long as I live, I close [here]

The old *Washington* Steam Boat came along side of us Took 70 barrels of salt rais[d] steam and made herself fast about 2 miles Below—

FRIDAY November 17th 1820

We left early—I took the Skiff and Went to the Mouth of the Ohio, and round the point up the Misisipi—

Eleven Years ago on the 2 of January I ascended that Stream to St. Genevieve Ferdinand Rozier of *Nantes* my partner in a Large Keel Boat Loaded with Sundries to a Large Amount *our* property

The 10th of May 1819 I passed this place in an open Skiff Bound to New Orleans with two of My Slaves—

Now I enter it *poor* in fact *Destitute* of all things and reliing only on that providential Hope the Comforter of this Wearied Mind—in a flat Boat a Passenger—

The meeting of the Two Streams reminds me a little of the [Young,] Gentle, [man] Youth who Comes in the World, spotles he presents himself, he is gradually drawn in to thousands of Dificulties that Makes him wish to keep [to himself] apart, but at Last he is over done [and] mixed, and lost in the Vortex—

The Beautifull & Transparent Watter of the Ohio when first entering the Misisipi is taken in small Drafts and Looks the More aquable to the Eye as it goes down surrounded by the Muddy Current, it keeps off as much as possible by running down on the Kentucky side for several miles but reduced to a narrow strip & is lost—I saw here two Indians in a Canoe they spoke some French, had Bear traps, uncomonly clean kept, a few Venaison hams a gun and Looked so Independent, free & unconcerned with the World that I Gazed on them, admired their Spirits, & wished for their Condition = here the Traveller enters a New World, the current of the stream about 4 miles per hour, puts the steersman on the alert and awakes him to troubles and difficulties unknown on the Ohio, the Passenger feels a different atmosphere, a very diferent prospect—the Curling stream & its hue are the first objects—the caving in of the Banks and the Thick Set Growth of the Young Cotton Wood is the next = the Watter's dencity reduced the thermoter from 62 to 20 degrees = We Landed Very Early, Cap[e] C & I Walked Through the Woods, and remark[d] the Great Diference of Temperature so suddenly felt—

I bid my farewell to the Ohio at 2 o'clock P.M. and felt a fear gathering involuntarily, every moment draws me from all that is Dear to Me My Beloved Wife & Children—

The Boats separated on Entering the Mississipi, as being safer to navigate it singly—We felt the better for this and Hope good Cheer will revive again—

Although I hunted a good Deal this Day I saw but Little and Nothing New, a few King Fishers, some Divers, Geese Ducks—some Gold Finches the Notes of Which reminded Me of the *Canary Birds*, a few Blue Geays, Now & then the Plaintive farewell note of the *Blue Bird:* rather sorry that the Strong Current we are in Will Not permit Me to go a Shore unless Landed by the force of Contrary winds—

The Mississipi is a good Midle Stage at present

SATURDAY November 18th 1820

floated within about 2 miles of the Iron Banks and Land in Kentucky at about ½ past 3 oclock—

I took a sketch of the River below us comprehending on our Left the Iron Banks, the Chalk Bank on our Right in the Back ground Wolff Island and Part of the Missoury shore.

My drawing finished took a Walk in the Woods, the Country full of Ponds of

Stagnant Watters, Shot 2 Malards While Dash was bringing out the Last one a White headed Eagle *Dashed* at the Duck the Bitch brought it—

Killed an Oppossum Many Blackbirds the Thermometer at 64. Bats in the Evening and Butterflies seen to day in quantity as well as many other Insects—

The Game Not so plenty as on the Ohio and Much Shier—

SUNDAY 19th November 1820

When We Left Cincinnati, we agreed to shave & Clean completely every Sunday = and often have been anxious to see the day come for certainly a shirt worn one week, hunting every day and sleeping in Buffalo robes at night soon becomes soiled and Desagreable. We pass this morning the famous Wolf Island, the history of which is amply given in the Ohio Navigator—here said Mr. Lovelace a Man [of] called *White* having become a Lunatic as I was going up the River, jumped over Board in the Night made the shore although he could not swim and I was then at anchor in 7 feet of watter and never seen afterward, I sent several *hands* to Look for him, his tracks only Led to the Top of the Bank—the Muskitoes being then remarkly Bad—he must have Died in the Same Bend of the River found 2 dead men shot through the head could not burrie them their stench was too great—

We floated to day almost 20 Miles Killed Nothing—saw Mr *James Asler* who told Me that Mr *Thomas Litton* & himself Lived with a Mile of *Chalk Banks*—

Ivory Billed Wood Peckers are Now Plenty, Bears, Wolf, &c but the Country extremely Difficult of Access, the *caves* Extending in Many Places several Miles from the River—

On Sundays I look at My Drawings and particularly at that of My Beloved Wife—& like to spend about one hour in thoughts devoted to My family—

Landed opposit the Head of N° 8. at the foot of N° 7—in Missoury—put out our Lines caught a Cat fish—Weather agreeable,

Saw Many Gulls—unknown—

the Woods deserted almost by small Birds—Saw four Partriges—

The Wild Geese here sits on the Banks Many feet above Watter and feed on the seeds of a small grape somewhat resembling *Fol Avoine*—but are extremely shy—

MONDAY 20th November 1820

The winds on this River are Contrary to our Wishes as that of an Ole Rich *Maid* to the wishes of a Lover of Wealth, We are anxious to Make progress on account of our Situation—but it is disposed off Diferently by a *Superior Power*

We Came but a few Miles and Landed about Noon in Such a Dreary place that Neither the Woods Nor the Stream would afford us any benefit—it rain[d] in the evening—

Killed to day a *Red Tailed Awk* a great distance with a Ball—a Red Owl *Strix Asio*—one goose—Nothing in the Woods that are rendered almost Impenitrab by the Caves—being on a Very Loose Muddy Ground caught no fish—

TUESDAY November 21st 1820

The Wind High all Day Landed at New Madrid at 3 o'clock P.M.—

This allmost deserted Village is one of the poorest that is seen on this River having a Name; the Country Back was represented to us as being good, but the Looks of the Inhabitants contradicted strongly their assertions—they are Clad in Bukskin pantaloons and a sort of Shirt of the same, this is seldom put aside unless so ragged or so Blooded & Greased, that it will become disagreable even to the poor Wrecks that have it on—

The Indian is More decent, better off, and a thousand time more happy—here familly dicensions are at their Zenith, and to Kill a [man] Neighbour is but More than a Kill Deer or a Racoon—

A M[rs] Maddis formerly the Lawfull Wife of M[r] Reignier of St. *Genevieve* resides and keeps a small store in Company with a French Gentleman. We where told that the Partnership was rendered agreable to both by a mutual wish of Nature—Went to this Lady's house who knew me first and exibited much of the french Manners

Felt dull, this evening for every object that brings *forward* the *Background* of My Life's Picture shew too often with poignancy the diference of situation—

Made some inquiries about the regulations of the Post office, now suficiently encouraging to enable my writing a few Lines to My Beloved Wife & Son—

Saw some geese, Killed one, & Golden Plovers and tw Shore Larks—Caught No Fish—

A Black Hawk passed within a few Yards of me to day when I had a Rifle and could Not kill it on the Wing, these Birds become more aboundant as we descend; a few swans are now & then seen very High,

All our Hands Playing Cards untill bed Tim about 9. oclock = The *Swamp Sparrows* and Snow Birds are plenty in the High Dry Grass that Lines the Banks of this River, but the Woods have Nothing in More than the *Pait Pait Pait* of the Monogamous—Wood Peckers—

WEDNESDAY November 22^d 1820

We Left New Madrid at day Break. We had gone but a very Short Distance when The Wind Rose, yet we made as the Best days' run since in this River and Landed after Dark 3 Miles above the *Little* Prairie on the Missoury Shore—

Joseph & Myself floated before the Boats almost all day in hunt of a White headed Eagle, in Vain, I shot at & Missed a Beautifull *Black Hawk* and a *Brown Eagle* having a very Long *Bend* to run through, We helped Mr. Lovelac's; While We when rowing, a White headed Eagle Dashed off from the Top of a High Cypress Tree, after a Little *Duck* the *Spirit* and was on the point of being in Possession of it—when I sent him Two heavy Loads from My Gun that Wounded him badly—

I called this Morning at *Belle Vue* Wood Yard to see Mr Dela Roderic, he was absent to a Cypress Swamp, I saw his Wife & Sister, and Left my Respects

the Weather quite Pleasant, although We have Light White frosts every Night—

Some minutes after Landed, Dash started an oppossum that took Every Man to Shore thinking it was a Bear, the Poor Oppossum came on Board with us—

fishing lines all out lost one, and a large *Cat Fish* in tringig to hawl him in—Geese, Sandy Hill Cranes Plentifull on the Sand Barr opposite us, they regularly resort to these for roost.

THURSDAY November 23 1820

As soon as we had eat our *Common Breakfast* fried Bacon and Soaked Biscuits—Joseph went to his station and I to Mine, i. e., he rowed the skiff and I steering it—Went to the *Little Prairie* shot at a Brown Eagle probably 250 yards and yet cut one of its legs—

at this Place We Saw a great number of Birds, Mostly *Red Breasted Thrushes*—

the sungs of Which revived our Spirits and Imparted within us the Sweet sensation that Spring brings to Minds of *our Kind.*

the Rusty Grakles extremely Plenty—Snow Birds & Many Sparows

I shot a Beautifull *White headed Eagle Falco Leucocephalus*—probably 150 yards off, My Ball Went through its body—

Returned to our Boats immediately and began My Drawings—it is a Handsome Male—

Many Shots at Geese, but We find them so shy that We Loose Much Ammunition in Contending With them—

Floated 23 miles Landed opposite Island No. 20 according to the Old Navigator, some Indians camped on it, Made us Load all our pieces—

I saw Two Eagles Nest, One of them I remembered seeing as I went to New Orleans 18 months ago it had being worked upon and no doubt young Where raised in it, it is in a Large Cypress Tree, Not very high, Made of Very Large Dead sticks and about 8 feet in Diameter—

Since I killed the one before me I am Convinced that the *Bald Eagle* and the *Brown Eagle* are Two Diferent Species—

FRIDAY November 24th 1820

high Winds, remained at Our Last Night's harbor all day—at Day Break saw a Deer crossing the River below us, ran him down and Brought him to the Boats; Cleaned, it Weighed [1/12 ?] had 9 points to its horns and so much run down that its Neck Was swolen ½ the Size of its Body—

I spent the greater part of the day drawing; All hands hunting—Killed Two Geese, 1 Racoon & 1 Oppossum—

the Woods here are so Dreadfully tangled with *Bull rushes*, Green Briars and Canes that the Travelling through them is extremely Irksome =

Saw some Carrion Crows and some *Turkey Buzards* that were attracted by the scent?? of the Deer We had hung in the Woods??

Immediately below us is a family of Three people in Two Skiffs a Woman & 2 men; they are Too Lazy to Make themselves Comfortable, and Lie on the Damp Earth, near the Edge of the Watter, have *Racoons* to Eat and Muddy Watter to help that food down, are from the Mouth of Cumberland and moving to a Worst Part of the Worst Without Doubt—

Saw some Ivory Billed Wood Peckers, these Birds allways go in Paires and when they Leave a Tree to fly to another they Sail and Look Not unlike a *Raven*. I shot and Killed a Turkey Buzzard a great Distance, Mistaking it for a Carrion Crow—unfortunately we are in bad part of the River for Fish—

SATURDAY November 25th 1820

I spent the whole of this day drawing the *White* headed *Eagle,* the Weather exceedingly Warm, the thermometer rose to 70°. The wind Blowing Strong a head We remained still; in the Course of the Afternoon a small Steam Boat, the *Independence* passed us, I saw with the spy glass old Cape Nasson of Louisville; Butterflies, Wasps & Bees plenty all day about us—the Skiff family of yesterday a few hundred yards below us, the Woman washed for us

At Sunset the Wind Shifted, A heavy Clouds came over us and Made a great difference in the atmosphere—

Killed 2 Geese and Two Racoons—Saw a few *Carrion Crows* attracted by the Scent of the Deer We Caught yesterday

Our Hands sailing all day

SUNDAY November 26th 1820

Drawed all day, floated 18 Miles—the family in the Skiffs came on Board this Morning. Nearly frozen, the Thermometer down at 22—the ground very hard, and my being without a Shirt—Made Me feel rather unpleasant.

the Woman of the Skiffs Mending My Good Brown Breeches—

to Look on those people, and consider cooly their Condition, then; compare it to mine; they certainly are more Miserable to Common Eyes—but, it is all a Mistaken Idea, for poverty & Independance are the only friend that will travel together through this World.

Shot at an Eagle With a White head and Brown Tail.

Ducks, Geese, Swans, & other Birds all going southwardly—

MONDAY November 27th 1820

the weather raw and Cloudy. Finished my drawing of the White headed Eagle, having been 4 days at it—

That Noble Bird weighed 8½^{lb}, Measured 6 feet 7/2/12 his Total Length 2:7/2/12—it proved a Male, the heart extremely large, My Ball having passed through his Gizzard I could not see any of the Contents—

Those Birds are becoming very Numerous, hunt in pairs, and roost on the Tall trees above their Nests—One this morning took up the head of a Wild Goose thrown over board, with as much ease as a man could with the hand—they chase Ducks and if they force one from the Flock he is undoubtedly taken, carried on a Sand Bank and eat by Both Eagles—they are more shy in the afternoon than in the morning—they seldom sail High at this season, Watch from the tops of trees and Dash at any thing that comes near them—to secure a Goose, the Male & Femelle, Dive alternatively after it and give it so little time to breath that the poor fellow is forced in a few Minutes.

We are all unwell having eat too freely of the Buck. Mr Shaw went off this morning to Mr Lovelace's boat—Made a good run—saw a Large Flock of White Gulls—but not a Land Bird—Much to My surprise I have not yet seen a Pelican, nor a Swan on the Barrs or in the River—Malards are the only Ducks we now see—No Game, to be procured Not able to hunt on the shores—We are Landed at the foot of Flour Island, opposite the first Chicasaw Bluff—the First High Ground since the Chalk banks—

While Looking at My Beloved Wife's Likeness this day I thought it was altered and Looked sorrowfull, it produced an Immediate sensation of Dread of her being in Want—yet I cannot hear from her for Weeks to Come—but Hope she and our Children are Well—

The Eagles along the Banks of this River, retire in bad weather to the Inner parts of High Cypress Woods and remain on Low Limbs for whole day, I had an opportunity of seeing several from our Landing place, with my spy glass.—

TUESDAY November 28th 1820

As it is a rainy morning, I cannot, hunt, and will take this opportunity to retaling to you such incidents relative to my Life as I think you may at some future period be glad to know—

My Father John Audubon, was born at *Sables D'Olorme* in France; the son of a man who had a very Large familly, being 20 males & one femelle. his Father started him at a very early age *Cabin Boy* on Board a Whaleing Ship—of course [his youth his] by education he was [no more than what is call here w] nothing; but he naturally was quick, Industrious and soberly Inclined; his voyage was a hard one but he often assured me that he never regretted it—it rendered him Robust, active and fit to go through the World's rugged paths. He soon became able to command a Fishing Smack, to purchase it, and so rapidly did he proceed on the road of Fortune, that when of Age, he commanded a small Vessel belonging to him, trading to St. Domingo—

A Man of Such Natural Talents and enterprise could not be confined to the common drudgery of the Money Making Annimal, and entered an officer in the French Navy's Service under Louis the 16th was fortunate and Employed an Agent at St. Domingo to Carry the trade—Every movement was a Happy hit, he became Wealthy = the American Revolution brought him to this Country Commander of a Frigate under the Count Rochambeau, he had the honor of being presented to the Great Washington, and Major Croghan of Kentucky who has told me often that he then Looked Much Like me was particularly well acquainted with him. My Father was in several Engagements in the American service and at the taking of Lord Cornwallis—

Before his Return to Europe he purchased a Beautifull Farm on the *Schuillkill* and *Perkioming* Creek in Pennsylvania; the Civil Wars of France and St. Domingo, brought such heavy ravages of Fortune on his head, that it was with the utmost Dificulty that his Life Was Spared—

he along with thousands now saw his Wealth Torn from him, and had Little More left than was Necessary to Live and Educate Two Children Left out of five—having 3 older Brothers killed in the Wars—

he remained in France reentered in the Service under Bonaparte; but the French Navy prospered not and he retired to a Small [but] beautifull Country Seat, Three Leagues from *Nantes* in Sight of the Loire and ended his Life happy = Most Men have faults, he had one that never Left him untill sobered by a Long Life common to Many Individual, but this was Counter balanced by Many qualities—his Generosity was often too great—as a Father I never complained of him and the many Durable Friends he had prove him to have been a *good* Man—

[Two lines blotted out here]

My Mother, who I have been told was an Extraordinary beautifull Woman, died shortly after my Birth and My Father having remaried in France I was removed thereto when only Two Years Old and receive by that Best of Women, raised and cherished [by her] to the utmost of her Means—My Father gave me and My Sister *Rosa* an education appropriate to his purse. I studied Mathematicks at an early Age, and had many Teachers of Agreable Talents. I perhaps would have much stored up, if the Continual Wars in Which France Was engaged had not forced me away when only Fourteen Years Old = I entered in the Navy and was Rec^{d} a Midshipman at Rochefort Much against my Inclinations—the Short Peace of 1802 between England & France ended My Military Carreer [——]. & [but] the Conscription determined My Father on sending me to *America* and Live on the *Mill Grove* Farm I have mentoned above—he sent me to the care of *Miers Fisher,* Es^{qr} a rich and honest Quaker of Philadelphia who had been his agent for many years, and who received me so Politelly that I was sure he Esteemed My Name—

A Young Man of *Seventeen* sent to America to *Make Money* (for such was My Father's Wish) brought up in France in easy Circumstance who had never thought on the Want of an article I had had at Discretion, was but ill fitted for it—

I spent much Money and One Year of My as Happy as the Young Bird; that having Left the Parents sight carolls Merily, While Hawks of All Species are Watching him for an easy prey

I had a Partner with whom I did not agreed, he [——] [tried] waited his opportunity [——] [that] We parted forever.

Here it is well I should Mentioned, that I Landed in *New York,* took the Yellow Fever and did not reach Philadelphia for Three Months—

Shortly after My Arrival on My Farm, Your Mother *Lucy Bakewell* came with her Father's Familly to a Farm Called *Fatland Ford* and divided from mine only by the [road going from] Philadelphia Road.

We soon became acquainted and *I* attached to her. I went to France to Obtain My Father's Consent to Marry her, and returned with a Partner, Ferdinand Rozier of Nantes entered in Business for the thoughts of Marriage brought Ideas so new to me that I [entered] began with pleasure in the [Business] War to secure my Future Wife and Familly the Comforts We had both been used to—I travelled through the Western Country and Made Louisville my Choice for a residence—On my return and being of age I married your Beloved Mother on the 5th of April 1808 and removed to Kentucky—Louisville did not suit our Plans and we left that place with a View to Visit St. Louis on the Mississipi; but it is so seldom that our Wishes are favored that we did not reach that Place, for My Partner not being on good Terms with My Wife, I left her and You *Victor* at Henderson, you when there a babe, having reached St. Genevieve through Many Dificulties, Ice, &c I parted from Mr Rozier and Walked to Henderson in Four Days 165 Miles.

Your present Uncle T. W. Bakewell Joined me in opening a House at New Orleans that the War with England Made us Remove to Henderson.—

This Place saw My best days, My Happiest, My Wife having blessed me with Your Brother Woodhouse and a sweet Daughter I Calculated, to Live and died in Comfort, Our Business Was good of course We agreed. but I was intended to meet Many Events of a Disagreable Nature; A Third Partner Was taken in and the Building of a Large Steam Mill, the Purchasing of Too Many goods sold on Credit of course Lost, reduced us—Divided us—

Your Uncle who had maried a Short time previous removed to Louisville—Men with whom I had Long been connected offered me a Partnership. I accepted and a small ray of Light reappeared in My Business but a *Revolution* occassioned by a Numberless quantities of Failures, put all to an end; the Loss of My Darling Daughter affected Me Much; My Wife apparently had Lost her spirits. I felt no wish to try the Mercantile Business. I paid all I could and Left Henderson, Poor & Miserable of thoughts.

My Intention to go to France to see My Mother and Sister was frustrated, and at Last I resorted to My Poor Talents to Maintain, You and Your Dear Mother, who fortunately [now] apparently became easy at her Change of Condition, and gave me a Spirit such as I really Needed, to Meet the surly Looks and Cold receptions of those Who so shortly before where pleased to Call me Their Friend.

in Attempting the Likeness of James Berthoud, Esqr a Particularly good Man and I believed the Only *Sincere* Friend of Myself and Wife We ever had—to please his Son & Lady I discovered such Talents that I was engaged to proceed and succeeded in a Few Weeks beyond my Expectations.

Your Mother who had remained at Henderson to come by Watter, was at Last obliged to come in a Carriage, and for the second time You had a sweet sister born. How I have dwelt on her Lovely features, when sucking the nutritious food from her Dear Mother—Yet she was torn away from us when only 7 months old = having taken all the likeness Louisville could afford I removed to Cincinnati, leaving you all behind untill satisfied of some Means of Making something for a Maintanance—Through Talents in stuffing Fishes I entered in the service of the Western Museum at One hundred and Twenty five Dollars per Month, and raised a Drawing School of 25 Pupils, Made some Likeness, and had You around Me Once More—but small towns do not afford a support for any time.

Ever since a Boy I have had an astonishing desire to see Much of the World & particularly to Acquire a true Knowledge of the Birds of North America, consequently, I hunted when Ever I had an Opportunity, and Drew every New Speciman as I could, or dared *steel time* from my Business and having a tolerably Large Number of Drawings that have been generally admired, I Concluded that perhaps I Could Not do better than to Travel, and finish My Collection or so nearly [so] that it would be a Valuable Acquisition—My Wife Hoped it might do Well, and I Left her Once More with an intention of returning in Seven or Eight Months; I wrote to Henry Clay Es[qr] with Whom I Was acquainted and he Enclosed Me in a Very Polite & Friendly Letter One of General Introduction = I received Many from Others—General Harrison, &[c]—

from the day I left Cincinnati untill the present My Journal gives you a rough Idea of My Way of Spending the tedious Passage in a Flat Boat to New Orleans—

We moved from our Landing of Last Night and only crossed the River for the rain Lowered the *Smoake* so Much that it was impossible to see, beyond 20 or 30 Yards; played great deal on the flutes, Looked at My Drawings, read as Much as I Could and yet found the day very Long and heavy for Although I am Naturally of light spirits and have often tried to Keep [my spirits] these good, when off from my Home, I have often dull Moments of Anguish—[it stopped] the rain abated for a few Minutes. Cap[e] C. Joseph & I took a Walk to a Sand Barr Where Joseph Killed a Large blue Crane, unfortunately a Young one—saw few Geese, many *Cardinals*, some *Carolina Wrens*—We are better to day—[fortunatly] Luckily our Boat does not Leek—Saw a few Purple Finches =

WEDNESDAY November 29th 1820

the rain that begun two days since, accompanied us the whole of this day, yet We Left our Harbour at about 7 this morning and removed 20 miles—We passed the second Chicasaw Bluff, raining so much that I could not draw them; they are Much More Interesting than the Chalk Banks indeed they Look grand and Imposing, they are from 150 to 200 feet High Irregularly Caving down and Variegated in stratas of Red, Yellow, Black, and deep Lead Colors, the Whole of Such Soapy and Washing Nature as to give an interesting contrast by the Dashing of the Wash down to the edge of the Watter which here is very deep the upper Strata (the whole run horizontal) is perferated with Thousands of holes, the Nests of the *Bank Swallow*—these Bluffs are about two Miles ½ Long, the Country back, Barren & Poor—

Confined to the Inside of the Boat nearly the day, saw but a few Gulls, apparently all White, some White headed Eagles, & a few Cranes—a Large flock of Gold Finches—a few Blue Geays, & *Cardinals,* the Ivory Bill Wood Peckers heard from time to time,—

We are Landed at the foot of N° 35 a few miles above what the Navigators call the Devil's Raceground—but the whole of the Mississipi being so much of the same nature, it feels quite immatereal to follow the Devil's tracks any where along its Muddy Course—

THURSDAY Novembr 30th 1820

We found the race path of the Devil well cleared and beaten, and went through it with great Ease, Many places on this River are rendered More terrible in Idea by their Extraordinary Names than real dificulties—

We run a Race with Mr Lovelace's Boat of several miles that was well nigh terminating in a dispute—it reminded me of Gamblers that although playing for Nothing are allways grieved by Lossing—

We came 25 Miles and Landed a little below the Twelve Outlets, passed the third Chicasaw Bluffs, the view of those was intercepted by our Running on the Right of an Island—the Weather Cold, & very Disagreable, Wind blowing all day mostly ahead = Some Men came on Board Mr L. Boat who said had Killed Three Bears a few days' before—saw a few Indians at a Small Encampment—this morning I remarked Two Flocks of *American Teals* flying up the River; the Parokeets Numerous in the Woods—a Large flock of Sand Hill Cranes Sailed over us for some time, sounding & Elevating themselves to a Considerable Hight took a southwardly course—One Swan was seen on a Barr, but so shy that he flew several times at shore as he perceived the Boat. Whenever We Land a Number of *Swamp Sparrows* are seen [Sku] Scullking through the High Grass that borders the Banks, the seeds of Which are the heath or more properly Mud; the Geese feed freely on these Whilst the Grass affords them an agreeable place during the day—Saw a few Gulls, a Large flock of Sprig Tail Ducks going all southwardly—the Cedar Birds Ampellis Americana fly northeast.

We passed this afternoon 19 Flat Boats Lying at the Shore, some of which had Left the Falls 10 days previous to our Departure—

I saw to day 2 Common Crows the only I have seen on the Mississipi—

FRIDAY December 1st 1820

This Morning was Cold and Cloudy, Flocks of Ducks Geese, &c flying High plentifully, all bound to the Southwest—I remarked Large Flocks of Mergan-zers, I Mean, The Large. These Birds seldom Leave the Clear Ohio or its tributary streams untill Compell^d by the Ice Closing most parts of them; their passing southwardly so early indicates a severe winter above us—A their Flight is Direct, regularly formed in Acute Angles and so swift that one might suppose the noise over head as proceeded from a violent storm of Wind—Saw Early to day several Hundreds of Gull, playing over a Large Barr—When We attempted to Close on them, they rose high and off South—4 White headed Eagles where at the same time regalling themselves on the Carcass of a Deer—The sworms of Grakles that are passing us is astonishing—the Purple Finches also very Numerous saw several hundreds in one flock—Where the Weather is Intensely Cold, scarcely a Fowl is to be seen along the Banks, the ponds offering themselves at that time food & Shelter Passed a Large Settlement of Wood Cutters—Mr Shaw killed 5 Geese 4 at one shot saw two *Slate Colored Hawks, Falco Pennsylvanicus* and a *Winter Hawk.* I went a shore to a House about 5 Miles above *Wolf River* in a sharp running Bend, saw 2 beautifull Trees the *Pride of China* here the High Land is within 2 miles of the River and the spot on which the Plantation stands never overflows, these are remarkable spots—

We are Landed immediately at the foot of Old *Fort Pickering.* We Walked up to it through a very narrow crooked path, and found in a very decayed situation; the Position a Beautifull one the Land Rich about it—and were told that the Spaniards own it it was an agreable spot to Live at—about 2 Miles above this, the Mouth of Wolf river came in from the East, and is the Landing place of a Town Called *Memphis* = have runned 24 miles—Saw some Towe Buntings and Many Sparrows—
[The following paragraph is written across page]

At New Orleans on the 25th of February same year saw an a/c in a New York Paper saying that the Weather had been Intensely severe that the Mercury had been so Low as 24 Degrees below 0 = that the Port of New York was completely closed by Ice and that all the Streams in that state & Pennsylvania where closed—Much pleased to see that particularly the Migrations of a swift Moving Bird such as the Merganser can so truly be considered as the herald of Weather in its Movements when going Northward & Southwardly at the Watter Edge (then about midle Stage) there is a Bed of Coals running Orisontally about 2 feet Deep above the surface—this and the Eligeability of the Situation May become Valuable—

I saw this afternoon Two Eagles Coatiting—the femelle was on a Very high Limb of a Tree and squated at the approach of the Male, who came Like a Torrent, alighted on her and quakled shrill untill he sailed off the femelle following him . and zig zaging herself through the air—this is a scarce proof I have had the pleasure of witnessing of these and all of the *Falco Genus* breeding much Earlier than any Other Land Birds—

We lost one of our Fishing Lines, a Large Fish Must have Carried it, its being very Strong and Well fastened to a Strong Willow Pole—

The Beard of the Turkey shews about one Inch Long the 1st Year, and one of the Male in full growth and plumage must be 3 Years old =

So well exercise[d] are the Geese at extricating themselves from danger when a Flat Boat Comes Near them that they will Walk off from the Edge of the Watter out, into the Young Willows & Cotton Wood several 100 yards—but when in a Bend where the Shore is Steep they May be reached With more Ease.

SATURDAY December 2nd 1820

Cloudy & Cold, took the Skiff and went ahead of the Boats, the only way I have now of hunting, and when any game is on Sight Lay down in the Bow and float untill distance—Shot at a Large W. H. Eagle and a *Black Hawk* Missed both, this Latter in going off *flapped* his Wings Like a Pigeon, they more Swifter than any bird in their Common flight. I could Not Well account My Missing these Birds, they were not More than 100 y[ds] off—I shot 3 Turkeys at 2 shots—their Crops completely filled with *winter Grapes* [their] Gizzards of the seeds of the same and large gravel. They were extremely gentle [and] I floated immediately to them and Came within 25 y[ds] Cold days force the geese away from shore—the Woods Literally filled with Parokeets great Many Squirels—and Many *Snow Birds*—begun raining at about one—Now & then a Wood Cutter's Hut is seen in a Small Parcel of clear[d] Land between Two Thick Cane Brakes—

The Lights or *Pulmon* of the Turkeys I Examined This Day, had much the appearance of *Waddles* connected by a Thin Skin, the *Wadles* about ½ Inch Long = the Thin Skin about ¼ of an Inch—

Many *Golden Wings* Woodpeckers—a few Sparrow Hawks

We Landed on What is Named a *Tow head* a little above the Island N° 51—a Tow head is a small Willow Island overflown in High Watter—

in was Not Dark and We Walk Round it—Saw Many Geese, and a Young *Bear*, but its being so late & dark Was Called out, and gave it up.—

Raining & very Disagreable—Many Gulls flew about us to day and Picked up the Lights & fat of the Turkey, & Geese that Were thrown over Board

SUNDAY December 3d 1820

It rained Heavy all night, this morning it was concluded the boats would not Leave, yet our Captains went hunting the Wind Blowing at Intervals and then ceasing, made some diference of the force of the rain = Saw several Crows Many *Brown Larks*, Many Geese, and Malards = Killed to day 3 Geese, 2 Malards and 2 Mergansers, both femelles measuring 25 Inches—the Bill & Legs Not so bright a cealing Wax Color as usual—tongues sharp Triangular and Toothed think them young—one of them had Caught a Fish about 9 Inches of the *Sucker* Kind and had only partly down its throat when Killed—there was 5 together we drove them several times out of a Pond in the inner part of an Island and at Last Shot these on the Wing = the Geese very Shy—Three Keel Boats passed us about 2 o'clock P.M.—they had left the Falls of Ohio 3 week ago—Left our Harbour and floated about 4 Miles to the foot of Buck Island—here I saw with the setting sun hundreds of Malards travelling *South* and the *Finest* rainbow I ever beheld, the Clouds were also beautifull apposite it = Looked at my Beloved Wife's Likeness Shaved and Cleaned One of the few enjoyments Flat Boats Can afford—the Goose we eat at Dinner extremely fishy.

Joseph who now is obliged to officiate as Cook does not appear to relish the thing—the more I see Capе C the more I Like him = Wish that we could say the same of all the World—

The Towhead on Little Island on Which Laid Last Night and to day had *Vast Many dry* Nests of Thrushes on the small Willow Trees—the Tall Grass with many Sparows—Saw 2 Flocks of Partridges, Many Parokeets.

Saw a Blue Crane.

MONDAY December 4th 1820

We had a dreadfull night of Wind, the hands obliged to move the boats. I did not sleep, the Knocking of the Boats against the Sand Barr very desagreable—This Morning the Wind still blowing hard. Went to a Small Lake to Shoot—there I saw a *Killdeer plover,* a *King Fisher;* Many Geese & Ducks—but no *Swans* as we had been led to expect from reports of a Squater—Killed 3 Geese, and 3—*American Teals*—Saw a few Turkeys—this Lake about 2 Miles from the River, contains some of the Largest Muscles I ever saw, and Vast Many perewinkles that appear to be of a Peculiar Species and I put some in My pockets—I found these in round Parcels of about 20 to 30 close together—My Slut Dash brought out of the Watter very well—When returned to our Boats, 9 of the Boats we had passed some days previous went by, and we pushed off immediately—We floated only about 4 miles—and Landed on the Tennessee Shore—

Saw to day great Many *Autumnal Warblers,* the first I have seen since on the Mississippi, a few *Crows* some *Winter* Wren, some *Eagles W. H^{d}*—it is very Seldom that brown Eagles are seen;—the others are now *courting* seeing them every day chasing off the Batchelors—

I doubt not that the Migrations of the *Autumnal Warblers* is the latest of all that Genus—

I was taken off Suddenly a few Minutes ago to take a Cat Fish of our Line. I had some trouble for a few minutes but having drowned him put my Left hand in his Geels and hauled it in the Skiff—it weighed 64½ lb and Looked fat—Killed it by stabbing it about the Center of its head, this was so Effectual that in a few Second it was quite Motionless—

I would be Inclined that, from *Shape, Size, Color* & Habits so diferent to that of the Cat Fish Caught in the River Ohio the present one is a diferent *Species.*

TUESDAY December 5th 1820

Skinning the Cat Fish was the first Job this morning—this was done by cutting through the Skin (which is very Tough) in Narrow Long Strips and tearing those off with a Strong Pair of Pincers While at this saw Several hundred of those Black Birds yet unknown to me that I denominate *Black Pelicans* flying South forming a very obtuse Angle, without uttering any Noise—have some Hopes therefore to see some of them on the Watters of Red River or Washita—Sand Hill Cranes were also fling and We saw More Geese than usual—Joseph Killed 4 American Teals—those fly up stream—saw 3 Swans—While Geese are flying in a Travelling order the Young or Smallest are about the center of the *Lines* and the Larger Gander Lead the Van, the Oldest Goose Drives the Rear—the Weather beautifull but cold, and No Doubt that the Frogs that Wistled so merily Yesterday are well buried in the Mud this Morning—

We made an Awkward Landing, Lodged in the Mud for about ½ an hour, and our Commander had a good opportunity of Exercising his Powers at Swearing—More particularly when Anthony broke his Sweep Oar. this Took Place about 30 Miles from our starting place of this Morning, [at the foot of No] opposite the head of 57 & 58—"fine Weather but No Fish" says Cap[e] Cummings—

WEDNESDAY December 6th 1820

Light frost, Rich Clouds of Purple & Light Green Indicates Wind—extremely anxious to overtake the fleet ahead our Commanders have yesterday and this morning exerted themselves more than usual and have left our harbours as soon as Day Light would permit—how beneficial a fleet constantly a head Would be!—

Saw 2 Large White Cranes with Black Tips—too Shy to get in floating distance;—Many A[n] Teals and as Many Geese as Ever—

Passed the *St. Francis River* the mouth of which at this time appeared Closed by a Mud Barr—but the people who Lived on the Point formed by that stream and the Mississippi told us that there was plenty, and that Keel Boats go up it 400 Miles, many settlements on the banks the first about 15 miles—the Same people told us that they had seen many *Pelicans* a few days previous in Passing—saw some *Old Blue Cranes* on the Trees but could not go within 150 *yds* of any of them—a Little before we passed the Place called the *Big Prairie* shot a Monstrous Turkey Cock, I think the Largest I every saw; it appears considerably Larger than one I weighed that was over 31 *lb*—My anxiety to have it Made Me Miss—the *Big Prairie* is a Tolerable sized Plantation rather higher than usual on this River, about ¼ of a Mile Back—the Land rises in Gentle Hills, and when told is extremely Rich there, I first saw the *Mississippi Kites*[a]: ascending in the Steam Boat Paragon in June 1819—Bought some Delicious *Sweet Potatoes* at ½ Dollar pr Bushel the Squater assured me that a few Weeks previous, the Pelicans were so numerous that Hundreds where often in sight on a Barr below this Place—people very sickly—Landed on the Tennessee Shore about 7 Miles from the *Settlement of the Hills*, the Wonderfull fleet still about 4 miles a head, our Comodores had a Meeting, the result was that We should *Start* one hour before day and run down the D*d* Rascals—

[The following lines in the diary are put on a blank space of the preceding page as an explanation.]

[a]The Mississippi Kite Were Busily Employed in Catching small Lizards off the Bark of Dead Cypress Trees, this effected by Sliding beautifully by the Trees and suddenly Turning on their Side and Graple the prey—having At that time no Crayons or Paper, did not Draw one, and determined Never to Draw from a Stuffed Specimen, Carried No Skins—

THURSDAY December 7th 1820

Caught a Nice *Cat Fish* weighing 29 lb. at 3 o'clock this Morning—stabd him as we had the former but it did not die for One Hour—at Day Break the Wind stiff a head, a Couple of Light Showers Lulled it, and we put off—Mr Aumack Winged a *White headed Eagle,* brought it a live on board, the Noble Fellow Looked at his Ennemies with a Contemptible Eye. I tied a String on one of its Legs this Made him Jump over Board. My Surprise at Seeing it Swim well Was very great, it used its Wings with great Effect and Would have Made the Shore distant there about 200 *yds* Dragging a Pole Weighing at Least 151*bs*—Joseph went after it with a Skiff, the Eagle Defended itself—I am glad to find that its Eyes were Coresponding with My Drawing—this Specimen rather less than the one I draw—the femelle hovered over us and shrieked for some time, exibiting the *true sorrow* of the *Constant Mate*—Prepared a Bed for My Slut *Dash* expecting her to be delivered from her Burthen every Day—

Our Eagle Eat of Fish freely about one hour after we had him, by fixing a piece on a stick and putting it to its Mouth—however while I was friendly Indian toward it it Lanced one of its feet and caught hold of My right thum, made it feel very sore—

Went to an Eagle's Nest; busily Employed Building Shot at the femelle, which is at all times distinguished by her size, the Male was also asitting—Killd *One* Goose—as We Reachd the head of the Sand Barr close to Island N° 62 We passed the *Fleet* at anchor but they all pushed off when they saw us go through this Place—One Boat suffered Much, by being runned down by another in Landing—Came 25 Miles—the Evening Looking Stormy, the Current strong as our Landing Could not put out our Lines—Our Comodores Much Elated—

FRIDAY December 8th 1820

Not satisfied about our Landing in a strong Current I slept but little. Whenever I walk*d* on the deck the Eagle Hissed at Me, and ruffed itself in the Manner that Owls do generally—the Weather Warm, Cloudy, & Windy—put off late and only run about 3½ miles forced by the storm to land at the foot of what I supposed the Island N° 63—leaving all the fleet behind us—

with Some Hopes of Shortly being at the Arkansas Fork I feel Inclined to Copy a few of the Letters I had for that Place Particularly those of Generals *Harrison* & Lytle.

Cincinnati Sept 7th 1820

Dear General

Mr. Audubon who will have the honor to hand you this, is upon a Tour through the Extensive forests of *Western America* for a Scientific Purpose—that of Completing a Collection of American Birds—I beg leave to Introduce him to you & to request your aid and [assistance] Countenance in the [highly] accomplishment of his highly laudable project—

Mrs Harrison is well & my daughter Lucy & Son Syomms both Maried since you were at my House, the latter to a Daughter of Genl Pikes—

Your Friend
W. H. Harrison

Governor I. Miller
Arkansas T—y.—

Cincinnati Ohio Octr 9th 1820

Dear Sir—

Permit me to Introduce to your Acquaintance John J. Audubon es*qr* who is on a Visit to the Territory of Arkansas and the Norwest as an ornithologist—for the taking drawings of Birds—Fowls &c—for a Work he has on hand. Any facilities You May have it in Your Power to offer him toward promoting the Object in view will be thankfully received by him and duly appreciated by

Yours friend and Hble Servt
W[am] Lytle

Cincinnati Octr 10th 1820

Revr Gentlemen—

Permit me to Introduce to your Kind regards John J. Audubon Es*qr* who proposes traversing Louisiana for the purpose of Compleating a Collection of Drawings of the Birds of the U. States which he proposes to publish at some Future Period—he has been engaged in our Museum for 3 or 4 Months & his performances do honor to his Pencil—

I regreat to hear that you have been visited by Sickness. I Hope you May get Safe to your journey end, & be prospered in the great & glorious work on Which you have Entered—I should be Pleased to hear from you frequently

I Remain Your Sincere Friend & Brother
in the *Lord*

Elijah *Slack*

Reverend Veil & Chapman—

Elijah Slack was then President of the Cincinnati College

Bess[i]des the above I received several Letters from Docr *Drake* Directed to the Reverend Mr *Chapman* Osage Mission—*Coll Breasly* Indian agent & *Governor Miller*—

I Will give You here the Copy of the Letters I Received from the Honorable Henry Clay to the one I Wrote from Cincinnati the Copy of Which is annexed at the Beginning of the Part of My Journal

Sir—

I received your letter of the 12th inst and now do myself the pleasure to transmit to you inclosed such a letter as I presume you want—I suppose a general letter would answer all the purposes of special introduction, which I should have been at a loss to give as I do not know the particular points which you may Visit—and even if I did, I might not have there any personal acquaintances—

Will it not be well for you before you commit yourself to any great Expense in the preperation and publication of your Contemplated Work to ascertain the success which attended a similar undertaking of Mr Wilson?

With Great Respect
I am Yours
H. Clay

Lexington—25 Augt 1820

I have had the satisfaction of a personal acquaintance with Mr. John J. Audubon; and I have learnd from others who have Known him longer and better, that his Character and Conduct have been uniformly good: being about to take a Journey, through the Southwestern portion of our Country with a laudable object connected with its Natural History, I take great Pleasure in recommending him to the Kind Offices of the Officers, and Agents of Government and other Citizens whom he may meet, as a Gentleman of amiable and Excellent qualities, well qualified, as I believe, to execute the object which he has undertaken

H. Clay

H. Clay Was then the Speaker of the House of Representatives, and I Hope to rece some benefits from this Letter—

Mr Aumack killed a Goose and Joseph an *Intrepid Hawk*, Swans extremely Plenty fired at them Many Times with Balls without Success—

Drifted at 3 o'clock P. M.—about 4 Miles and Landed at foot of N° 64—

I began a letter to My Beloved Lucy with some Hopes of reaching the Fort of Arkansas Tomorrow, but Hopes are Shy Birds fling at a great Distance [and it is] seldom [that] reached by the best of Guns [can reach them]

SATURDAY 9th December 1820

I have nothing to say for this day. I drew a little [today,] seeing a Green Briar with seeds on—Wrote to My Lucy and Lived on Sweet Potatoes—how Surly the Looks of Ill fortune are to the poor. I Hope to see the fort of Arkansas tomorow and Hope to Leave the Boat I am now in if there is What the Kentuckians Term a "*half Chance*". Our Commanders Looks and acting are so strange that I have become quite Sickened—

the Weather quite rough, all day, cleared at night, the Flat Boats passed us this evening—We have made a bad Landing according to my Ideals—

SUNDAY 10th December 1820

We floated down to the *Caledonian point* or Petite Landing about 4 Mile above the *real* mouth of *White River*

here it was Concluded that Mr Aumack should *walk* to the old *Post of Arkansas* of course I & Joseph prepared and having made Enquiries concerning the road we determined to go by Watter to the mouth of the *Cut off* and then walk the remainder; Anthony joined us, and the Skiff doubled oared was taken; We left at 10 o'clock with Light hearts. Small Bottle of Whiskey a few Biscuits, and the determination of Reaching the Post that Night—

At the Entrance of White River we discovered that that stream Was full and Run Violently, the Watter a Dull Red Clay Color; We soon found ourselves forced to Land to Make a Natural Cordel of several *Grape Vines* and pull up by it—the distance to the Cutt off is Seven Miles that appeared at Least 10 to us here We Met 2 Canoes of Indians from the *Osage Nation*, Landed our Skiff on the opposite side of White River Which we here found a *beautifull Clear Stream* and Backed by the Watters of Arkansas running through the Cut off; We Walked through a Narrow Path often so thickly beset with green Briars that We Would be forced to give back and go round—this followed through *Cypress Swamps* and round *Pounds* and Cane Breaks untill We reached the first Settlement owned by a Frenchman Called Monsr Duval, this friendly Man about going to bed offered us his assistance put on shoes & clothing and Lead us 7 Miles through Mud & Watter to the Post; and at 9 o'clock P. M. We Entered the Only Tavern in the Country—*Wearied*, Muddy, Wet, & hungry—the Supper Was soon calld for, and soon served, and to see 4 Wolfs taring an old Carcass would not give you a bad Idea of our Manners while helping *Ourselves* the *Bright Staring Eyes* of the Land Ladies Notwithstanding

however I found Mrs Montgomery a handsome Woman of good Manners and rather superior to those in her rank of Life—to Bed and to sleep sound was the next Wish for 32 Miles in such a Country May be Calculated as a full dose for any *Pedestrian per day*—Led into a Large Building that formerly perhaps saw the great *Concils of Spanish Dons* we saw 3 Beds containing 5 men, Yet, all was arrangd in a few moments and as the Breaches were Coming off our Legs, Mr Aumack & Anthony slided by into one and Joseph & myself into Another, to force Acquaintance with the strangers being of course necessary a Conversation ensued that Lulled Me a Sleep, and Nothing but the Want of *Blankets* Kept Me. from Resting Well, for I soon found a Place between the *Tugs* that Supported about 10 lbs of Wild Turkey Feathers to save (?), My roundest Parts from the Sharp Edges of An Homespun Bedstead—

The Morning broke and with it, Mirth *all about us*, the *Cardinals*, the Towe Buntings, the Meadow Larks and Many Spcies of Sparrows, chearing the approach of a Benevolent sun Shining day—dressed and about to take a View of *all things* in this Place, Met a Mr *Thomas* known formerly when in the Paragon Steam Boat—he introduced Me generally to the Medley Circle, around, and from thence took Me to a Keel Boat to receive the Information I Wanted about the Upper Countries through Which this Noble Stream Meanders—think of My Surprise at seeing here a Man who 13 years ago gave me Letters of Introduction at Pittsburgh (Penn) for Men in Kentucky—this Was Mr. *Barbour* the former Partner of Cromwell—he Met Me with great Cordiality, told me of the absence of the *Governor*, the Indian agent and also that the Osage Missionaries had proceeded about 150 miles up to a Place called the *Rocky* Point.

the Cadsaw is [hi] beyond that where a New town, the seat of Govérnt was expected to be situated.

disapointed to the utmost in Not Meeting those who I supposed Would of Course give me the best Information I requested of Mr Thomas to give the Governor My Letters and beg of him to Write Me a few Lines at New Orleans to the Care of Governor Robertson—[the Gentleman—] Mr Barbour told Me that he had for Several years past gone up to the Osage Nation about 900 Miles and that his Last Voyage he fell in with *Nutall the Botanist* and had him on board for 4 Months—that Many species of *Birds* were in that Country unknown in this and that the Navigation Was an agreable One, at the same time that it was rendered profitable by the enormous profits derived from the Trade with the Indians, whom he represented as friendly and Honorouble in all there dealings—that he would be extremely Happy of My Company and that of My Companions and that if I did not go with him at present that he Hoped I would Meet him when coming down the Arkansas Next Spring or Summer for he is about 6 Months employd each Voyage—The Post of Arkansas is Now a poor, Nearly deserted Village, it flourished in the time that the Spaniards & French kept it, and One 100 years passed it could have been called and agreable Small Town—at present, the decripid Visages of the Worn out Indian Traders and a few American famillies are all that gives it Life, the Natural situation is a handsome One, on a high Bank formerly the Edge of a *Prairie*, but rendered extremely sickly by the Back Neighborhood of Many Overflowing Lakes & Swamps.—

I was assured that only Two frosts had been felt here this Season and that the Ice in the River never Stopped the Navigation—

the Town now Prospering at *Point Rock* is high healthy and in the Center of a Rich tract of [Land] Wood & Prairie Lands—and probably may flourish—the *Arkansas River* flows a Thick Current of red Clay & Sand, and if not for its coloring would have much of the appearance of the Mississippi—Cotton is raised here With some advantage—Corn grows Well, game & Fish are plenty—

I here feel Inclined to tell you that an oportunit of Good; Fresh Flour Whiskey, Candles, Cheese, Apples, Porter, Cider, Butter Onions, Tow Linen and Blankets would meet with advantageous Sales during Winter, accompanied by *Powder Lead, Flint*, [and] Butchers Knives, Rifles, and *blue Shrouds* for the Indians.

After Breakfast We Left the Post of Arkansas with a Wish to see the Country above, and so *Strong* is My Anthusiasm to Enlarge the Ornithological Knowledge of My Country that I felt as if I wish Myself *Rich again* and thereby able to Leave My familly for a Couple of Years—here I saw a French Gentleman who but a few Weeks passed had Killed a *Hawk* of a Large size *perfectly White* except the Tail Which Was a *bright red*. Unfortunately, no remains of its Skin Legs or Bill were to be found—We travelled fast—reachd the Cutt off and Landd our Skiff, having Killed 5 *Crows* for their *Quills*, Never before did I see these Birds so easily approachd and in fact all the Birds We saw, 2 Hawks I did not know hovered high over us—the Indians still at their Canoes, We Hailed, and gave them a Drachem of Whiskey, and as they could not speak either french or English, I *Drew* a *Deer* with a stroke across its hind parts, [——] and thereby Made them Know our Wants of Venaison hams—

they brought 2 We gave them 50[cts] and a Couple Loads of Gun Powder to each, brought out smiles, and a Cordial Shaking of Hands—a Squaw with them a *Handsome Woman* waded to us as Well as the Men and drank freely—[Never do I] Whenever I meet *Indians* I feel the greatness of our Creator in all its Splendor, for there I see the Man Naked from his Hand and yet free from Acquired Sorrow =

in White River We saw a great number of Geese Malards and Some Blue Cranes—also Two Large Flocks of these *unknown Divers* or *Pelicans*—

reached our Boats about 6 in the afternoon fatigued but Contented a good Supper, Merry Chat—and good Looks all round—Went to bed all Well—

before I leave the Trip to the *Arkansas Post* I think I will give you More of it—We saw there a *Velocipede* Judge how fast the Arts & Sciences Improved in this Southwestern Country—I want also to tell you that the Squaw on White River While Wading out to us Craked a Large *Louse* taken from under her arm—

The *Intrepid Hawks* are extremely plenty along the Banks of the Mississipi where the feed aboundantly on the *Swamp Sparrows* as also on the *Sturnus depradatorius;* some of these are so strong and daring that they Will attack some Ducks on the Wing and often carry them off several hundreds of yards to the Sand Bars—

The Brown Eagles that were so plenty on the Ohio have entirely disapeared and nothing by White Headed Ones are to be seen—

The Lakes found in the Interior are stored with the finest of Fishes Such as Pikes, Salmons—Rock, Bass Sun Perches &c and the bottom covered with Thousands of Muskle Shells and Perrywinkles of many species—those Latter of Course find their Way while the Spring floods are so general—the Bottom of Most of these Lakes is firm and Level—

TUESDAY Decembr 12th 1820

This day Mr *Shaw* and *Anthony* Walked off to the Post and We floated down to the Mouth of that River, this We reached and Landed. I was so fortunate as to Meet the *Steam Boat* the Maid of [New] *Orleans* on board of Which I put a Letter for My Dearest Friend My Wife—with orders to put it in the St. *Louis* Post Office—Saw to Day Many Crows Mergansers and Geese, some *Dun Divers*—and a Large flock of My unknown Divers—

the Blue Jeays are now & then seen—the great *Carolina Wrens* are very Aboundant—but the Snow Birds have disappeared, the weather is so Warm that *Buterflies, Bats, Bees,* & Many Insects are flying about us and at the Arkansa I was assured that they had had but *Two* Light frosts—

Mr Shaw is Expected Tomorrow Night and perhaps Will Leave us here to Proceed up that River, for we are told the Orleans Market is Extremely dull—Killed a Gull precisely Such as I shot at *New Port Kentucky,* rather Fatter

The Wild Geese we now Shoot have Eggs swollen to the size of N° 3 Britt Shots—

An Indian Chief at the Mouth of the Arkansas Killed *Three Swans* one of which

I was told Measured 9 feet from Tip to Tip—Those Indians had Left when We arrived,—a View of Such Noble Specimen Would have been very agreable—

The *Prairie Hawk* that I see here is not the *Marsh Hawk* of Willson it is Much Less-Lighter Color, the Tip of the Wings Black and only One Large Bend of Dark ending the Tail—they fly Much Like the *Night Hawk* and Catch Small Birds on the Grass Without Stopping their course—

WEDNESDAY Decemb[r] 13th 1820

A Beautifull day, Walked up the Arkansas in Search of a Lake but the Cane so thick that we give it up Killed Two Geese. M[r] Aumack Shot at a *Prairie Hawk* but did not Kill it—I wrote Governor Miller the Copy of which is here annexed—

To his Excellency Gov[r] Miller of the Arkansas—Sir—having had the Honor to receive several letter of recommandation to your Excellency, from Gen[l] *Harrison*, Gen[l] *Lyttle* and other Gentlemen, I felicated myself with the pleasure of an Interview with you—I Reached the Post of Arkansas but was foiled by your absence; having only a few Moments to remain at that place I begged of M[r] Thomas to present you the Letters I was the Bearer of—these I even was not able to seal—

My ardent Wish to Compleat a collection of drawings of the *Birds* of our Country, from *Nature*, all of Natural *Size*, begun about 15 years since, and to Acquire either by *occular*, or reliable observations of others the Knowledge of their Habits & residence; makes me wish to travel as far at Least as the Osage Nations on the Arkansas as also along the Whole of our Frontiers—

Should your arduous avocations admit—I would Consider myself very Highly honored and under great obligations to receive from your Excellency a few lines of Information respecting the *Time*, the Manner of travelling and what might be necessary to render such a Journey fruitfull to my Views—as well as your Personal Information of the discoveries you have made in Ornithology, in that part of America—My Intention is to Visit the country around New Orleans as far East as the Florida Keys, then ascend Red River, and to go to the Hot Springs—thence across to the Arkansas and Come down [it] to its Mouth where on My return to Cincinnati at present My familly Residence—yet My Plans are alterable as advised by Gentlemen of more Experienced

Should your Excellency contemplate any expedition upon Your River and I was sufered to Join it, I would be anxious to Meet any Wishes my Humble talents

Could afford at any time—

With Hopes that your [Excellency] will not be displeased at the liberty I have here taken

I remain with high respect
Your V. H. S[t]
J. J. A.

P. S. if agreeable please
Direct to Governor
Robertson

Saw some Turkey Buzzard
Some Merganzers and Sand Hill Cranes—
Malards—Crows—Towe Buntings—
Winter Wrens—Meadow Larks—Partridges
Red Winged Starlings—and a Vast Number of Swamp Sparrows in the High Missisipi grass
Parokeets—Golden Crowned Wrens—

M[r] Shaw & Anthony [not] returned [yet] at eleven o'clock, but did not effect their Business, They returned in a canoe, 30 Miles down the Arkansas to the *cut off* through that about 6 (miles) down White River 7 and down the Mississipi 15—this makes 58 and the Distance from Arkansas Post to the Mouth of that River is 60—

about One Mile below the Mouth of Arkansas in Thick patch of *Cane* are *Two Women* the remainder of a party of Wandering Vagarounds that about 2 years ago Left some part of the Eastern State to proceed to the *Promised Land*—these Two Wretches, Never Wash Comb or Scarcely clad themselves, and subsist from the Scant generosity of the Neighbours—Now and then doing a little Sawing and Washing—

THURSDAY Decemb[r] 14th 1820

After Long consideration, our Gentlemen determine to do *Nothing*, and [Left our Landin] We *Cutt Loose* about 10 this Morning, the Weather quite Warm, Distant frequent Clapps of Thunder announcing a change—it soon begain to Rain, the Fog raised and We Landed again about 2 o'clock—saw here 5 Ivory Bill Wood

Peckers feeding on the Berries of Some Creeper they were gentle—Keeping a Constant Cry of *Pet Pet Pet*—Killed a Crow on the Drift wood it was not untill then that I discovered that a Crow Killed by M[r] Shaw while I was at the Arkansas was a *Fish Crow*—it rained all Night Watter raising fast—

FRIDAY Decemb[r] 15th 1820

Rainy & Cold, floated about 6 Miles—at One o'clock the Steam Boat, *James Ross* passed us—in the afternoon I had the good fortune to Kill a Beautifull *Marsh Hawk Falco Eulginosus* feeding on a Swamp Sparrow that I saw him Catch—When I approach him before I Shot he saw me first and flew a few yards where he sat tearing at his prey untill death reached him—the *Prairie Hawk* seen Yesterday is entirely diferent, in Size, Color, & Manner of flying and [h]as it is a Nondescript I Hope I may meet it again—Killed 4 Teals and 2 Geese—the Watter rose 20 Inches this day—

I have seen the Marsh Hawks about September flying down the Ohio River, and several times about that season I have seen *flocks* of them travelling high and southwesterly—finding them Now plenty on these Shores Where great deal of rich food is afforded them by the Numerous quantity of Swamp Sparrows the mildness of the Weather, no doubt assures them a good Winter Residence—

the One I Killed Was a Male, in good order, Weighing Only ¾[lb]—Measured 18 Inches full Length, breadth 3½ feet—the Insides of the Mouth Black—

SATURDAY December 16th 1820

the weather much the same, heard this Morning that the Ohio has raised Immensely—Spent the day Drawing my Hawk but so dark and disagreable that I could not finish it—having Landed for 3 or 4 hours at the Head of the *Cypress Bend* We floated but a Short Distance—

Vast Many Geese seen to day—Joseph Killed a *Pewee fly catcher* close by the Boat—the little fellow Was very active, and in very good order—a femelle—

this Evening about *100 Pelicans* were in view on a Sand Barr, and although I had no expectations of reaching them We put off in the Skiff, when about 300 yards off from them, they flew and I sent them My Ball, without effect—these are the first I have seen this Journey. I Hope I will have one to draw before I reach N. Orleans—

The generality of the geese We Kill are very poor—and scarcely fit to eat—the Steam Boat *Govr Shelby* pass^d us heavily Laden—

When I was trying to approach the Pelicans, they rose from their slumber one at a time, and Shook their Wings as if to try if able to fly in Case of need—the Nearer We Came the faster they gethered and walked off untill they all flew without uttering any Noise—

SUNDAY December 17th 1820

Raining all day. I finished my Drawing—Landed at *Pointe Chico* a few Miles before this the *Spanish Beard* is seen—Pointe Chico is a handsome spot on this river that never overflows—and answers well for the Growth of Cotton Corn &^c—Peach & Aple Trees flourish well here, but Sugar will not grow—

a Man of good Manner assured me that the *Marsh Hawks* were very plenty here all winter, but not to be seen in Summer—and that the *Pelicans* disapear at this season for the South and return early in April with young—Many Red Winged Starlings—and Many *Bleu Birds* these were pleasing to me, the poorest note of these is allways wellcome to Mine ears—

the Watter raising very fast—

Last Night Mr Aumack who was *rather Merry*—Went to Shoot a Pelican about 10 o'clock P. M. to Cool himself return^d without a Shot—

Saw Winter Wrens and a beautifull plant in full Bloom—*Ivory Billed Wood Peckers* becoming more plenty—

MONDAY Decr 18th 1820

Raining all day, floated but a few Miles—Landed at a Place where Geese & Ducks abounded—Killed a Crow, a *Great Horned Owl* and a Winter *Falcon.*

TUESDAY Decemb^r 19th 1820

Rain and fog all day—Landed within 7 Miles of Last Nights—Killed a *Carrion Crow*, a *Winter Wren* and *16 Parokeets*, I heard and saw once a *Thrush* unknown to me but could not get a Shot at it—Immense flocks of Parokeets and Swamp *Blackbirds*—the *Carrion Crow* I shot at Would Not suffer us to go near than about 100 yards and forced me to draw My Shot and Put in a Ball this brought him down Lifeless—

This Morning I Shot at a Bird unknown to Me and no doubt a Non Descript—it was of the Sparrow Genus—

Saw several Thrushes, very Shy. they sung Sweetly and also Constantly. Took for the Golden *Crown*[d] *Thrush Turdus Auracapillus;* Also a *French Mocking Bird Turdus Rufus*—the Trees in Many places in the Thick Canes full of Leaves—and during this Rainy *Spell* the Weather as felt Much Like the Beginning of May—

WEDNESDAY Decemb[r] 20th 1820

The Weather as desagreable as one could wish it—Raining and so foggy that we could not see 50 yards—drawing all day—in the Morning the *Winter Wren Sylvia troglodites* and afterwards the *Carrion Crow, Vultur Atratus*—at Twelve o'clock a Short Clearing taking place We floated about 4 Miles and Landed opposite side of the River—Cap[e] Cummings Shot at an *Ivory Billed Wood Pecker Picus Principallis* broke his Wing and When he Went to take it up it Jump up and claimed a tree, as fast as a Squirel to the Very Top, he gave it up having but a few Loads of Shot—Joseph Came and saw it—Shot at it and brought him down—

We Boiled 10 Parokeets to night for Dash who has had 10 Welps—purposely to try the effect of the Poisoning effect of their hearts on animals. Yesterday We Were told that 7 Cats had been Killed Last Summer by Eating as Many Parokeets—Killed Two Geese—Several Boats Landed along us to night

THURSDAY Decemb[r] 21th 1820

We at last had fine weather, floated about 35 miles this brought us to the upper Part of *Stack Island* Now only a Barr the former having being sunk by the Earthquakes—Drawing nearly all day I finished the Carion Crow, it stunk so intolerably, and Looked so disgusting that I was very glad when I through it over Board—

Saw in the afternoon a *Black Hawk*, a flock of *Pelicans* at which I shot at about 200 yards as near as I could approach, Without effect. the Sand Barr where they were was Literally covered with excrements and their Feathers. Vast Many Geese seen all day, these Birds Now *Pairing.* Spanish Moss very aboundant on all the Cypress trees—Large flocks of *American Teals* and the constant Cry of Ivory

Billed Wood Peckers about us—scarcely any other except a few *Peleated* & Golden Wings—have not seen a red head Wood Pecker for some time—the Carolina Wrens and *Cardinals* exercising their Vocal powers all day—

We received the visits of 2 Men, Wood Cutters from the Shore—they assured me that *Pelicans* were here at all seasons but that when the Weather is bad they keep in the Lakes in Great Flocks along with the Geese, Swans, Ducks, Cranes—and there find an aboundance of food—they spoke of a *Black Hawk* that Lives on Fish but I could not ascertain Much about size of anything else concerning it—

they reported that a few Weeks passed a Youth of about 12 Years having Met a Large *Brown Tiger*, or *Cougar* called here a *Painter* was so frightened that he died after reaching his Parents' house—those animals are now scarce, but Dear, Bears and Wolfs are plenty—Anxious to Know if Alligators were seen during this season, they answered in the affirmative. They could be seen every few days—some Keeping in Small ponds too Shallow to cover their backs and there Catch the *Garr Fishes*, root for frogs &c that one Killed Lately had a Large Quantity of Black Walnut and Hickorys—they are Killed here for the Skin, that when tanned gives a fine Leather preserving the Lozanges of the Scales. One of the men said he owed them a Grudge for Killing An Excellent hunting Dog—while following a wounded Deer across a Lake, and that he retaliats on the Whole Species Whenever an opportunity offers—a Boy told us that one that had dugged a hole about 20 feet under the ground to resort to, in bad weather, was taken a few days ago; the earth having fallen on him during a heavy Dash of rain—they are easily Killed with a Clubb—the usual way of destroying them—they Move slow on the ground—but swiftly in the Watter—this is all hear say and put it here to Compare it with My Own future Observations

FRIDAY December 22d 1820

Started at 5 o'clock this Morning, and certainly deserves Noting—after breakfast Joseph & I push off as usual in the Skiff in which we remained untill near Sun Down—Saw *Three Black Hawks*, Shot at this Twice, but these Birds are So Shy that I dared not advance nearer than Rifle Shot and Missed them—

Went to a house, to Warm our fingers, the wind blowing rather Sharp this morning—found a Handsome familly of Young Brats who as well as their Mamma Looked Clean and healthy—here We saw the *Pameta Plants* along the Fences—

in the afternoon the sun Shown Warm, the Geese where in Thousands on the Willow Bar, fighting and Mating. Malards, Teals and Wood Ducks aboundant—Killed One Malard and Two Wood Ducks—saw One Swan—One Redtailed *Hawk*, several *Sparrow Hawks*—Many Crested Titmouse—Autumnal Warblers all through the Shaggy *Beards*—The *Carrion Crow* plenty, and their relation the Buzzard.

the *Pewee Fly Catchers* very busy diving at Insects and Singing Merely—saw several Bald Eagles that I Might have Shot at—

a little before Sun down a Steam Boat Called the *Mars* passed us, a poor running Machine—apparently an Old Barge—

Our Commanders spoke of *Cutting a Stick* at 12 o'clock, but the Axes were dull and We did not get up untill 3.—

SATURDAY Decemb^r 23^d 1820

the Moon shining beautifully Clear, the Weather calm a heavy White frost—started at 3 o'clock—

as soon as the fog disapeared, J. & I putt off for the Mouth of the *Yazoo River*, seeing some Geese Made for them and Killed One—in the Mouth of that River I perceived a Large flock of My unknown *Blackbirds* that I suppose Brown Pelicans—Landed below them, and after crawling on My belly for about 300 yards I arrived within about 45 yards I fired at 3 that were perched Close together on a dead Stick about 7 feet above the Watter, at my shot they all fell as so many stones. I expected them to be all dead but to My surprise, those and about 20 swimming under them had dove, they soon rose and took Wing after running on the Watter about 50 Y^ds at the exception of the One I had taken aim on—it would not raise, the Skiff brought up We rowed after it, diving below us up the Yazoo Nearly one Mile, Yet I could not give it up, it became Warier, & remained Less under Watter the Nearer We approach when at Last Joseph Shot at its *head* [the] & Neck (the only part in view Looking much Like a Snake) and Keeled it over—I took it up with great pleasure and anxiety—but I could Not ascertain its Genus—for I could not Make it an [d] *Albatros* the only Bird I can discover any relation to—

We had to exert ourselves to reach our Boats—this done, I began Drawing—We passed to day the *Walnut Hills* a handsome situation on the Mississipi covered with Cotton plantations—We also passed the Small Village of *Warren* commonly Called "Warington" opposite this place (Not Improving) Met the *St. Elca S.—B.*

The *Yazoo River* flowed a Beautifull Stream of transparent Watter, Covered with 1000[ds] of Geese & Ducks and filled With Fish—the Entrance Low Willows & Cotton Trees—We run to day 49 Miles—the Weather rather too Warm—

MONDAY December 25th 1820 Chrismas day

We passed, the Petit Gulf—early this morning—the Steam Boat Comet passed by from Louisville 9 Days—

I had the pleasure of seeing M[r] Aumack Killed a *Great Footed Hawk*, the *Bird* Alexander Willson heard so Many wonderfull Tales of—these Birds are plenty on this River at this season every Year according to all the accounts I have Collected but allways extremely Shy, and I Believe few Men Can Boast of having Killed Many of them, for 15 years, that I have hunted and seen probably one hundred [of] I Never had the satisfaction of bringing one to the Ground—I often have seen them after hearing their Canon Ball Like wissling Noise through the Air seize their Prey on the Wing particularly at *Henderson* Kentucky, where I watched for Weeks near a Pigeon House, that furnished one of those daring Robbers, with food & Exercise—No doubt that the Clouds Ducks of some many Species as are found on this River, renders it a pleasant and fruitfull Winter Residence—

We have seen about 50 since a few Weeks—they fly fast, with quick motions of their Wings, seldom Sailing except when about alighting:—the Specimen before me is a very Old Bird and a beautifull one, When on the Wing they appear Black and are often Mistaken for the *Falco Niger*—Killed 3 Geese;—

Saw a *Tell Tale Godwit*, the Only one seen since I Left the Ohio—but understanding that that River is now considerably raised, I expect they are forced to abandon it—

Cap[e] Cummings saw 4 Deer this Evening—

We are Now Landed about 15 Miles above Natchez, and if No head Wind takes place Must reach that City To morrow

I Hope that My Familly wishes me as good a Christmas as I do them—Could I have spent it with My Beloved Wife & Children, the exchange of situation would have been most Agreable—I hope to have some tidings of them Tomorrow = the Shores are now Lined with Green Willows the Weather much Like May—at *Henderson*—The Thermometer is allmost every day from 60 to 65—

TUESDAY Decembr 26th 1820

Beautifull Morning, Light frost—I began my drawing as soon as I could see—drawing all day—

We saw to day probably *Millions* of those *Irish Geese* or Cormorants, flying Southwest—they flew in Single Lines for several Hours extremely high—

At half Past 11 o'clock The Boats Landed at the Natchez Bluffs amongst about 100 More, several S.B.ts were also at this place = the Carrion Crows first attracted my Attention, hundreds of them flying constantly Low over the Shores and alighting on the Houses—

I Recd Two Letters from my Beloved Wife, dated 7th & 14th of Novr the Last date contained one from My Brother *G. Loyen Dupuygandeau* dated July 24 1820

So Busy I have been all day drawing, that I did not even go to the Shore—a Little before Dusk I saw from our Boat Roof the Magnolia & Pines that ornament the Hills above this Place—

I Wrote a Long Letter to My Lucy with Hopes that it would be in time for the Mail of Tomorrow—

Our Commanders and Mr Shaw found every article of Produce Low, perhaps too Low to resound themselves—

I found the Stomack of the Great footed Hawk filled With Bones, feathers, and the Gizzard of a Teal, also the Eyes of a Fish and Many Scales—it was a femelle Egg numerous and 4 of them the size of Green Peas—

As we approach Natchez I [saw] remarked in several places—Saw Mills, placed over ditches cut from the River and running to the Swamps which in time of floods afford a Good Current—these ditches also serve to furnish the Mills with timbers floated through them from the Interior =

We have also seen very Large Rafts of Long Logs Intended for Mr Livingston's Warf at New Orleans—a Rafter assured us of having recd 6000$ for the Last Parcel he stole from the Government's Land—

WEDNESDAY Decembr 27th 1820

As soon as my drawing was finished, I cleaned and Went to Natchez properly speaking—there to my uttmost surprise I met Nicholas Berthoud, who accosted Me Kindly, and askd me to go to New Orleans in his Boat—I accepted his offer—

from the River opposite Natchez, that place presents a Most Romantick scenery, the Shore Lined by Steam vessels [and f] Barges & flat Boats, seconded by the Lower town, consisting of Ware Houses, Grogg, Chops, Decayed Boats proper for the uses of Washer Women, and the sidling Road raising along the Caving Hills [about 200] on an [Inclinat] oblique of a quarter of a Mile and about 200 feet High covered with Goats feeding peaceally on its declivities, while hundreds of Carts, Horses and foot travellers are constantly, meeting and Crossing each Other reduced to Miniature by the distance renders the whole really picturesque [the Carrion Crows Looking on the Whole with Wistfull Eye]; on the Top of this the Traveller comes in sight of the town as he enters [and] avenues of regularly planted Trees Leading to the diferent Streets running at right Angles towards the River; on the left the *Theater* a poor framed Building and a New and Elegant Mansion the property of M[r] Postlewait attract the Anxious eye—on the right the rollings of the hearth [and left in the St] thinly diversified by poor habitations soon close the prospect—advancing, [Two feet] he is Led into Main Street; this as well as the generality of the place [is] too Narrow to be Handsome, is rendered Less Interesting by the poorness & Iregularity of the Houses, few [Built] of which are Bricks—and at this season very much encumbered by Bales of Cotton—the Jail, Court House are New and tolerable in their form the Lower part of the former [is] a Boarding House of some Note. there are Two Miserable Looking Churches; [and already the] I dare not say unattended but [I] think so—

the Natchez's Hotel is a good House built on the Spanish plan, i.e. with Large Piazas and Many Doors and Windows—Well Kept by M[r] John Garnier and is the rendez vous of all Gentile Travellers and Boarders—Several Large tavern which I did not Visit furnish Amply the Wants of the Strangers that at all times abound from [all] different parts of the Union—this place now Contain about 2000—inhabitants and Houses, has a Bank in good Credit—a Post Office receiving the Diferent Mails Thrice per Week, a Public reading Room and 2 printing offices—

the Naturalist will immediatly remark the general Mildness of the temperature on seeing at this season the [Vegitation in] premature Growth of Lettuces, Radishes and other vegetables that in our Eastern Latitudes are Carefully nursed in April and sometimes in May—

the *Pewee* fly Catcher, the Notable Mocking Bird, constant residents [often] assure him that if frosts are few they must be of short duration, and the Numberless prostrated Carrion Crows in the less frequented Streets [and Yards] prove to him the unhalhiness of the atmosphere—those certainly may be considered as necessary Evils, for no Birds are more disagreable at the same time that few are More Valuable in Climates Like this—

I saw here a Gentleman with Whom I travelled some distance down the Mississipi My first Voyage but as he did not or Would Not recognize my features I spoke not to him—

the Country back of Natchez was represented as Good and fitted by rich planters who once raised a Large [annual] quantity of Cotton the principal article of Export—Opposite this the Lands are extremely Low and overflow to a great Extent and Depth the Mail in Times of flood goes by Watter through the Woods nearly 40 Miles Toward Natchitoches on Red River.

Indians are Daily seen here with diferent sorts of Game—for which they receive high Prices, I saw Small Wild Turkey sold by them for One Dollar each, Malards at 50^{cts}

Although the Weather is Comparatively Mild, the Orange trees will not bear the Winters in open air = and sometimes the frosts for a day or two are severely felt—the remains of an Ancient Spanish fort are perceivable, the Center is now Honored by the Gallows and the Ditch serves as buriing ground for Slaves—the Cemetiere Lies at the extremity of the Town—about 2 Years ago a Large part of the Hill gave Way, Sunk probably 150 feet and Carried Many Houses into the River—this was occasioned by the quick Sand Running Springs that flows into the Strata of Clay and pebles of Which the Hill is Composed—

this sunken part is Now used as the depot of Dead Carcasses, and often times during the Summer emits such Exalaisons as attract hundreds, Nay I was told Thousands of Carrion Crows = an Engine is now Nearly in Operation Intended to raise the Watter of one of the springs or *ecoulement* or drains to suply the City—This indeed is much wanted, Watter hauled from the River is sold at 50^{cts} p^{r} Barrel [and] taken out of the Eddy very impure = I found few Men Interested towards Ornithology except those who had heard or pleased to Invent Wonderfull Stories respecting a few Species—

Mr Garnier on whom I can rely told me that he had given Liberty to a Mocking Bird after several Years confinement and that for several Years afterward the Bird came daily in the House as if to thank him for his Generosity and Past Kind Attentions = Mr James Willkins to whom A. Willson had Letters of Introduction assured me that his Work was far from Completed, that through his Mere transient observations he had discovered several New Specimens but being a *Man of Business* he Never had Noted any

a Bird Much resembling the femelle humming Bird is often seen (it is said) during Summer, feeding by Suction amongst the Magnolias She is about Twice the Size of a Wren—

The Carrion Crows Never breed in or Near the Town—having Not one Cent when I Landed here I imediatly Looked for something to do in the Likeness Way for our Support ([for] unfortunately Naturalists are Obliged to eat and have some sort of Garb) I entered the room of a Portrait Painter Naming himself *Cook* but I assure you he was scarcely fit for a Scullion, Yet the *Gentleman* had some politeness and procured me the drawing of two Sketches for 5$ each, this was fine sauce to our empty stomacks

One was imediately paid for, the other a very excellent resemblance of M[r] *Mathewson* probably never will be, for that Gentleman absented the same Evening and never Left orders to any body to pay—I merely put this down to give you the Best advice a Father Can present you with. Never to Sell or Buy without imediately paying for the same—a constant adherence of this Maxim will Keep your Mind and person all times free, & Happy

M[r] Cook much pleased with My Drawings and quickness of performance, desired to travel with us if suitable Mutual arrangements could be Made. I Asked him to pay me Two Dollars per day Monthly in advance and furnish besides, One Third of the Whole Expenses, [furnishing] providing himself with Whatever Materials might be necessary—

He spoke of Joining us in a Couple of Weeks; I thought it very uncertain—the awkwardness I felt when I sat to Dinner at the Hotel was really justified to me; having not used a fork and Scarcely even a Plate since I left Louisville, I involuntarily took Meet and Vegetables with My fingers several times; on Board the flat Boats We seldom eat together and very often the hungry Cooked, this I perform[d] [often] when in need by Plucking & Cleaning a Duck, or a Partridge and throwing it on the Hot embers; few Men have eat a Teal with better appetite than I have dressed in this manner—

Others prefering Bacon would Cut a Slice from the *Side* that hung by the Chimney and Chew that raw with a hard Biscuit—Such Life is well intended to drill men Gradually to hardships to go to Sleep with Wet Muddy Clothing on a Buffalo skin stretch on a Board—to hunt through Woods filled with fallen trees,

Entengled with Vines, Briars, Canes, high Rushes and at the same time giving under foot; produces heavy sweats [and] strong Appetite, & Keeps the Imagination free from Worldly thoughts, [and] I Would advise Many *Citisens* particularly our Eastern *Dandys* to try the experiment—leaving their high heeled Boots, but not their *Corsets*, for, this would no doubt [would] be Serviceable whenever food giving way, they might wish to depress their stomacks for the occasion—

THURSDAY December 28th 1820

Weather sultry. Saw some Mocking Birds and was assured that they remained during the Winter here—

Nicholas having invited me to stay at his Lodgings I Breakfasted at the Hotel of Mr Garnier a French Gentleman of Agreable Manners who kindly procured me Willson's Ornithology from Mr James Wilkins to whom I was introduced to by Nicholas.

FRIDAY December 29th 1820

The weather this Morning had taken a remarquable Change, the Thermometer had fallen from 72 to 36—it snowed and blew hard from the Northwest—Last night the Musquitoes, were quite troublesome

I made Two Sketches to day for 5$ each; after Many Inquiries for the 9th Volume of Willson I was disapointed in my wish of examining it none of the subscribers have recd it—

Joseph and Cape Cummings still remaining on Board of Mr Aumack's Boat—I had the satisfaction of ransacking the *Fables* of Lafontaine, with Engravings—Wrote to Dr. Drake and Mr. Robt Best—

SATURDAY Decembr 30th 1820

the Weather very Cold, the Thermometer a 25—

Spent all day Writting the Name and Such Descriptions of the Watter [s] Birds in Willson as would enable me to Judge whenever a New Specimen falls my Praise—

Mr Aumack Left this Morning in our Boat taking with him Cape Cummings—I felt Sorry at parting With that really agreable Compagnion [and the More so When I thought that Like Me he had to meet all the ——.] I Wrote to My Beloved Wife—

SUNDAY Decemb^r^ 31st 1820

Early this Morning We prepared for our Departure, our things were Collected, and Carried to the Keel Boat—however it Was not untill One o'clock that the Steam Boat Columbus hauled off the Landing—

We Made fast to her Stern with Two Ropes and went very Swiftly the Moment She was under full headway—

I drew this afternoon—and here I have to tell a sad Misfortune that took place this Morning—having Carried under My Arm My Smallest Port Folio and Some other articles I Laid the Whole on the Ground and ordered Mr Berthow's Servant to take them on Board

I unfortunately Went off to Natchez again to breakfast the Servant forgot My Folio on the Shore and Now I am Without, any Silver paper, to preserve my Drawings, have Lost some very Valuable Drawings, and My Beloved Wife's Likeness = the greatest Exercions I now Must Make to try to find it again, but so dull do I feel about it that I am nearly Made Sick

I Wrote to Mr John Garnier, requesting him to advertise and procure some one to try to find My Port Folio [for] but no Hopes can I have of ever seeing it when Lost amongst 150 or 160 flat Boats and Houses filled with the Lowest of Caracters—No doubt My Drawings will serve to ornement their Parlours or will be Nailed on Some of The Steering Oars—

We passed to day A Long Line of Bluffs exquisiting grand to the sight—

My Port Folio Contained 15 Drawings Three of Which were Non Descripts—One a Duck extremely Curious and rare that I had Named the *Fintail:* Should I not get it again, it may retard My return home very Considerable—

MONDAY January 1st 1821

This day 21 Years since I was at *Rochefort* in France. I spent most of that day at Copying Letters of My Father to the Minister of the Navy—

What I have seen, and felt since, would fill a Large Volume—the Whole of Which Would end at *this Day January 1st 1821. I am on Board a Keel Boat going down to New Orleans the poorest Man on it*—& What I have seen and felt has brought some very dearly purchased Experience, and Yet Yesterday I forgot that No servant could do for Me What I might do Myself; had I acted accordingly; My Port Folio Would now have been safe in my possession—

Not Willing to dwell on Ideal futurity, I do not at present attempt to forsee where My Poor Body may be this day 12 Months

at 12 o'clock to day the *Columbus* Came too at Bayou *Sarah*—a Small Village at the Mouth of that Inlet—Many flat Boats the Steam Boat and 2 Briggs waiting for Cotton—the Steam Boat *Alabama* put off as we came to, and about half an hour after; the Columbus Left us to ourselves to try to reach *Baton Rouge* before her, to procure the freight there—promising to wait 3 hours.

The Lands are flatening fast—the Orange trees are now and then seen near the Rich Planter's habitation—and the Verdure along all the shore is very Luxuriant and agreeable—the Thermometer at—68 at 12 o'clock in the Shade, the Day Beautifully fair—Expected to see some Alligators = Many *Irish Geese* in the Eddys—Malards but few geese = at half past 6 o'clock P.M. we came opposite *Baton Rouge* but the Steam Boat had left and of Course we proceeded on our Way floating—this Last place is a Thrifty Villege on the New Orleans State—from some distance above *Levees* have made their appearance—I saw a Negro Man fishing by deeping a Scoup Net [constantly] every moment in the Watter immediately at a point Where the current ran swift forming an eddy below, he had taken several tolerably Large Cat Fishes—

TUESDAY January 2[d] 1821

We floated all night without accidents, the river since Natchez is much deeper, and free of Sawers and Snaggs—at day breake found ourselves about 50 Miles below Baton Rouge; the day Cloudy, raw, and some Wind a head—

the Plantations increase in number, and the Shores have Much the Appearance of those on Some of the Large rivers of France, their Lowness Excepted, the points are quite diferent to those on the River above, One May see the River below them by Looking across in Many places—and from the Boat we can only have a View of the upper windows, Roofs and Tops of the Trees about them, the Whole is backed by a dark Curtain of Thickly Moss covered Cypresses—flat Boats are Landed at nearly every Plantation, this being a Sure method of disposing of their produce to a better profit, travellers on horse Back or Gigg go by us full Gallop as if their Life depended on the accelerity of their movements—I have Seen More Common Crows since Natchez than I ever saw in My Whole Life before, the Shores and Trees are Covered with them but yet very few fish Crows have been Seen—saw some Pelicans, Many Gulls, Buzzards & C. Crows—

our Situation in this Boat is quite Comfortable—We have a good Servant to wait on us, are served with regular Meals, Clean and in Plates—Move much faster than With Mess[rs] Aumack & Lovelace, having here 8 Roaers who dare not contradict orders—

it rained and blowed hard a head, about One Mile below *Bayou Lafourche*. We Came to—the weather did not stop Joseph and Myself from taking a walk to the Swamp back of Plantation in front of which the Boat was moored—after Chasing the Note of What I supposed a New Bird for a considerable time, I found the deceiving Mocking Bird close by me and Exulting with the Towe Bunting's *cheep*—Joseph was more fortunate he Killed Two Warblers, one the *Red Poll* (of this We saw about a dozen) the other I have Not yet ascertained—although in Beautiful plumage; Both Male—how Sweet for me to find Myself the 1st of January in a Country where the woods are filled with Warblers, Thrushes, and at the same time see the Rivers and Lakes covered with all Kinds of Watter Birds—

the Pewees are quite gaily, I have seen this day 3 Cat Birds—if this is not the winter retreat of all our Summer Birds it is at all Events that of very many—how happy would I feel to see some future January surrounded by the diferent species of Swallows Skeeping about, with the Whippoorwill & Night Hawk

—I drew the Likeness of M[r] Dickerson the Master of the Boat—he paid me in Gold—took the Outlines of Both the *Warblers by Candle Light to afford Me time to morow to finish both*—

—3[d]—

Raining & Blowing hard all Night the weather cooled Considerably, much Like some of our April day at Henderson—took a Walk Early, while waiting for the Light to Increase and enable Me to Work—passed through a Large Cotton Plantation yet unpicked Looked Like if a Heavy Snow had fell and frose on every Pod—

the great regularity with which this is [Planted] sowed and raised attracts the Eye imediatly; it Lays in rows I believe allways runing at right Angle to the

River, about 6 feet distant from each other and the plants about 3 and so straight that your Eye is Carried to the farthest extremity of the field without the Lease obstruction—even at this time that the Cotton has Ceased to be attended for Many weeks, it is quite free from Weeds of any sort—

the Woods here have a new and very romantic appearance—the Plant Called Pamitta raises promiscuously through them the Moss on every tree darkens the under growth and affords to the [Mind] melancholy [fashioned] Mind a retreat thooted by the Chirpings of hundreds of [——] Beautifully Plumed inhabitants—

the flocks of *Blackbirds* taking the Species En Masse, feel the air, they pass Southwest constantly; forming a Line Like disbanded Soldiers all anxious to reach the point of destination each hurring to pass the companion before him—

Doves are plenty, the Cardinal Gros Beaks very numerous and all Species of Sparrow inabiting the Interior are here—I remark[d] great Many *Brown Larks* busy feeding on the Drift Wood that feels Many Eddys—

I drew both My Birds, the first on a plant in full bloom that I plucked Near the Boat—saw about 50 Mocking Birds some of them extremely Gentle, and holding their tail Leaning back allmost over their Heads—

We were Visited by several *french Creoles* this is a Breed of animals that Neither speak French English nor Spanish correctly but have a Jargon composed of the Impure parts of these three—

they Stared at My Drawing, and when a litle Composed Gazed and Complimented Me very Highly—on asking them the names of about a dozen diferent Birds then lying on the Table they Made at once and without hesitating a Solid Mass of *Yellow Birds* of the Whole—One of them a young Man told Me that he could procure 3 or 4 dozen of them every Night by hunting the Orange trees with a Lantern—I can said he "see the Rascalls White belies and Knok them down with a Stick very handy"—few of these good Natured Souls could answer any valuable account of the Country—

Some toads were hoping about this evening, and on turning a Dead Tree over, we found several Lizards, who moved with great Vigour—at Sunset the Wind Lulled, the Captain, Sailors and passingers all anxious to reach New Orleans, it was determined that after a good supper the oars would be used untill day break tomorow, if so We May see the City early

I Shot this Evening at a *Sparrow Hawk* that being badly Wounded sailed directly for a hole (probably that a Wood pecker) and [escaped us] no doubt died there—a few moments previous he Was teazing an American Buzzard—

Joseph Killed One Teal—with several Goldfinches and Warblers, some Sparrows, but Nothing New—

THURSDAY January 4th 1821

at 4 o'clock this morning the wind was so high, that it forced us to Come to a litle above *Bonne Care's* Church—the weather was rather cold, as soon as day broke walk[d] over to the Swamp

Saw some Birds that I took for Large non described *Cukoos* as they flew high over us, they had a new Note to me—Many Warblers, Robins, blue Birds, Cardinals, Grakles, Sparows Goldfinches, doves, Golden Wing Woodpeckers, One Red headed one, Many, Carolina & Winter Wrens—Sparrow Hawks, and a Large one unknown, on our return to the boat we Started with Hopes that We could make *some* headway—but were forced to about a Mile below the Church—I paid my Respects to the Pastor, to make some Inquiries respecting M[r] *Lecorgne, George Crogham,* and the country; but I found only a tall thin dirty Creole who could not say much besides the prayer for the prosperity of the Brick Church now erecting—from this pennsionary of Bigots I went to a School House; there I had the pleasure of meeting an Old, Polite, and well Instructed French Gentleman in charge of about a dozen of Pupils of Both Sexes—he told me that *George Crogham* resided about 3 Miles from this Place across the Mississipi, that he was not acquainted with Lecorgne's name—that this Country was a fine field for my Wishes; [he] walked to the Boats, examined attentively My Drawings and told Me that having Left Europe and the World of Talents for so many Years such a Sight was very gratifying—We hunted again and Saw more of the *Cuckoos* these [by] this time I saw on the Ground and Knew them at once for some of the same Birds I had Shot my preceeding Voyage and had taken for *Boat Tailed Grakles*—I Killed 3, Two femelles and one Male and had the pleasure of examinng their Manners very Closely—their Voice is Loud and Sweet and their Movements elegantly airy—a Beautifull male was very busily Engage[d] in carrying some straws to a Large Live Oak, but loosing the Bird every time through the Spanish Moss, I Could Not see any appearance of Nest, and so early in the season could not presume it was for the purpose of building one—I Shot him—and Joseph Killed another femelle—these Birds are considerably More Shy than any other Grakle—fly very Loosely when in flocks, uttering constantly a Chuck [very] diferent to that of the *purple Grakle—Gracula quiscula*—and their flight resembles that of our Cuckoos and that of the

Cukoo of Europe—While on the ground their walk is Elegant and Stately carrying their Long concave tails rather high—feed Closer to each other than Swamp Black Birds: Turton speaks of their Shortish Bills and gives for total length 13 Inches—this was taken no doubt from a Young femelle—the Male now before me Measures 15¾ Inches—my drawing Shews you Male & femelle, and tomorow it being finished I will give you a description—the French here call them starlings but on all questions respecting them or any other birds their answer is a constant *Oh Oui*—the country is here richly adorned by handsome dwelling Houses, Many Sugar and Cotton Plantations running about One Mile and half to the Swamp, free from Old trees and Stumps—every house a *L' Espagnole*—orange trees, now hanging with their golden fruits forming avenues and Edges—the fields Well fenced in and dreaned by ditches running to the Swamps—

the Mocking Birds are so Gentle that I followed one along a fence this morning for nearly one Mile Keeping only one panel between us the whole of the distance—I have Not heard one sing yet; but imitating Many Birds—

about 5 o'clock we again ventured off and again the Wind drove us in Shore and now we are Landed on a point where our Boat rolls merily—raining hard—thermometer to day at 52—this morning the French Gentlemen Wrap^d up in their Cloaks Kept their Handerkerchiefs to their Noses—What would become of them on the Rocky Mountains at this Season—our Captain Exchanged some Apples for Oranges receiving 2 for 1—

FRIDAY January ^5th^ 1821

We had some Light Snow this morning, drawing nearly all day, the wind blowing violently—Shortly after Breakfast I saw some *Terns Winowing* in the Eddy below us. Killed Two on the Wing—on the Falling of the first, the Second approach^d as if to see What was the Matter. I shot it dead, when the remaining Two that where coming fast wheeled and flew out of our Sight imediatly—these Birds flew Lightly with their Bills perpendicular over the Watter on which they appeared to Keep a close attention, Now & then falling to it and taking up Small fragments of Buiscuits thrown from our Boat—I finished my *Gracula Barita* but not the drawing; the rocking of the Boat quite disagreable—took a Long Walk towards Evening, saw Many Warblers particularly the *Maryland Yellow throat.* Shot an *Hermit Thrush;* paid a Visit to a Cottager a French Creole, handsome Children who were all afraid of me—the Lady *remarkably handsome* their Little Garden was adorned with a few orange Trees some fine Lettuces filled the Borders; Gren Peas nearly in bloom, Artichaux, reminding me of the Happy days spent in France—bought some delightfull radishes, and Enquired of Birds of Course; One League distant is a fine Lake, now the rendez vous of Ducks, Geese, &c but could not obtain Valuable remarks—the transient Cool Weather as rendered the Mocking Birds so gentle that they Scarcely would move out of the Way—

at Night I drew the Outline of the Tern I had shot and ransacked Turton's but all without effect. Yet I do not Consider this as a New Speci[men], until I See Willson's 9th Volume—the Gracule Male I draw Measured in Breadth 22 Inches—Tongue bifid—and I found to day that the Carrying Straws up in the Trees was simply to pick the rice contained in the heads—I saw this day thousands of them—particularly found of Catle pens, alighting close to them and hunting in the fresh Dung in the manner of the Uropean Starling

SATURDAY January 6th 1821

the Thermother fell by Sun rise to 30. We had some ice on the running Boards; after So much Warm weather experienced since the Latitude 33 this felt very uncomfortable, and our Litle Stove was good Company, the Wind blew hard ahead Yet—I drew slowly, the Tern I Killed Yesterday; Joseph Made his first attempt from Nature on the femelle—I was very much pleased with his assay—the wind falling aloft at 8 o'clock We pushd off—and rowed 12 Miles with much Dificulty, I did not expect our Commander would Leave our harbour with as dull a prospect—We were blowed a Shore opposite *Monsieur St Armand's* Sugar Plantation—

Out on Shore with Guns imediately—the Swamps about 3 Miles back we gave up going to them fearing the departure of our boat while absent that far—here Was the finest Plantations we have seen Mr S. Ad own 70 Negroes and Makes about 400 Hogshead of Sugar—besides raising, Corn Hay, Rice &c—this Gentleman, apparently Young was Shooting Red Winged Starlings, on the Wing for his amusement, had a richly ornamented Doubled Barelled Gun of which he Made excellent use—the Slaves employed at Cutting the Sugar Cane—this they perform with Large heavy Knifes not unlike those used by Butchers to Chop—some cutting the Head of the plants and others the Cane itself—tying the Last in small fagots with the Tops. Carts with Entire Wooden Wheels drawn by 4 oxen haul it to the

House where it is, bruised, pressed [and] Boiled & Made into Sugar—the Miserable Wretches at Work begged a Winter Falcon We had Killed, saying *it Was a great treat for them;* the Overseer a Good Looking Black Man, told us of his being in the same Employ for 8 Years and had obtain so much of his Master's Confidence, as to have the Entire Care of the Plantation—he Spoke roughly to his under servants but had a good indulgent Eye, and no doubt does what he Can to Accomodate, Master and All—those Immense Sugar Plantations Looked Like Prairies early in Summer for Scarce a Tree is to be seen, and particularly here where the Horizon was bounded by Cleared Land—

We saw Many Catle, Horses, and Sheep, but all poor and Slack, the Latters have but Litle Wool and that only on the back the Rams wear a Long Kind of hairy beard Like Goats—

the Gardens were beautifull. Roses in full bloom revive the Eye of the Traveller—who for lengthy Days has been Confined to the Smoky inside of a Dark flat Bottomed Boat—the Wind entirely Lulled away at Sun Set—the Moon's Disk assured us of a fine Night and We Left our Station to drop within a few Miles of the City—Tomorow perhaps May take us there, yet so uncertain is this World that I should not be Surprised if I never Was to reach it—the further removed, the Stronger My anxiety to see My family again presses on My Mind—and Nothing but the astonishing desire I have of Compleating my work Keeps My Spirits at par—

saw a few Fish Crows; a Marsh Hawk—a Red Shouldered Hawk, one the Boatman Killed a *Barred Owl,* this and the Winter Falcon Much Lighter Colored than usual—

Several Steam Boats passed us going up & down Length of the Tern 13 Inches—to the end of the tail—Wings extending 2 Inches beyond, tail 12 feathers—tongue Longue Slender and bifid. Mouth orange Color—Breadth 2:7¼ 12—Eye dark brown—Legs & feet red Orange—

SUNDAY January 7th 1821

at New Orleans at Last—We arrived here about 8 o'clock this Morning; hundreds of Fish Crows hovering near the shipping and dashing down to the Watter Like Gulls for food—uttering a cry very much like the young of the *Common crow* when they first Leave the Nests—I saw Mr Prentice, who directed me to the House of Messrs *Gordon & Grant* where he told me N. Berthoud Was; I saw him and was

Introduced to Mr Gordon of whom I shall have opportunity of speaking probably frequently hereafter and the *British Consul* Mr Davisson I heard that my familly Was Well, and saw a Note from My Wife to N. Berthoud, that accompanied a p^{re} of Gloves made by her for him—

We walked out, Met Col^{l} George Croghan, our former acquaintance saw Many of the Louisville Gentry too tedious to mention names—

Arrived at the House of Mr Arnauld an old friend of N. B^{ds} father. we were invited to take dinner, and although we had engaged previously to Mr Gordon we staid, We had a good dinner and great deal of Mirth that I call *french Gayety* that really sickened me. I thought Myself in Bedlam, every body talk^{d} Loud at once and the topics dry Jokes—Yet every one appeared good, well disposed, Gentlemen, and were very polite to us—a Monkey amused the Company a good deal by his Gambols and pranks—formerly I would have been able as well as anxious to go to the Theatre but now I can only partake of the Last, and after having paid a Short Visit to Mr Gordon I retired to the Keel Boat; with a bad head Hake occasioned by drinking some Wine—and very sorry that I probably Could not have Letters from the Post Office until Tuesday, Tomorow being a grand French Fete the aniversary of the Memorable Batle of Orleans

Joseph had spent his day visiting the Town and was not prepossessed in its favor—

I saw at Mr Arnauld 2 American Doves who have been Caged for Two Years, they *Layd* Eggs Last Spring, and incubated them for four days, but they were broke by Accident—

MONDAY January 8th 1821

at Day breake, went to Market having received information that Much and great variety of game was brought to it—We found Vast Many Malards, some teals, some American widgeons, Canada Geese Snow Geese, Mergansers, Robins; Blue Birds; Red wing Starlings—Tell Tale Godwits—every thing selling extremely high \$1.25 for one pa^{re} of Ducks, 1.50 for a Goose &^{c} Much surprised and diverted on finding a *Barred Owl* Clean^{d} and Exposed for sale Value 25^{cts}—some fine Fish; Indiferent Meat—some Vegetables both of this Country and West Indies = these Latters are put up in Small parcels on the ground opposite the owner, who has fixed prices for each Lot—I went to the review and will remember it and the 8th of

January forever—My Pocket was rifled of my pocket Book taken in this Morning with an intention of going to the Governor with the Letters I had received for him, and to Mr Wheeler's brother in Law—when I mention^{d} My Loss to N. Berthoud he called me a *Green Horn*, I do not Know the Color of My Horns but well, *those* of some Neighbours of Mine—

Not blaming fortune as is generally the Case I peaceably pack the whole to Myself and will try to grow Wiser if possible—I think the Knave who took it is now good deal disapointed and probably wishes I had it—the Parade was only tolerable I had a view of the Governor that is now no doubt all I May expect, he Looked about 60 a french face of good Countenance—We Walked to Bayou St John absolutly to Kill the time, the whole City taken with the festivals of the day—Joseph rec^{d} a Letter from his Parents—

this evening one of our Men Called *Smith* fell over board drunk and Would have drowned if Providence had not interfered a Women heard the Noise and the Yawl of the S. B^{t} U. States saw him—

—9th—

Breakfasted with J. B. Gilly. Rec^{d} a letter from My Wife—My Spirits very Low—Weather Cloudy & Sultry—begun raining paid a Visit to Jarvis the P. Painter—Saw W^{am} Croghan,—Wrote to My Wife—Wished I had remained at Natchez—having found No Work to do remained on Board the Keel Boat opposite the Market, the Dirtiest place in all the Cities of the United States—Wrote to John Garnier about My Port Folio—

WEDNESDAY Jany 10th 1821

Raining hard all day wrote My Brother G. L. Dupuygandeau and to My Mother—in the afternoon Cap^{n} Cummings arrived and dined with us—his appearance much Worst—the Weather so bad that I had no opportunity of doing any thing toward procuring Work—Strong thoughts of returning to Natchez Saw Cap^{n} Penniston who rec^{d} me very Politely

—11—

Spent the Day Walking about trying to find some work Shewed My Drawings to Mr Gordon & the British Consul Mr Davisson—spoke good of the Publication—the former raised My Expectations of their value—

Remarked in Market, Blue Cranes, great Many Coots Caldwall Ducks, some Geese, Keeldeers—1 White Crane—or Herons and one Sand Hill Crane—

Was sometime with Mr Prentice who gave me a letter to Doctor Hunter, whom I wished to see, to procure the Information I so much Need about the Red River, What life &c—Joseph Employed in Making Enquiries about the Lost Port Folio from every Boat Landing from Natchez—No Work yet—rain, Warm, the Frogs all piping—

—12th—

Early this Morning I Met an Italian, painter at the Theatre. I took him to N. Bd Room and Shewed me the Drawing of the White Headed Eagle, he was much pleased, took me to his painting appartement at the Theatre, then to the Directors who very roughly offered me 100$ per Month to paint with Mons L'Italian.

I believe really now that my talents must be poor or the Country—Dined with Mr Gordon, conversation Birds & Drawings, Must exibit some again and again as New Guests came in—

I Recd to day a Letter from My Beloved Wife Dated Novr 28th 1820—gave My Letter for Mr Garnier to the Columbus—No work yet—paid a Visit to Monsieur *Pamar* but Audubon was poor to day and he Knew it when I made my bow—

Wrote this Evening to Henry Clay Esqr for another Letter of Recommandation—

Weather, Warm, rainy, foggy, and altogether Disagreable Saw One Wood Cock in Market—

SATURDAY 13th January 1821

I rose early tormented by many disagreable thoughts, nearly again without a cent, in a Busling City where no one cares a fig for a Man in my situation—I walked to *Jarvis* the Painter and shewed him some of my Drawings—he over-looked [at] them, said nothing then Leaned down and examined them minutely but never said they Were good or bad—Merely that when he drew an Eagle for Instance, he made it resemble a Lyon, and covered it with Yellow hair and not Feathers—some fools who entered the room, were so pleased at seeing my Eagle that theh prised it, and Jarvis wistled—I called him aside, While Joseph Rolled up our Papers and asked him if he needed assistance to finish his Portraits i.e. the Clothing and Grounds—he stared, I repeated my question and told him I would not turn my Back to any one for Such employment and that I had received good Lessons from good Masters—he then asked me to come the following day and Would think about it—

in following N. B. through the street while nothing better could be done, We entered the Warehouse of Mr Pamar and at once was surprised to hear *him* ask what I Charged for my Drawings of Faces; 25$;—but said he I have 3 Children and you May put them all on one piece; then I must have 100$—

N. B^{d} requested me to make a Sketch of the Litle Girl then present; a sheet of Paper was procured, My Pencil Sharped and Sitting on a Crate was soon at Work and soon finished; the Likeness was Striking; the Father Smiled, the Clerks stared me emased and the servant was dispatched to shew My Success (as it Was Called) to Mistress—Monsieur Pamar *Civilly* told me that I Must do my Best for him and Left it to Myself as to the Price—I would have Liked to earn the half of the Money that day, but the Eldest Daughter could not be ready perhaps for several Days—Yet here is found Hopes—how I Calculated on 100 Dollars; What relief to My Dear Wife and Children for Said I if I get this, I may send it her and no doubt I will soon procure some more Work—

I spent the remainder of this day in better Spirits took a Long Walk With Joseph toward the Lake—Saw an Aligator—

Wrote this Evening to Doct^{r} D^{l} Drake—and read some Interesting tales, borrowed from Mr Prentice—

SUNDAY 14th January 1821

I dispatch Joseph and *Simon* (N. B. Servant) across the river for some Life Oak to draw, brought some not fit—

Dressed I Went to Jarvis—he took me imediately in his painting room, and asked me many questions, until *I thought* that he *feared* my assistance [might become troublesome to him]; he very Simply told me he could not believe that I might help him in the Least—I rose, bowed, and Walk[d] out without one Word, and No doubt he Looked on Me as I did on him as an Original, and a Craked Man—

The Levee early was Crowded by people of all Sorts as well as Colors, the Market, very aboundant the Church Bell ringing [and] the Billiard Balls Knocking, the Guns heard all around. What a Display this is for a Steady Quaker of Philad[a] or Cincinnati—the day was beautifull and the crowd Increased considerably—I saw however no handsome Woman and the Citron hue of allmost all is very disgusting to one who Likes the [——] rosy Yankee or English Cheeks—

I took My Gun, rowed out to the edge of the Eddy and Killed a Fish Crow, these Birds are plenty on the River every fair day—(when otherwise, the food is plenty in the Swamps, the Crabs, Young Frogs, Watter Snakes, &[c] Shewing out in great Numbers)—When the one I Killed fell, hundreds flew to him and appeared as if about to Carry him off, but they soon found it their Interest to let me have him. I drew Near and Loaded for another; they all rose in Circling Like Hawks extremely High and then flew down the Stream, out of sight calling aloud all the times they suffer their Legs & feet to hang down as if broken—

I brought it on Board and began to Work imediately—at Dark took a Walk to M[r] Gordon, from there on to M[r] *Laville* where We saw some *White Ladies* and Good Looking ones—returning on Board the Quartroon Ball attracted My View but as it cost 1$ Entrance I Merely Listened a Short time to the Noise and came Home as We are pleased to Call it—

MONDAY 15th January 1821
TUESDAY 16th " "
WEDNESDAY 17th " "
THURSDAY 18th " "

This is a Way of Cutting Matter Short, but indeed the time has been so Long and dull during these days that I think it a good ridince to use them thus [——] [I was sad when they passed heavily by] I spent them running about to procure Work, being sadly disapointed by M[r] Pamar who said the Lady Wanted Oil Colors—

Yesterday I Made My Long Acquaintance's Likeness John B. Gilly purposely to expose it to the Public—it is considered by every one who Knows him to be perfect—and to shew it this Morning (for I made it in a few hours) to Pamar procured Me the Making of that of his heldest Daughter; by the time We receive the pay for it, We will be penny Less—

to day I rec[d] a letter from My Beloved Wife who rufled My Spirits Sadly it Was dated Cincinnati Dec[r] 31. 1820 I answered it. [——]

Saw in Market—2 White Herons—one New Species of Snipe, but could not Draw any of them, being partly pluck[d]—Joseph Who hunted all day Yesterday, Killed Nothing New—saw Many Warblers—

	British Consul's Likeness		$25.00
19th Friday—	Euphraim Pamar's Likeness		25.00
20 Saturday—	Another Sister	"	25.00
21 Sunday			
22 Monday	M[rs] Pamar		25.00
23 Tuesday	Litle Daughter of	D[o]	25.00
24 Wednesday	M[r] Forestall	D[o]	25.00
25 Thursday	Young Lucin	D[o]	25.00
26 Friday	M[rs] Lucin	D[o]	20.00
27 Saturday	Mr [——] Carabie	D[o]	25.00
			$220.00

28 Sunday Drawing a Brown Pelican—

fatigued, Wearied of Body but in good Spirits having plenty to do at good Prices, and my Work much admired—only sorry that the Sun Sets—

MONDAY January 29*th* 1821

Drawing all day the Brown Pelican, Collected My Earnings purchased a Crate of queens Ware for My Beloved Wife, Wrote to her, W*m* Bakewell and Charles Briggs forwarding her by Letter and Parcel care of M*r* Buckamain of Louisville 270 Dollars—the Crate Cost 36: 33$—

TUESDAY 30*th* Mr Duchamp's Likeness 25$
WEDNESDAY 31 Nothing disapointed by M. Laville

[THURSDAY "Drew a common gull]
THURSDAY—February 1st 1821—

Began a Likeness of Mr Louallier and Drew a Common Gull—

FRIDAY 2d Mr Smith began hunting for Me at 25$ per Month stopped Thursday Morning—and the Girl *Began Cooking for us at 10$ per Month with washing*
SATURDAY 3d Wrote to My Beloved Wife—

SUNDAY 4*th* Disapointed again by M. Laville, returned to the Boat and Drew a *Grey Snipe*—

Joseph & Mr Smith out hunting all day, brought a few Red Breasted Thrushes, some Pewee fly Catchers and Many Swamp Sparrows and Savanah Finches—

Complaining very Much of the Hardships of hunting in the Cypress Swamps—Remarked in the Market many *Purple Galinules* but all so Mangled that I could not see one fit to draw—Saw also several *Rails or Soras*

MONDAY February 5th 1821

Running about pretty much all day trying to procure some More Work and also Enquiring about Willson's Ornithology, but in Vain—the high Value set on that work now particularly Lately as rendered it extremely rare, and the few who possess it will not Lend it—the Weather extremely sultry and Damp heavy Rains and thunder—saw to day at Mr Pamar, where I often Breakfast or Dine Mr Delaroderie

having been so extremely engaged the Two Days that I Drew the *Brown Pelican* I had not time to make such Memoranda as I wished—it was given me by Mr Aumack, the Bird was Killed on a Lake in this Vicinity and are rare—it was a Male, in tolerable order—Mr Gordon who before he had seen My Drawing thought it a Pelican Common in some of the West Indies, was much pleased at finding it a different Species—I was assured that these Birds were seen in Immense flocks in the Neighbourhood of *Buenos-Aires* by a Scotch Gentleman, of respectability—here the hunters call them *Grand Gozier* and say that seldom more than two are seen together—and only for Short Periods after heavy Gales from the Sea—

the Common Gulls pay us regular Visits With their usual Companions the Fish Crows every Morning about Sun Rise—Comming across the Land from the

Lakes Barataria where I am told they resort to Roost—I was not a litle surprised at finding the Stomack of several We Killed a few days since filled with *Beetles* of different Kinds; Joseph examined the River and found Vast Many of these Last floating dead on the surface, the Crows also feed freely on the Same—the Gulls often chase the Crows for some Considerable distance but Never Successfully, the Crow being much Swifter on the Wing—the quantity of Robins or Red breasted Thrushes Killed here is astonishing, the Market abound with them and yet they bring 6¼ cents each—they are at present the principal Game to be found. Birds of all description are destroyed and eat, our Men Cooked the Gulls and found them excellent food—

I saw 12 of the Grey Snipes Similar to the one I Made My Drawing from, all alike in size and Marking, but the Stupid Ass who sold me one Knew Nothing, Not even where *he* had Killed them—

FEBRUARY 9th—1821—

In walking this morning about a mile below this city I had the pleasure of remarking thousands of purple martins travelling eastwardly they flew high and circling feeding on [incectes] insects as they went they moved onwardly about ¼ miles an hour Thermometer at 68 weather Drisly.

Hundreds of Coots where in the market this morning

FEBRUARY 15th 1821

Wrote to My Beloved Wife this Day per N. Berthoud List of Drawings Sent My Beloved Wife February 17th by Nicholas Berthoud of Shipping port Esq[re]

1 Common gallinule—Not Described by Willson
2 D°—Gull— D° D° D°
3 Marsh Hawk
4 Boat Tailed Grakles Male & femelle—Not Describd by Willson
5 Common Crow
6 Fish Crow
7 Rail or Sora
8 Marsh Tern—

9 Snipe Not Described by Willson
10 Hermit Thrush
11 Yellow Red Pole Warbler
12 Savannah Finch
13 Bath Ground Warbler Not Described by Willson
14 Brown Pelican Not Described by Willson
15 Great footed Hawk
16 Turkey Hen—Not Described by Willson
17 Cormorant— D° D° D°
18 Carrion Crow or Black Vulture
19 Imber Diver
20 White Headed or Bald Eagle

May I have the Satisfaction of Looking at these and Many More in good Order on My return the fruits of a Long Journey—

MONDAY 19th February 1821

the Weather beautifull, Clear & Warm, the Wind having blown hard from the Southwest for 2 days & nights—

Saw this Morning Three Immense flocks of *Bank Swallows* that past over Me with the Rapidity of a Storm, going Northeast, their Cry was heard distinctly, and I knew them first by the Noise they made in the air coming from behind Me; the falling of their Dung resembled a heavy but thinly falling Snow; No appearance of any feeding While in our Sight—Which Lasted but a few Minutes—

I was much pleased to see these arbingers of Spring but Where could they be moving so rapidly at this early Season I am quite at a Loss to think & yet their Passage here was about as long after the Purple Martins that Went By on the 9th Instant as is their arrival in Kentucky a Month hence—perhaps Were they forced by the last Winds and now Enticed to proceed by the Mildness of the Weather the Thermometer being at 68—

how far More south Must I go Next January & February to see these Millions of Swallows Spending their Winter as Thousands of Warblers, fly Catchers, Thrushes and Myriads of Ducks, Geese, Snipes &c Do here?

the Market is regularly furnished with the *English Snipe* Which the french Call

Cache Cache. Robins Blue Wingd Teals, Common Teals, Spoon Bill Ducks, Malards, Snow Geese, Canada Geese, Many Cormorants. Coots, Watter Hens, Tell Tale (Godwits, calld here *Clou Clou*) Yellow Shank Snipes, Some Sand Hill Cranes, Strings of Bleu Warblers, Cardinal Grosbeaks, Common Turtle Doves, Golden Wingd Wood Peckers &c

WEDNESDAY 21st February 1821

I Met this morning with one of those slight discouraging Incidents connected with the life of the artists; I had a Likeness [——] Spoken of in very rude terms by the fair Lady it was Made for, and perhaps will Loose My time and the reward expected for My Labour,—Mrs *Andre* I here mention the Name as I May Speak More of the Likeness as the occasion Will require—

Saw Many Green Baked White Belied Swallows to day and Also four *Martins* Hiroundo Purpurea—All of them very Lively and not exibiting much of the Muddy Appearance that immersion in the Swamps about this City would undoutedly give them, [if they] had they remained buried in it since Last December at Which time late in that Month they were plenty and remarked passing by [and] ariving from North & East [Courses (?)] moving South Weswardly—here they Must Make a Long Pause or Move Eastwardly very Slowly as Seldom do they arrive in Pennsylvania before the 25 of [April] March and more frequently in the first days of [May] April—they find nowhere an abondance of Insects, and the Millions of Musquitoes that raise from the Swamps Would Sufice to feed the Swallows of all the World—

Saw Many Brown Larks.

the Fish Crows are remarkably fond of alighting in flocks on some Pacan trees about 12 Mile below this City about 9 o'clock in the Morning when they retire on these to rest from their Fishing excursions and remain Croaking untill the Midle of the day—

THURSDAY New Orleans February 22d 1821

We at Last have the Keel Boat off and have moved on the hearth again—Our present situation is quite a Curious one to Me, the room we are in and for which We pay $10 per Month is situated in Barraks Street near the Corner of that &

Royal Street—between Two Shops of Grocers and divided from them and our Yellow Landlady by Mere Board Partitions, receiving at once all the new Matter that Issues from the thundering Mouths of all these groupes—the *Honest Woman* spoke much of honesty in Strangers and required one Month paid in Advance, this however I could not do, and satisfied her with one half Not taking a Receipt although She appeared very urgent—

I walked a good deal about the City in search of Work & Willson's Ornithology but was not favored with any success—extremely anxious to receive some news from My familly—am very much fatigued of New Orleans Where I cannot *Shoot Two Birds with one Stone* I retired to our Lodgings at Dusk—

Saw Cape Barton's of Henderson who said he would Not have known me if I had not Spoke within his hearing (one half page blotted out)

SATURDAY 24th February 1821

Idle, and the weather fair, took a *Wade* in the Woods the vegetation forwarded beyond my expectations. Saw some handsome Plants in Bloom that made me regret of having sent My Drawings home—Birds extremely Shy and nothing New;—in the afternoon Paid a Visit to M*r* John F. Miller from Whom (I was told) I might expect an uncommon Share of Politeness, perhaps I walked in at a bad time but be this as it May, I Was received and Dispatched as promptly as the case would permit; the subject of Course During the few Moments I Left then rolled on Birds. I was asked if Many Ducks such as the *Canvas Back* &c (fine food) where residents of this Part of America, had I put those queries I might have expected a No from One that had resided 10 Years here, and yet this No I Was obliged to give—No doubt for want of True Knowledge for the Birds in question are said to be plenty here during Winter—and are Most Likely are the Sheldrakes Dun Divers and Mergansers having Many of these latters offered for Sale [——]

This Morning the Market was well Stocked with Green Backed Swallows *Hiroundo Veridis*, the Whole very fat and in beautifull plumage; if these Dear Litle Cherubs have preferred their coats and these flesh so fresh during the pretended Torpor Occasioned by Winter' frost how much more fortunate they are than the Pork Beef & Butter of Kentucky that sowers however well Salted.

but I have been assured by Men on whom I can rely that some Winters are so Mild that Swallows are seen from time to time during every Months—

the Swallows in Markets Were caught in the holes about houses their resorting places during the Nights—this Morning the Weather is Quite Cold, and yet the Swallows are flying about the Street, over the River &c twitering very Lively—

SUNDAY 25th February 1821

Killed Some Green backed White bellied Swallows *Hirondo Veridis*—extremely fat, the Gizards completely filled with the remains of Wingd Insects—could not perceive any outward diference between the sexes, the femelles however were well Stored With eggs and the Males strongly marked—the Brother of Mr Pamar Killed a Beautifull White Robin, but his Dog Mangled it so much that I did not draw it—this extraordinary change of Color appeared as the Cause of old age, the Bill of the Bird much worn and the Legs were cicatrised in Several parts, the Bird however was very fat as well as Nearly all the others he killed this Day—

Saw a few Partridges these Birds are here much Sought and hunted down without Mercy, not even do the Sportsmen permit a few Paires to remain untouch & thereby the race is nearly extinguished Near the City—We Waded to day through an extensive Swamp with hopes of Meeting Some New Species, but Saw Nothing of the Kind to my astounishment—

March SATURDAY 10th 1821

Sent My Letter to My Lucy, Victor; Wam Bakewell & N. Berthoud by the steam Boat Car of Commerce—

Saw in Market this Morning some *Ampellis Americana* Vast Number of *Common Snipes* but the Robbins have nearly disapeared and are not even to be seen in the Woods—Was assured to Day that the *Baltimore Oriole* Wintered on the Island of *Cuba* and that the south west side of this island was the resort of Millions of Swallows during November, December & January; So strongly was this afirmed that it has determined me to go there Next Winter and to the bay of hondurass, where it is said they also are plenty during these Months Only—

Recd this Morning a Letter from My Beloved Wife dated Shippingport written a few days previous to the one I red by the S. B. James Ross—

This Evening Capn Cummings saw a Night Hawk or a *Chuck Will's Widow* flying about the Street Near our Lodgings—I presume it was the Latter Bird having been assured that in the first days of April they are seen in numbers at the Bay of St. Louis—

SUNDAY March 11th 1821

Walkd out this Morning with Joseph to try my Souvenir Gun and found it an excellent One. Shot Many Green Backed Swallows on the Wing—Some red Wing Starlings, Savannah Finches, one fish Crow, & c—but Nothing New Yet in the Woods, which unfortunately are Now very deep with Watter, the River being about 4 or 5 feet higher than the ground behind the *Levee*—

During a Walk this afternoon a Beautifull Mississipi Kite sailed by me, unfortunately I had no Gun.

Near our House a Mocking Bird regularly resorts to the South Angle of a Chimney top and salutes us with Sweetest Notes from the rising of the Moon until about Midnight, and every Morning from about 8 o'clock untill 11, when it flys to the Convent Garden to feed—I have remarked that Bird allways in the Same Spot and Same Position, and have been particularly pleased at hearing him [try to] Imitate the Watchman's Cry of *All's Well* that Issues from the fort about 3 Squares Distant, and so well, has he sometimes performed that I Would have been mistaken if he had not repeated too often in the Space of a 10 minutes

March 15th THURSDAY 1821

Wrote to My Beloved Wife and Mr Rob Best Last Night—Saw Many *Chuck Will's Widow* flying about the streets and some *Night Hawks*

I Made a Likeness to day for a Lady's Sadle a thing I had not the Leass use for, but the Man I had Made a portrait for, Wanted his Wife's Very Much and Could not Spare Money, and Not to disapoint him I Sufered Myself to be Sadled—

While at Dinner We were all surprised at the astounishing Leaps that Some *Maggots* took about our Table.

they Issued out of a Very good piece of Cheese to perform this I remarked them drawing up their heads toward the Tail untill Nearly runing both half of the Body Parell and The Suddenly striking one of the ends Could Not see Which they through themselves about 50 or 60 Times their Length some time One Way sometimes another apparently in Search of the Cheese—

March 16th FRIDAY 1821

I had the pleasure of receiving a Letter this Morning from Mr. A. P. Bodley dated Natchez 8th Inst Informing me of my Port Folio having been found and Deposited at the office of the *Mississippi Republican* and that I could have it by writing—

I Acknowledge with a very sensible pleasure the Kindness of Mr. P. who worked his Passage down in Mr Aumack Boats—and at the same time cannot Conceive how the Book had escaped the researches of Mr Garnier—

Mr Gordon had the goodness to write to a friend to have it forwarded imediately and pay whatever charges there might be, the Politeness of that Gentleman is remarkable to a Man who is no more than a Stranger to him, but No doubt it would be impossible for a Good heart to act otherwise—

I took a Walk with my Gun this afternoon to see the Passage of Millions of *Golden Plovers* Coming from the North Est and going Nearly South—the distruction of these innocent fugitives from a Winter Storm above us was really astonishing—the Sportsmen are here more numerous and at the same time more expert at shooting on the Wing than any where in the U. States [and] on the first sight of these birds Early this Morning assembled in Parties of from 20 to 100 at Diferent places where they Knew by experience they told me the birds pass and arranged themselves at equal distances squatted on their hams, as a flock Came Near every man Called in a Masterly astonishing Manner, the Birds Imediately Lowered and Wheeled and coming about 40 or 50 yards run the Gantlet every Gun goes off in Rotation, and so well aimed that I saw several times a flock of 100 or More Plovers destroyed at the exception of 5 or 6—the Dogs after each Voleys While the Shooters charged their Pieces brought the Same to each Individuals—this continued all day, When I Left One of those Lines of Sharp Shooters then the Sun Setting, they appeared as Intent on Killing More as when I arrived at the spot at 4 o'clock—

a Man Near [me] where I was seated had Killed 63 dozens—from the firing before & behind us I would suppose that 400 Gunners where out. Supposing each Man to have Killed 30 Dozen that day 144,000 must have been destroyed—On Enquiring if these Passages where frequent I was told that Six Years ago there was about such an Instance, imediatly after 2 or 3 days of Very Warm Weather a blow from the Northeast brought them, Which Was Nearly the Same to day—some few Were fat but the Greatest Number Lean, and all that I opened showed no food—the femelles Eggs extremely small

SATURDAY 17th March 1821

This Morning the Market Was plentifully supplied with Golden Plovers and *Pures*—I also saw a White Crane. Spent the day Walking about at the exception of an hour Drawing at a Likeness—

SUNDAY March 18th 1821

This Morning I was Witness and in some measure contributed to the presserving of a Tara of a new Kind at Least to Me.—Walking along the Levee to Mr Pamar Where I had an Appointment for a Likeness. I Was Invited to breakfast by Mr A. Liautaud—I Walked in and Met a Large Party, Well engaged round an old Gentleman at Pleasing him by the most extravagant round of praises—I understood the Caracter was rather Moony, and very gay When well managed, productive of Much Mirth to his hearers—During the Breakfast that Certainly Was a good one & on Which One *Prince Guest* touchd heavily, We were several times struck by unexpected Voleys of Verses, composed for the Occasion, and that could not Indeed have had a better Effect than that produced—Every One enjoyed himself Very Much, particularly the Compositors who were highly Clapped, sometimes to be sure to put an end to his Loquacity.

Breakfast over I was told to remain and see the best part. Mr Liautaud the Learned Guest Was about being recd a *Mason* and My being a Brother entitled me at once to a seat—this was conducted in the most Ludicrous Manner any one can conceive, and I really pitied the Newly Initiated When all Ceremonies Were over the Man Was Burned in several parts, baptised in a Large Bucket of Watter, Tossed in a Blanket, and Make to Crawl Over about 50 Casks of Wine, on his belly and Knees, and When at Last given up for want of Invention, the Poor Devils Who had being praying for Mercy during all this was Left in the Necessary—

To this Man this Might be done perhaps again, but few could bear such treatment and I expected several times that his Cries or a Change of sensation from Cowardice to Courage would shew a very diferent scene but all however was Ended as Intended and the poor fellow took it for Granted that he [was] really Was a Mason—

I left and Made Mr Doucillier, Guesnore Likeness—

We purchased this Morning in Market a Beautifull *Blue Crane* the choice of 5 that were nearly alike—those I was told were come earlier than usual. Are extremely dificult of access

I Drew it and its coresponding so well with A. Willson Description stop*d* me from writing it Myself—

Total Length	30½	Inches
to end of tail	23½	"
Breadth	39¼	"
Tail Feathers	12	"

Weight 10 Onces—Midle Claw Serrate inwardly The Cotony Substance on the Breast followed the Breast Bones only—and reapeared on each side of the rump—I was assured it a very beautifull Male—

WEDNESDAY March 21st 1821

In reading the Papers this morning at Mr Pamar I saw the Treaty between Spain and our Country.

the 4th Article Speaking of an expedition to run the Line of Division formed by both Parties and to leave Natchitoches during the Course of this Year; I imediatly went to Mr Gordon to know from him What steps Would be necessary to procure an Appointment as Draftsman for this So Long Wished for Journey he advised me to see Mr Hawkins Who would introduce me to Governor Robinson—

I saw Mr Hawkins who very politly promised to See the Governor and Mention to him my Wishes and to Call at his office on the 23d

to Join in Such an enterprise and to leave all I am attachd to, perhaps for ever, produced Many diferent sensations & thought, but all are Counterbalanced When persuaded as I am that My Labours [will all be of use to them] are all for their use & benefit.

I did not wait late on the 23d but My Spirits were sadly dampened when Mr Hawkins told Me that it was the Governor's opinion that nothing more [than common] would be done than to run the Line in question and that none but the surveyors would be Necessary;

Disapointed but not less anxious to try further I Calld on Mr Gordon, he Joined me in the Idea of My Adressing the President Directly and that he could Not think that a Journey so Interesting Would be performd only to say that Men had gone & Come back—in leaving this truly Kind Gentleman, I Met Mr Grayson of Louisville spoke to him to my thoughts and Wishes.

"I Can render a Service I believe Mr A and I Will do it. I will give you some Letters to diferent Members of Congress With whom I am Well acquainted and that will be glad to Meet Your Views, but Write to the President—
This sounded better to my Ears—

full of My plans I went home & Wrote to N Berthoud to request his Imediate Assistance—Walkd out in the afternoon seeing Nothing but hundreds of New Birds, in Imagination and supposed Myself often on the Journey—

on the 24th I called again on Mr Hawkins. Mr Gordon had spoke of me to him and the former again to the Governor, I spoke of adressing the President, he acquiestd and promissed to give a Letter for the same and procure one from the Governor—

going through the Street Not unlike I dare say a Wild Man thinking too much to think at all My Eyes Were attracted by a handsome faced Man. I Knew him it was My Old Acquaintance & friend George Croghan. We Met freely and I was eased, he Knew what I was going to say having dined the day previous at the Governor's with Mr Hawkins, he said he had spoke of Me but Would do more and promised to find Some Letters to Mr Hawkins for my use, and Invited Me with such forcible Kindness to go and spend Some time at his Plantation that I Accepted his offer—see me again Walking fast and Looking Wild, but recollecting the high price of time I hunted M[r] D[d] Prentice, and asked him if he would form a letter for Me—he answered Yes but told Me that I would do better by writing Myself and that he would freely give his advice and help if Needed—I Was then reduced to My poor thoughts to Express My Wishes—[but] Anxious, and Determined to leave no power of Mine untried, I sat to the Paper & Wrote in as Great a Hurry as I am Now doing, a Letter, that M[r] Prentice to My utmost astonishment pronounced *all suficient*—he spoke much about the Journey and anticipated he said the pleasure of reading My Journal on My return—feeling a great Weight off My Shoulders I returned to My Room, took Gun Ammunition & Joseph & to the Woods Went in Search of New Species—

My Life has been strewed with Many thorns but could I see Myself & the fruits of my Labour safe, with My Beloved familly *all Well* on a return from Such an expedition, how gratefull Would I feel to My Country and [Thank]full of the Greatness of My Author—

In Market this Morning I saw Three of What Willson *Calls Bartrams's* Snipes, they Where very fat are called here *Papacots*—Saw a Beautifull White Crane but without legs—Vast Many Green Wing[d] & blue winged Teals—hundreds of Snipes, *pures*, Solitary Snipes—Green back[d] Swallows—but robbins have disapeared—

The Migration of Birds does Not go a pace with the Vegetable Kingdom in this Part of America When, The Trees are as much in Leaves in Pensylvania, or Ohio, or Kentucky or even the upper Part of Tennessee [Birds] (*and this is about the Midle of May*) Birds back to the 25 of April have reach[d] these Parts and are preparing to answer the Calls of Nature—

to My Astonishment, the Many Species of Warblers, Thrushes &[c] that Were numerous during the Winter have all Moved on Eastwardly and None of the Species that resort this Part have yet reach[d] at the exception of Swallows and a Few Watter Birds—

this Would tempt me to believe that Most of Our Migratory Birds Leave their Winter resort [When Influenced by their Natural habits] with such Certainty [in the Weather] of Knowledge about the Weather and Such Swiftness of Movement over the Country as does not give us even the opportunity of remarking their Passage and the greater number [all] at once at the time appointed by the Strength of Passion—

SUNDAY 25th March 1821

Bought a beautifull Specimen of the *Great White Heron* [grus Americana] in perfect order it had been sent me by a hunter with Whom I had formed acquaintance a few days ago—Worked on it the Whole day and found it the most dificult to Imitate of any bird I have yet undertaken—Took a Walk in the afternoon and heard the Voice of a Warbler new to Me, but could Not reach it—

MONDAY 26th Walked early this Morning in search of Plants to form the back ground of My Drawing—Left Joseph out who Killed a beautifull, *Blue Yellow back Warbler Male Sylvia Pusilla.* I had seen some in the Swamp but the litle fellows where so Nimble and brisk that they had all escaped Me—Worked Nearly all Day Not having time to go and Purchase some articles I wanted to send My Lucy—

Saw M[r] Gordon, who on reading My adress to the President Tura told Me he would Make some Alternation to its form—I left it with him—

TUESDAY 27th March 1821

Drawing at My Heron yet, it Smelt so dreadfully bad that When I opened it I could only take time to See *how plainly it proved a Male*—the Cottony Substance Was round the breast bone, on each Side of the Vent and on each Side the romp—the Midle Claws very pectinated

I Drew to Day Male & femelle of the *Blue Yellow back Warbler Sylvia Pusilla*—and in a Short Walk taken this Evening saw perhaps thousands of the same Species Killed several all alike in their sexes—

forwarded My Beloved Lucy pr Mr Prentice who left at 12 o'clock in the James Ross a piece of Linen, some Stockings, a piece of goods for our Boys and the Woman's Sadle I was almost forced to take some time since—

Gave Mr Forettale who also went in the same boat an order to get My Port Folio at Natchez and begged of him to forward it to Mr Gordon—

SATURDAY March 31st 1821

I spent my time these 3 Last Days More at thinking than any thing else—and often indeed have I thought My Head very heavy—

This Morning I Waited on Mr Gordon with a Wish to receive from him an amendment to my Letter to the President, for all in my head is the Pacific Expedition he Wrote, I read it, but was Not altogether Satisfied—I Called on Mr Vanderlyn the Historical Painter With my Port Folio—to shew him some of my Birds with a View to Ask him for a few Lines of recommandation—he examined them attentively and *Called* them *handsomely* done, but being *far* from possessing any Knowledge of Ornithology or Natural History, I was quite satisfied *he* Was No Judge but of their being better or Worst *Shaded* Yet he spoke of the beautifull Coloring and Good Positions and told Me that he would With pleasure give me a Certificate of his having *Inspected* them—Are all Men of Talents fools and Rude purposely or Naturally? I cannot assert, but have often thought that they were one or the other.

When I arrived at Mr V's Room, he spoke to me as if I had been an abject slave, and told Me in Walking Away to Lay my Drawings down *there the Dirty* [——] that he would return *presently* and Look them over.—I felt so vexed that My first Intention Was to *Pack* off, but the Expedition Was in View, I thought how Long

Kempbell the Actor Waitted Once at the theatre in England, and stood patiently *although* not Laying My Drawing *Down there*

About 30 Minutes Elapsed, he returned with an officer and with an air More becoming a Man Who *Once Was Much* in My situation ask me in his private room. Yet I could plainly see in his Eye that selfish Confidence that allways destroy in some degree the Greatest Man's Worth.

the Swet ran down My face as I hastily open[d] My Drawings and Laid them on the floor; I Look[d] up to him. he Was looking at them, the officer's *By God* that's handsome, struck my eyars Vanderlyn took up a Bird Look[d] at it closely put it down and said they Were *handsomely done.*

I breathed, Not because I thought him a Man of the Most Superior Talents, for to come to such a pitch one Must have no faults, and I With My Eyes *half Closed* (as you know the pretended Juges of our Day Look at Painting [Look]) saw a great Deffect in One of his figures of Women (the deffect that had being Corrected by the Lady I drew Lately.) but because this Gentleman had *some Talents*, that he Was Look[d] on as a Very Excellent Judge and that I had been Told that a few Words from him Might be serviceable—of My Likeness he spoke very diferently, the one I had Was fair, hard, and Without Effect, although he Acknowledged it Must have been a Strong one—

he sat, he Wrote, and I, thinking More of Journeying to the Pacific Ocean, than of Likenesses, Cared Not [Not] a *Pecayon* about these Later Observations—

as I Was Walking away from his house corner of S[t] Louis and Royal Streets—the *Corner of Events* the officer who had followed me, ask[d] me, the price of My *Black Chalk* Likeness and where I resided—all answered; I thought how Strange it was that a poor Devil Like me Could Steal the Custom of the Great Vanderlein—but fortune if not *blind* certainly Most have his Lunatic Moments—the officer said he would Call on Me Liking My Style Very Much—

M[r] Hawkings saw this afternoon some of My Drawings and I gave him My Letter to the President. he was apparently Much pleased With both—and told me he Would do all in his Power for Me—

I Put My Letters to My Beloved Wife—N. Berthoud & Judge Fowles in the [Office] Mail that Leaves every Sunday at 8 o'clock, and return[d] to our Lodging with a compound of Ideas Not Easily to be described—

I had Shot during a Short Walk I took Yesterday afternoon a *White Eyed fly Catcher* that the Rats having eat Last Night I could Not of course draw to day—

Joseph to day [Drew] Shot a *Tyrant* Fly *Catcher* and a *Yellow belied Wood Pecker*

The Politeness of M[r] Vanderlyn Will be remembered—a long time by me; and When ever I Look over these Scrawls it will do me good to have a Litle of the same feelings—the following is the Copy of the Lines he handed me—

M[r] John J. Audubon has shewed me several Specimens of his Drawings in Natural History—such as Birds, with their Natural Colors, & other Drawings in plain Black & White Which Appear to be done With great [accuracy] truth & Accuracy of representation as much so as any I have seen in any Country—the Above Gentleman wishes Me to give this as My Opinion in Writting believing it may Serve as a recommendation to his being employed as a Draftsman in any Expedition to the interiors of our Country.

J. Vanderlyn

NEW ORLEANS 20th March 1821

April 5th 1821

I have just now recovered My Lost Port Folio M[r] Garnier sent it me a fortnight ago to the Care of his son the trouble this gave me I will mention hereafter—I have to thank M[r] Garnier but More *he* that found it on the River Bank and took Such very remarkable good Care of it—for on opening it I found the Contents in as good order as the day it Was Lost and *Only* One Plate Missing—

Blue Yellow Back Warblers
Orchard Orioles
Cardinal Grosbeaks
Yellow Eyed Cuckoos
Large Crested Fly Catchers
White Eyed —" —"
Night Hawks at Dawn of Day Plenty
Turkey Buzards
Carrion Crows
Common Gulls
Carolina Wrens by Vast Number
Partridges a few Very Shy

to see these in hunt of others I Was out since half past 2 o'clock this Morning untill 4 this afternoon Wading often to our Midles through the Swamps and then Walking through the Thickest Woods I believe I had yet seen—

Cap[e] Cummings Left us on the 10th for Philadelphia the Poor Man had Not *One Cent* with him—saw M[r] Hoyteura who arrived in the Columbus—Rec[d] by that Vessel Letters from Lucy William Bakewell, Charles Briggs, N. Berthoud—Car of Commerce brought another from My Wife and one from M[r] Mataleon the great flute player—answered

Painted Bunting

Although these Birds are taken now and quite tamed in a few days, so much so as to sing as if at Large, when Caught Next Month they Die in a few hours, and shew dejection from the Instant they are caged—they are fond of Nesting in Live Oaks, Wild plum trees, Briars, Orange Groves—when domesticated are fed on Rice—*Breed Twice*—femelle opened on the 15 April had Eggs the size of N° 5 Shot—

I had from reliable source that One Made his Escape from a Cage and returned Thirty odd Days after to the house, went to the Cage, and remained Many Years—a femelle Was seen in Cage to carry and Arrange the Material given her to form a Nest and Compleated it—but Never Laid—

Saw Some Young Mocking Birds in Market this Morning able to fly—these Birds are said by People here to breed as often as 4 Times during one season I also was told that Young Ones sufered to be approach[d] by the Parents after a separation of Several days are often Poisonned by them, this *unatural* conduct demands [more] *Self* Confirmation—

MONDAY April 16th 1821

having received a Visit from D[r] heermann and Lady While hunting Last Saturday, I called on him at his House, he wished I would give a few Lessons of Drawing to M[r]s H—I acquiested and am to begin tomorow;—Joseph Hunting nearly all day Killed a Red Headed Wood Pecker, a Red bellied one—[Baltimore] Orchard Orioles, Black headed Titmouse—Green Painted Buntings—drew one With a blue head—Joseph Saw a *Yellow Breasted Chat* and some Baltimore Orioles—

Spent part of the Day with M[r] Hoyteura & *Young Towns* of Henderson on Board—

Was sorry as well as Surprised Not to have a Word by the S.B[t] Manattan from My Beloved Lucy—

THURSDAY 19th—April 1821

Low in funds again Left home with my *Port Folio* My *Best friends* as to Introduction and travelled as far as D[rs] *Hunter* the renowned *Man* of Jefferson—We Came on him rather *unaware* the good Man Was P X X X g We Waited and I gave him My Letter of D[d] *Prentice*—

This *Phisician* May have been a Great *Doctor formerly* but Now deprived of all that I Call Mind I found it *Necessary* to leave to his Mill's Drudgery—

Called on a New Phenomena in Painting M[r] *Earl* I believe & there Saw M[r] *Earl Jackson*—*Great God* forgive Me if My Jugment is Erroneous—I Never Saw A Worst painted Sign *in the Street of Paris*—

APRIL Sunday 22[d] 1821

Rec[d] Yesterday a letter from My Beloved Lucy, answered it Last Night and this Morning Wrote to N. Berthoud & W[an] Bakewell sent Home One Box & 1 One Bag the Whole by the Steam Boat Car *of Commerce* that Left at 9 o'clock P.M.

Dinned at M[r] Pamar as I usually do on Sundays. My Pupils all religiously inclined did not give them Lessons—the great Ease and of Course Comfort that I find in the Company of M[rs] Pamar's familly renders my visits at that house quite what I wish and often need for a relief of exersions—Finished to day a Drawing of a *Snowy Heron Ardea Candidissima* a beautifull Male—Joseph Drawing Flowers all day—

MONDAY 23[d]

found in Market a Gallinule that differs much from What I call the *Purple* one—the Yellow Legs & feet, their Stoutness—the Blue baby Top—& all Coloring—the hunters assured me they Never saw *One* with red Legs—but I cannot depend on their Memory—We also found another Male of Blue Grosbeak *Loxia Purpurea*—

Saw My Old Acquaintance John Gwathway of Louisville, he was *a la guettée* on the Levée the appearance of My Clothes, did Not please him we talk but Litle together.

TUESDAY April 24[d] 1821

Much in want of Cash Walked to the Columbus Steam Boat and Made *Baxter Town's* Portrait for 25$—gave My Likeness and Drew the *Black Poll* Warbler Male—*Sylvia Striata*, pleased with my days Work, rec[d] a letter from My Son Victor. —Much pleased also at his improved hand write—

WEDNESDAY 25—

Went again on board Columbus rec[d] My Pay from Towns—Made M[r] Hall's *Portrait* one of the best I believe I ever have taken—Met Gwathway & Thompson from Louisville—Dined at M[r] Pamar had a great Wish to see General Jackson but no time to spare yet—Paid our Board Rent & Washing to day—15$ for the 2 first Items and 5 for the Latter—raining hard—M[r] Hoyteura still in the Columbus—HIS THOUGHTS—

I am forced here to Complain of the bad figure that My friend Willson has given of the Warbler I drew yesterday, in the Bill only the length exceed that of Nature 1/8 of an Inch—an enormous diference—and he has runned a broad White line round over the Eye that does not exist—

April FRIDAY 27th 1821

Walked on Board of the Columbus at 6 o'clock this Morning. Made John De Hart's Portrait M[r] Hoyteura came a litle before breakfast, had been absent all Night. Much taken by the other passengers

General Jackson Left the City about Twelve I saw him *thrice* found Vanderlyn's Likeness the Only good One I have seen, *Tully's* Plate *Miserable*—John De Hart's Likeness being Intended for M[rs] Hall took it there. Spent a few Hours with her extremely agreably—Wrote to Charles Briggs—

SATURDAY 28th 1821

Rose early and went on Board the Heda to take M[r] Bossien Portrait. Made it good Young *Guesnore* and Hetchberger the Painter spent the Evening with Me, Joseph Sick Wrote to My Beloved Lucy

SUNDAY 29th 1821

Rose purposely to go and collect some Money to forward home, received 105$ sent 100 in a U. S. Note N° 152 Mark A dated Philad[a] April 5th 1817, payable at N. O. to B. Morgan's order—I gave it to John De Hart in an open[d] Letter to remit to My Wife—Columbus left at 12 o'clock—Dined at Hetchberger Made his Lady's Likeness—coming home Walked *in Painter of Feathers*' room, very Civil, asked for My Card—Hoyteura p[d] me a Short Visit Joseph better

MONDAY April 30th 1821

Steam Boat Paragon arrived No Letter for me—M[r] Gordon had one from M[r] Berthoud—sadly disappointed almost sicken, could not do any thing—

May 1st TUESDAY 1821

Walked some, Wrote to My Wife but the boat I Intended for, not going—extremely uneasy about My Wife's health or her Children—done Nothing—

May 2 WEDNESDAY—

Wrote again to Lucy give the Letter to Baxter Towns [Give Lesson to] drew a Long Legged Plover—Contents Shrimpes & Insects and at the Diference of Size only found it nearly coresponding to Bewick's description—it was a Male I received it from M[r] Duval the Miniature Painter who assured Me that he had Killed 6 or 7 he leaves here, all alike no diference whatever in the size or Coloring—they are often seen on Lake *Borgne* during the Summer Months—Was pleased with the Position in my drawing—

THURSDAY 3[d]

Bought 15 y[ds] Nankeen for Summer Clothes—found M[rs] H more aimiable if Possible than usual, talked freely to Me—became acquainted With an other Sister—Work[d] at M[rs] Hechburgers Portrait—Weather fine Monday Morning Very Disagreably Hot—Thermoter, at 88–89—and to day at 3 o'clock in the Shade at 90—

Cases of Yellow fever in the City I was told—

May SUNDAY 1821

Rec^{d} a Letter by the Cincinnati from My Wife. Not Very agreable to My feelings, surprised at having Nothing from N. Berthoud—appraised *Joseph* of his Father's Death. Bore it well—Saw Mr Jesse Embrie of the Cincinnati Museum = Spent all day very Dull dined at *Hetchberger*

—Counfounded Hot—Young *Guesnore* afronted and Ceased Speaking to me—

TUESDAY 7th Wrote to My Wife but did not Close my Letter—Mr Hoyteura paid us a Short Visit—Much esasperated against Mr Gordon—Mr Gordon sent to England for Me for 101^{b} *Italian Chalk*, 6 doz Black Lead Pencils, 2 Grosses Pastels—but did Not advance the Money—

WEDNESDAY 8th

finished My Letter to My Lucy and Wrote a Short One to N. Berthoud—put them on Board the Fayette, the Rose, the one I expected to Leave first Not being ready—*thought* to day that a Certain Gentleman to Whom I go to dayly felt *uncomfortable* While I was present, seldom before My coming to New Orleans did I think that I was Looked on so favorably by the *fair sex* as I have *Discovered* Lately—

Saw Hoyteura at Mr Hawkins in a High State of Spirits I dare not Call it Intoxication—he sailed this Evening for Liverpool—paid a Visit to the Amiable Vanderlyn, this Gentleman Like all substantial Men gained on acquaintance saw his Portrait of My Fair Pupil M^{rs} H—the Likeness good but roughly painted—he complimented me on My Drawings I thought too Much to be true—saw Gilly about settling for our Passage to Shippingport should I determine on going there—

Wrote to My Wife, N. Berthoud, Henry Clay, D^{l} Drake, on the 16th forwarded all by S.B.^{t} Paragon—

Dined at Governor Robertson, Polite reception, promessed Me a recommendatory letter to Mr Monroe

No news from Mr Berthoud yet.—

17th May 1821

begun lessons with			Young Mr Bollin	@1.50 per Lesson		
"	"	"	Miss Perry	2.00	"	"
"	"	"	Miss Dimitry	2.00	"	"

NEW ORLEANS May 20th 1821

Sent a few lines to My Wife by the S.B.t- Tamerlane—Last Week I Recd a letter from Mr J. Hawkins and one from Mr Robertson the Governor for the President of the U. States—favors from Men of High Stations are favors indeed—

The Governor a Man of Strong Information extremely Polite—

Since So long without any news from my familly, My Spirits have failed me, and it is With Dificulty that I sit to Write at all—My Journal Suffers through the same Cause that affects me—attention—

June 16th NEW ORLEANS—1821

Left this City at about ½ past 12 o'clock—in the Steam Boat Columbus Cape John de Hart—bound to Shippingport Kentucky—

pressed by much work on hand within some weeks passed, during every day, and too much incommodated by Musquitoes at evenings My poor Journal as been put a side [but] events and a wish not to discontinue to put down incidents that are of some Note and agreable to My Mind I come again to it—

a personnage who had some week ago boasted of his Interest towards me, and who on one occasion carried his attention quite too far and awkwardly must first take my attention—and here I will [take the Liberty] give you a Lesson, should you ever be Employed as a Teacher to any ostantatious oppolent person—*flatter*, Keep flattering and end in flattery or else expect No pay—

My Misfortunes often occur through a want of attention to that Maxim in similar Cases after having with assiduity attended on a Gentleman's Lady (Whose Name I will not at present Mention) for forty Days, I received the rudest of dismisal and My pride would not admit me to the House—to even ask any compensation—how agreable the first Lessons were I shall allways remember. *She thought* herself endowed with superior talents, and her Looking glass pleasing her Vanity I dare say made her believe She was a Star dropped from the heavens to ornament this Hearth—but dificulties augmented and of Course drawing seased to please, I could not well find time to finish every piece that I had began for her, and Constancy the Lady said was [not belonging] never to be found the Companion of Genius—toward the Last she Would be unwell when I walkd in, Yawnd and postponed to the morrow—I believe the Husband saw her Weakness, but the good Man Like *one* or *Two* More of My Acquaintances Was Weaker still—

I Knew well that My conduct had been correct and I felt a great pleasure in Leaving them, and, the One hundred Dollars I had hearned *with them*

The *Dimitry* familly on Whose's Daughters I had the pleasure of *attending* as a Drawing Master—had become peculiarly agreable and I left them with anxiety for their wellfare and the pleasure that anticipation produces, having some Hopes of seeing them Next Winter,—Young Dimitry I never will forget a Youth of More genuine Natural Ability I never have Met—his sarcastism had much the turn of D[r] Walcot's I Rec[d] from the Young Ladies Miss Aimee & Euphrosine Two handsome Plants for My Beloved Lucy that I forward[d] under the Care of Cap[e] De Hart—

My True friend R. Pamar and his most amiable and Kind Wife I have to thank for all that I can Call the pleasures I felt Wilss at New Orleans—I Eat there whenever I could find time, and I was so Loved by the Children that I felt as if I parted from Mine when I left them—I had found[d] a very slight acquaintance with M[r] P. some years ago as he descended the Ohio on his way home—I had [he said] been Polite to him when he called at My Poor Log House at Henderson and he said often that Kindness had not been profuted to him that he was Well able to remember the Instances and that if I did Not please to Make free with his house he Would be Sorry for it—

I Rec[d] Many Attentions from M[r] Laville and Lady—M[r] Hollander the Partner of My old but too rich Acquaintance Vincent Notte I had the pleasure of seing—he I believe saw that I had No wish to disgrace the Handsome Rich furniture of the Wealthy with My Intrusions when reduced to My Grey Breeches, and taking Me by both Hands One day as I was trying to Make Way from him, he said My Dear M[r] Audubon Come and see me. I promise you I shall Not have any one at table and I will try to Raise your Spirits. I have some fine Paintings, and please bring Your Birds that I am Anxious to see—then You see that although I lived extremely retired and general show[d] those that I thought I Would Incomodate I now & then stumbled on an Less Indiferent Member of this Life toward his fellows Who like Me have been rich and poor alternately—

I had attended a Miss Perrie to Enhance her Natural tallen for Drawing, for some days When her Mother Whom I intend Noticing in due time, asked Me to Think about My Spending the summer and fall at their farm Near Bayou Sarah; I Was glad of such an overture, but would have greatly preferred her Living in the Floridas—We Concluded the Bargain promissing Me 60 Dollars per Month for One half of My time to teach Miss Eliza all I could in Drawing Music Dancing &[c]

&^{c} [and the] furnishing us with a Room &^{c} for Joseph & Myself—so that after the One hundred Diferent Plans I had form^{d} as Opposite as Could be to this, I found Myself bound for several Months on a Farm in Louisiana.

We left our abode in Quartier Street and Old Miss Louise without the Least regret, the filthiness of her Manners, did not agree with our feeling; and by this time We had fully discovered that a Clean Sweet HouseKeeper is quite Necessary to a Naturalist—

Our Good Spanish Neighbour Mr Taek We Loved, His Nieces sung Well, and his own Jokes now and then amusing us—

We Came to our Landing at the Mouth of Bayou Sarah of a hot sultry day without any accidents; bid farewell to Mr Gordon and after Mounting the Hill at St Francis Ville Rested ourselves some Moments at a Mr Semple dinner Was set, but Not My Heart for it. I wished Myself on Board the Columbus, I Wished for My Beloved Lucy, My Dear Boys—I felt that I would be Awkward at table and a good opportunity having offered to go to Mr Perrie place, We Walked slowly on, guided by some of their servants dispatch^{d} with the News of our Coming and some Light Baggage—

the Aspect of the Country entirely New to us distracted My Mind from those objects that are the occupation of My Life—the Rich Magnolia covered with its Odoriferous Blossoms, the Holy, the Beech, the Tall Yellow Poplar, the Hilly ground, even the Red Clay I Looked at with amazement [and] such entire change in so Short a time appears often supernatural, and surrounded once More by thousands of Warblers & Thrushes, I enjoy^{d} Nature

My Eyes soon Met hovering over us the Long Wished for, Mississipi Kite and Swallow Tailed Hawk, but our Guns Were pack^{d} and We could only then anticipate the pleasure of procuring them shortly—the 5 Miles We Walked appear^{d} Short We Arrived and Met Mr Perrie at his House Anxious to Know him I Inspected his features by Lavaters directions,—*We Were received Kindly*—

July 4th 1821

during the diferent excursions We have made through the Woods here, and from Report of Such persons as we thought proper to Lessen to on the subject, I have made the following remarks—Viz—

The Blue Jeay Corvus Cristatus is seen in Lower Louisiana during the summer, not having met with More than a dozen Individuals in All our rambles—Last April Immense Numbers of these Birds, so annoyed the Corn rows in this Neighborhood as to force the planters to Poison them with Corn boiled with Arsenic which had a great Effect—Killed the Thiefs often instantly—

Yellow Bird fringilla Tristis

Saw a very few during Last Winter Near N. Orleans None at present—Kite do not breed here—

Baltimore Oriole, Oriolus Baltimore

Not an Individual to be met with at any season—

Wood Thrush Turdus Melodus—

Extremely plenty, in its usual haunts, i.e. deep Shady Woods—the first bird that Sings at the dawn of day Never Killed one Coresponding with Willson's figure

Robin—Turdus Migratorius—

Resorts here during Winter in vast Numbers, and becomes very fat—the sport to all the Gunners—leaves early in March—

White Breasted Black Cap[d] Nuthatch
Sitta Carolinensis—

Scarce, have Killed a few, Nestle here—Young quite Grown Midle of June, first brood—

Sitta Varia Red bellied Nutatch—

Not found—

Gold Wing[d] Wood Pecker—Picus Auratus—

Pentifull—

Black Throated Bunting, Emberiza Americana

Not an Individual Met With

Blue Bird Sylvia Sialis,

Scarce, about one pair to each Plantation—Yet Nesting in holes of Peach & Apple Trees—

Orchard Oriole, Oriolus Nutatus—

Very aboundant, this Country appears to be chosen by them I found seventeen Nests on M[r] Perrie's Plantation With Eggs or Young, during Two days Looking for them—the Young of Many already flying about Midle June—first brood—Was deceived one day by one imitating the Cry of the Loggerhead Shrieke and followed it a great distance before I found My Mistake; it Kept[d] on the tops of high trees in the forest, a very unusual circumstance, the figure of Willson has the bill Much too Large & Long, the figure of the Egg is also too large

Great American Shrieke, Lanius exubitor—

a few seen during Winter—

Ruby Crowned Wren—Sylvia Calendula

a few during Winter seen Near N. Orleans—

Shore Lark Alauda Apestris—

None at any season—

Pine Gros Beak—Loxia Enucleator

None—

Maryland Yellow throat. Sylvia Marilandica

Great Numbers during the Winter, Leaves early in March

Yellow Breasted Chat. Pipra Poliglotta

As many here as I have met in any other state—that is, One about each Plantation, Never have seen a femelle—

Summer Red Bird, Tanagra Aestiva

Tolerably plenty—

Indigo Bunting, Fringilla Cyanea

Tolerably plenty Not so much as in Kentucky or Pennsylvania, but more so than in Ohio—

American Red Start, Muscicapa Ruticilla

Very plenty. Young quite grown Midle of June

Cedar Bird Ampellis Americana—

Was remarked this Spring feeding on the Holy berries and remained to reap the fruit of the Wild Cherries—in Immense flocks, extremely fat—disappeared at once—

Red Bellied Wood Pecker, Picus Carolinus

pleantifull as any where else

Yellow throated fly Catcher Muscicapa Sylvicola—

Never Met With—

Purple finch—Fringilla purpurea

a few during Winter seen Near N. Orleans—allways in litle flocks of 4 to 6—

Brown Creeper—Certhia familiaris—

Not seen—

House Wren—

Not seen—

Black Cap[d] Titmouse, Parus atricapillus

Very plenty—Young quite grown Midle of June—

Crested Titmouse—Parus bicolor

the same—

Winter Wren—Sylvia trogloditis—

during Winter, Numerous in the Cypress swamps—

Red headed Wood Pecker, Picus Erythrocephalus

plenty—Young quite Grown 15 June—

Yellow Bellied [bellied] Wood Pecker—

a [few] during Winter

Hairy Wood Pecker—

Not seen—

Downy Wood Pecker, P. Pubescens

Scarce,—

Mocking Bird—Turdus polyglottus

Extremely plenty—Nestles in all sorts of situations having found Nest in the higher parts of Tall trees in, Small Bushes and even between fence rails garded only by the rail imediately over the Nest—the Egg represented by Willson very Litle Like any of the great Number I examined—these Birds Mock indiscriminately every Note of Birds—are very gentle With every thing but the Bird of Prey, these they give chase to and follow a great distance with much apparent Courage—here during Winter—

Humming Bird. Trochilus Colubris—

Plentifull—Was assured of their existence of Two Species by Many—represented as Much Larger—have Not Met with any individual yet and fear it to be a Mistake—these Birds are easily Caught by pouring sweetened Wine in the Calices of flowers—they fall intoxicated.

Willson erroneously says that this sweet Bird does not sing—I have many times lissened to its Low toned Melody with great pleasure and can assure you that if its Voice was as sonorous as it is varried and Musical it would be considered as surpassed by few other Species—

Towee Bunting—Emberiza Erythropthalma

Saw a very few Near New Orleans during Winter Not One found at present—

Cardinal Gros Beak. Loxia Cardinalis

Extremely plenty—during the Whole season increasing still by vast Numbers that come from the Eastward to spend their Winter here, very depredatory to Corn Cribs—Young very Large 15 June, Second brood Hatching—

Scarlet Tanager, Tanagra Rubra

Plentifull—but do not find them by any Means confined to the Interior of the forests, but to the contrary found on the bordering Tall trees of Plantations—

Rice Bunting—Emberiza Oryzivora

passes early in spring from further Southern parts going eastwardly, remarked some Last february and March—

Red Eyed Fly Catchers, Sylvia Ohvacea [Muscicapa]

plentifull—Young quite grown early in June

Marsh Wren—Certhia Palustris

Never saw one Like Willson's drawing—but have Killed Many Individual of the Marking & Shape of My Drawings—some of them a few Miles Above New Orleans in April—but never imediatly Near Rivers—

Great Carolina Wren—Certhia Caroliniana

[to be seen] almost constantly in sight or hearing about the field or the forest—young full grown now found of damp places—

Yellow Throat Warbler—Sylvia Flavicollis

Never Met With—

Tyrant Fly Catcher, Lanius Tirannus

plentifull—young full grown now—

Great Greasted Fly Catcher—Muscicapa Crinita—

Very Common—young full grown, very Timorous & Shy—

Small green Crested fly Catcher. Muscicapa Querula

Extremely plenty—found of the road sides from where it dashes at flies from Low Bushes—

Pewit Fly Catcher—Muscicapa Nonciala—

Plentifull during winter Near N. Orleans, a few remains in the Hilly parts of Louisiana during summer—

Wood Pewee—Muscicapa Rapax—

Plentifull in the Woods—this bird hunts latter than any of its Genus. I have heard it uttering its Note Long after Dark—

Feruginous Thrush. Turdus Rufus—

have a very few Individuals, that had more the appearance of Lost Strangers, than of happy residents, rarely seeing more than one at a time—

Golden Crown Thrush. Turdus Aurocapilla—

Not Met [Lower than] at all during summer many seen during the Winter Months—

Cat Bird, Turdus Lividus

Not One seen Since Last March When one Evening I saw Many on the Canal row of Willows, the public Walk Near N. Orleans—

Bay Breasted Warbler—Sylvia Castanea

None Seen—

Chesnut Sided Warbler. Sylvia Pensylva[a]

None seen

Mourning Warblers. Sylvia Philadelphia—

Never Met with—

Red Cockaded W. Pecker. P. querulus

have seen and Killed one Only, but it Lodged and I Lost it—Not seen Near plantations, unless the weather be very Cold—Mostly in Pine Woods—

Brown Headed Nuthatch—Sitta Pusilla

Never seen—

Pigeon Hawk falco Golumbarius—

Never Met with—

Blue Wing[d] Yellow Warbler Sylvia Solitaria

Not Met with here

Blue Eyed Yellow Warbler Sylvia Citrinella—

So many of these about New Orleans early in March Nimbly hunting for insects amongst the willows—but could not find one in the month of May, having past I suppose Eastwardly where it is so common in all our orchards—

Golden Winged Warblers Sylvia Chrysoptera—

Not seen here, a plentiuss bird in the Lower parts of Kentucky—

Black Throated Blue Warbler—Sylvia Canadensis

Never have seen these Birds since I left Pensylvania Very Numerous in the Lower part of that state in April and May—

American Sparrow Hawk Falco sparverius

very Common, Nestle allways in hole, mostly that of the Wood Peckers—Young quite grown, Midle of June here—

Field Sparrow Fringilla Pusilla

Not here—

Not a single Species of Sparrow to be met with at this season—

Yellow Rump Warbler Sylvia Coronata—

remains here during the whole winter, in aboundance—saw them every day every where there was a Tree even in the City of N. Orleans—in May Not one to be found—

Maryland Yellow throat—Sylvia Marilandica

plenty during Winter, and very Gentle—

Blue Small grey fly Catcher Muscicapa Coerulea—

plenty during summer Nestles in Willows, Keeps in Small parties of 6 or 7—has much the manner of the Long Tailed Titmouse of Europe—

White Eyed fly Catcher—Muscicapa Cantatrix—

the Commonest of all the Birds in our Woods—the young of which Two brood are raised in the season full grown [grown] Midle of July—

Saw 3 Red Ibisses pass over the plantation Yesterday—

Chuck-Will's Widow

Caprimulgus Carolinensis femelle

Yesterday the 25th of July 1821 an Indian of the Choctaw Nation, who habitually hunts for Mr Perrie—brought me a femelle of the Chuck Will's Widow in full and handsome plumage. it Measured One foot in length, 25 Inches in breadth, the tail composed of Ten Feathers is rounded, but has not the White in any degree that Willson Mentions on the inner vanes of the 2 exterior feathers—the Craw of this was filled with the heads of Many of that Kind of bugs commonly called *pinching* Bugs, One of them a very Curious Large One Armed with Two equal paires of Pincers—

these Birds generally aboundant in this part of Louisiana, and at present very scarce, Not having been able to Meet one in any of our Excursions that often are of Twenty Miles—a few Weeks previous to our arrival they where heard from all parts of the adjacent Woods loud during the day—a few have been seen and a few heard since, all of which have eluded My [attempts at seeing and killing then] pursuits—they remain here untill Late in September. I suppose them at this time so occupied in search of food for their young that it has put an end to their Crying—Many of the Planters think that this bird has the Power and Judment of removing its eggs when discovered sometimes several hundred yards—these are usually laid on the bare earth under a Small bush or by the side of a Log—

Saw 3 of these birds on the 20th of August one Evening while I was Watching the arrival of some Wood Ibisses—they flew Lightly in the manner of the Night Hawk but Close to the Cotton plant tops, passing & repassing by me untill I could see No more—Not one of this Birds have I heard since Early in June—

On the 22d a beautifull femelle of this Bird was brought to me by Mr Perrie's Overseer who had Shot it the preceding evening on a Small dead Tree, Where the

Chuck Will's Widow had taken a stand to Watch Beatles & Seiz[d] on them as they passed by, he saw it raise several times and Catch them in the Manner that Mocking Birds often do While enchanting the observer of his Melodies—this Man's Wife had seen it for Several evenings previous at the same Spot and sport—These Birds differ Much in plumage particularly in being darker or Lighter as they are Older or younger, this had [several] Many Winter eggs; very fat—finding [generally] Birds of Migratory habits generally in good order during this season it May be concluded that it is a preparatory occasion Incident to them to Make them to bear the fatigues and probably the unavoidable wants of foods during their travels—the Stomack of this had entire Locusts—Green Wood Lice, Ground Crikets and 2 of these Long Beetles Called usually by the french *Scarabees*—as Most Birds at present Many of the Tail & Wing feathers were tender from the Moult—

July 26th 1821

Rec[d] yesterday from New Orleans a Packet of Letters, 4 from My Wife, One from Benj. Bakewell & one from M[r] N. Berthoud. much afraid that My Wife has not rec[d] My Packet for the President forwarded her by the Cap[e] of the Steam boat Car of Commerce—

Remarked to day that a Male of the *Orchard* Oriole, that I had wounded in the tip of the Wing and Caged had Violent fits of Convultions, that lasted for as much as ten Minutes each, this I atributed to the uncommon exertions he Made to escape through the Wire of the upper part of the Cage—

Yet he eat freely of Fruits and Also rice—

Saw Yesterday an unknown Awk of a Large Size, that at first appearance I took for an *Ash Colored Awk*, but at a dash he Made at some Pigeons I was able to see him Well and could not recognize it for any of the 22 Species I am acquainted with—our pieces not being Loaded, I Lost him—

The Martins have for about Two Weeks Every day followed a Course of Conduct quite New to me & very remarkable, they are seen Every Morning about 50 in Number, the Whole of Which were inabitants of the Boxes put up for them, assembled on the top of a Dead tree Close by the House, from 8 o'clock until dinner time about 2 P.M. they amuse themselves over the Yard, after this the Whole disapear and Spends the Night, I know not where—going every turn they Leave due West—and returning from that point about sun rise—every Morning—do they spend their Night in Large trees at a great distance? or do they fly to & fro that While to try their power is uncertain; however, I suppose the first Case the true One—

Carrion Crows are extremely attached to their roosting Dead Trees and will spend the Whole of the Summer nights on the Same—Leaving it very late every morning When flying on a long Course they much resemble the True Turkey—flapping their wings 8 or 10 times then sailing about 50 yards, & again flap—

Allmost every Genius of Migratory Birds begain to depart as soon as the Young are fully able—saw to day Large flock of the Tyrant fly Catchers going due south—

July 29th 1821

I had the pleasure of Meeting with Several Red Cockaded Wood Peckers yesterday during a Walk We took to the Pine Woods and [to] procured Two beautifull Males, both alive, being Slightly wounded each in the Wing—the particular & very remarkable cry of this Bird can be heard at a very considerable distance of a Still day, in articulation which resembles that of the Hairy Wood Pecker, but is much more Shrill & *Loud.* the Tall pine trees are its Chosen haunts and seldom does it alight on any other Kind of Timber—its Motions are quick, gracefull and easy, its Move in all directions either on the Trunk or Limbs, Looking often very cunningly under the Loose pieces of bark for Insects: is more Shy than any of its Genus. Watching attentively all our Movements below, they Kept allways on the opposite side—peeping carefully at us—The second one Shot did not Loose a Moment to think of its Misfortune, the Moment it fell to the ground it Hoped briskly to the Nearest tree and Would soon have reach its top had I not secured it—it defended itself With courage and so powerfully did it peck at my fingers that I was forced to let him go—

Confined in My Hat on my head, they remained still and stubborn. I looked at them several times, when I found them trying to hide their Heads as if ashamed to have lost their Liberty—the report of my Gun alarmed them every time I shot when they both uttered a Plaintive Cry—

through pain of the Wound or the heat felt in My hat one died before We reachd Mr Perrie's house—the other I put in a Cage,—he imediatly reviewd the premises hoping about and hunting for a place to Work through, and used his chisel bill with great adroitness finding the Small Chips he cut to the right & left and having made his way to the floor, run to the Wall and Climb[d] up it as easily as if the back of his favorite Pine [bark]—had been his post hold, picking between the Bricks and Swallowing every Insect he found—remarking often his looking under Craks and the litle Shelves in the rough wall, I drew him in that position—Sorry I am to have to say that M[r] Willson's Drawing could not have been Made from the *Bird fresh Killed* or if so it was in very bad order about the head; he having put the Small striek of red feathers of the head imediatly over the Eye While there is a White Line, the red being placed far back of the [h]ear—and the whole of the wing not at all Marked Like that of the Bird—the sides of the breast is also badly represented, the Lines in Nature are Longitudinal only, and Show more of a body the appearance of this Bird when on the pine trees Would Make one suppose it to be Black all over and the Red Line is often covered by the Crown feathers in the living Bird—I first Met with this Species a few Miles from Nashville, when on My Way to Philadelphia in 1806 seeing them from time to time untill I left the first range of Mountains Called the Cumberland—of the Nest or time of Incubation I Cannot speak. I am told that during severe winters they will leave the pine Woods and approach Plantations—

the Length of Both those I attentively examined Was 8¼ Inches, Breadth 14½—

the Gizzard filled with heads of Small Ants and a few Minute Insects—the birds smells strongly of Pine; as I Hope to be soon able to procure the femelle, I May probably with her portrait give More information—

Shot also a Young of the Great White Heron, entirely destitute of the Pendant silky feathers of the Shoulders, but so well grown that it Might easily in that stage be taken for another Species—Killed Two Young of the Snowy Heron, that Where with an old One, Neither of these had the recumbent feathers of the back and their Legs & feet were all of a Yellowish Green instead of the former Black and feet bright Yellow—Saw some Killdeer plovers. Kind fisher, Green Heron, and a ferruginous Thrush—Low Land bordering Watter courses are usually preferred by these—

The Martins Hiroundo Purpurea—that leave this place daily, Congregate with a parcel raised Near Thompson's Creek, about 5 Miles from this and I have no doubt will take their flight from thence, for their Winter resort—

This afternoon having finished My Drawing of the Red Cockaded Wood pecker and satisfied of its Correctness by a Close comparison to the living original I gave it its Liberty, and was glad to think that it must Likely Would do well as it flew 40 to 50 Yards at times and seemed Much refreshed by its return to Liberty—

WEDNESDAY August 1st 1821

We were awakened Last night by a servant desiring that I should rise & Dress to accompany Mrs P. to a Dying Neighbour's house, about one Mile,—We Went, but arrived rather late for Mr James O'Conner Was Dead. I had the displeasure of Keeping his body's Company the remainder of the Night. On such Occasions time flys very slow indeed, so much so that it looked as if it Stood Still like the Hawk that Poises in the air over its prey—the Poor Man had drink himself Literally to an everlasting Sleep; peace to his Soul I Made a good Sketch of his head and Left the House and the Ladies engaged at preparing the ceremonial Dinner—the weather Sultry Thermometer at 93—it has Not raised yet this Season here above 96—

Our Sparrow Hawk Was Killed to day by a Hen engaged in guarding her Brood—Nero had become extremely Temeraire, Would Fall on a Grown duck as if thinking all Must answer his Wishes when hungry—he flew at liberty about the Place, caught Grasshopers with great ease and Would Catch in the Air any of the unfortunate Small birds Killed in our dayly walks when thrown toward him for food—he regularly refused all putrid flesh, Never Would touch [a] Woodpeckers, but dearly received Bats & Mice—he had grown handsome from an apparent parcel of Moving Cotton—sailed with the Wild Birds of his Species, returning every Night to the Inner upper part of a Sash in Mr P.'s Room—he seldom made use of the Note of the Old Birds but allmost constantly uttered his Cree, Cree, Cree—

Our Orchard Oriole I gave Liberty to seing that the Departure of every number of his tribe, had rendered him melancholy to excess; I have No doubt that that specie could be Kept in cages Without much trouble, and its Sprightly Songs certainly would well repay for the care employed to furnish them with food & Drink—

August 4th 1821—SATURDAY

Louisianna Warbler. Sylvia Ludovicianna. I Shot this morning the same Bird or one of the same Kind that I pursued yesterday [with such anxiety] so eagerly and then without success—and Was Much pleased to discover in it a New Species—

during My Chase of yesterday it flew briskly from one tree or small bush to an other Not as if afraid of me, but as if anxious for food, hanging its Wings very much like the Hooded fly Catcher and constantly Keeping its tail much spread like the American Red Start the Only Note it repeated every time it left a place for another Was a simple soft single *Tweet*, all its Movements extremely quick gave Me much trouble to Shot it—this Bird I Never have Met before, and of Course I Consider it as a Very Scarce One, its Note attracted me as that of all New species do; More of its habit I would Like to Know—Total Length 5 Inches Breadth 8—Whole upper part of a rich Olive Yellow—deeper on the shoulders & back, Wing feathers Black edged with bright Olive—tail Much rounded, composed of 12 Feathers the 3 first exterior on each side Outwarding edged with brownish black and Yellow inwardly these edges broadening more as they goes to the Midle feathers that are of a dark brown nearly black edged with Olive—Under wing Coverts rich yellow—under tail Coverts the same, very long—

Eyes, full, Irises deep brown, bill the true Warbler horn col^{d} above and Clay below, very Sharp with a few black bristle tongue forked & slender—Legs feet & Claws Yellowish Clay—it proved on dissection a Male, extremely fat—Gizzard containing, remains of Caterpillars, Small beetles and diferent Kinds of Small flies with a few fine Clean Sands—

August 29th* saw Two of those Birds to day a Male & femelle that I approached and Examined very attentively for some Minutes they were in a Low damp & Shady part of the wood. I Killed the femelle & have joined it to my drawing of a Male—I was anxious to procure her Mate but the discharge of my Gun so alarmed it that it flew off and I could not see it more—those Birds resemble the young of the Blue eyed Warbler of Willson in much of their plumage but not in Manners and are a scarce species—

* [This entry under August 29th is written across the entry of August 4th in a different ink.]

SUNDAY August 12th 1821

We left this morning after an early breakfast to go and explore a Famous Lake about 5½ Miles from this where we were to find (as told) great many Very fine Birds—the walk to it was pleasant being mostly through rich Magnolia Woods, We Killed Two Wood Ducks in a Small pond that we had to leave on a/c of the depth of the Hole, but that were excellently wellcome to Two *Red Shouldered Hawks* that Carried them off in our Sight—these Last are the only Birds of this Kind I have seen at this season in this part of Louisianna—We saw a singularly rich col[d] spider that finding a Horse fly just entangled in her Net move to it and covered it in a Moment with the silk of her bag, Shooting it out in a stream and at the same time rolling the fly untill the whole Likened the appearance of a Small oblong ball of White Silk, the Spider then return[d] to the center of its Net—No doubt this is a Way of preserving the flies when the Spider is Not hungry;—

When we left the ridges We at Once saw a diferent Country in aspect, the Tall White & Red Cypress being the principal Trees in Sight with their thousand Knees raising Like so Many Loafs of Sugar—Our eagerness to see the Lake engage us to force Our Way through Deep stiff Mud & Watter—We came to it and saw several Large Alligators Sluggishly Moving on the Surface, Not in the Least disturb[d] by our Approach—

Saw a White Ibis on a Log where it sat a Long time arranging its feathers using its scythe Shape[d] bill very dexterously; Could have Killed it but having No boat and afraid of Sending a Dog in the Lake Left it Setting peaceably—

Saw a great Number of Prothonotary Warblers on the Low Bushes of the Swamp—Many *Yellow throated Warblers* these have all the habits of a Creeper, Moving quickly round, up and down the Limbs and trunks of the Cypress trees, fly swiftly in the Manner of the brown Creeper alighting generally low on the truck assending it searching nimbly for small Insects; these birds [*I* have] have so much the appearance of the *White & Black Creeper* that had I not seen one fly directly toward me and discovering then the beautifull & Rich Yellow Throat I would not have shot one—This however I effected and found Myself in possession of a Beautifull Male that Measured 5½ Inches in Length & 8¼ in extent answering Willson Description, on Dissection the Bird Was very fat as all the Warblers We Shot Now Are, and had its Gizard filled with shells of Minute Insects—so found of the Cypress trees is this beautifull spirited litle bird that I Was Tempted to Call it the Cypress Swamp Warbler, where it is only to be Met in this part of the Country—

I Was also fortunate in Shooting a Male of the *Green blue Warbler*—One Week ago I had shot one but Never could find it, there was at this time five in Company, and Within only a few feet of Me M[r] Willson shot a femelle on the Cumberland River, and Never any more; about Two Months since I Discovered One in a small swamp Nearer M[r] Perrie—these birds sing sweetly, and No doubt breed here. Look

Much like the *blue Yellow Back* Warblers and hang downwards by the feet like these and the Titmouse. Saw only the One I shot to day and having as much as I Know I could Well draw before they would be Spoiled by the heat of the Wheather returned to the House.

This Male Measured 5 Inches in Length and 7¾ in breadth—All the Colors brighter & stronger than Willson's femelle, every Tail feather having White on their inner Vane except the Two Midle Ones—so fat was this bird and of so solid a Nature was that [grease] [?] fat that it Cut like Mutton fat; its gizard Was filled entirely with some Small brown Shelly Insects and the remains of the same Kind that are extremely plenty in the Cypresses of these Swamps—

Shot a Watter Trush, have found Many here.

Went to the Lake that We Visited Last Sunday and in Going there, I Was much pleased to observe that the *sound* that We heard on Sunday and had taken for the Plaintive Note of the *Wood Peewee* was in fact that of a Young Mississipi Kite, While waiting for the return of the Parents with food—this Young it seems had Actually remained [it appears] on the same tree Where we had heard it before but could not then discover it—this Morning perceiving that a long Vine reach^{d} near to the Top of the Tree and hearing the Noise without Knowing it nor where it actually issued from, I Walked toward it still looking up to the Topmost branches When I perceived something Like a dead stick Lodged Cross ways in a limb—I Eyed it particularly and saw it Moved, I Shot at it and the Noise stop^{d} but the Young M. Kite Closed her wings and destroyed the Dead Stik like appearance it had before my fire—I Waited for it to fall, it cried again shortly and I then saw the Old Bird bringing food and alight Close to the young with one of those Large Grass hopers that abound in the Mississipy flats—but the young was too far gone to relish food the Mother exibited much distress and after several trials to Make the young Bird take it it dropt it and taking old of her offspring by the feathers of the back carried it off with ease for about 25 yards to another tree where I follow^{d} and Killed both at One Shot—the Young instead of having the head of a Light blue ash color like its Mother, had it of a handsome buff and remainder of the body Was Nearly black. I intended drawing Both and I purposely hided them under a Log, but on my return some quadrupedes had discovered them and eat them both—I regret much the Loss the young Bird was nearly full grown [but probably the want of Knowledge to procure food] Saw several paires of Ivory Bill *Wood Peckers* Killed a handsome Male.—Louisianna affords all the *Picus Genus* of the U. States—

Arrived at the Swamp and there saw a great Number of Small Birds. Shot a beautifull *new* Species of Fly Catcher Muscicapa, which I Will give you Tomorrow when my Drawing of it Will be finished. I had the pleasure of seing Two that appeared Much alike, they were quarelling when I shot at them but fell only One—cannot say any More of this truly handsome bird having never seen any thing of them before to Day—

Saw within a few feet a beautifull *Mourning Warbler* but Was so situated Knee deep in the Mud that I could Not retrograde without alarming it I preferred gazing at it as it innocently gazed at Me hoping it would fly at a Short distance, but it Moved with a *Tweet* and out of sight in an instant. Much disapointed at My having lost the only oppr[y] I Ever have had of procuring this rare Bird.

Shot several of the *Yellow Throated Warbler* all alike and all Males the woods were full of them and yet Not a femelle could I Shot—they Move sideways on the small limbs of the Cypress in a Hoping Manner extremely quickly hang often to the ends of Limbs like the titmouse and run up or down the Large Trunks much like the Nuthatches—Killed Many *Blue Yellow back Warblers.* Saw Many *Prothonotarys,* several *Watter Thrushes* that I consider More Like Warblers the Habits of Which genus they exibit to a very great degree and the Bill of Which they [co] have—Alligators as numerous as before basking in the Sun that this day was more than ordinarily uncomfortable—saw several Ibiss at respectful Distances in their common dull postures—

My Litle fly Catcher had only one wing touch[d] When I presented Myself to pick it up, it spread its Tail & open its Wings and Snap its bill [20 or] about 20 times in the [——] Manner that Many of this Genus do when they seize a fly, particularly those that are Nearest the Standard of the Genus.—I seldom have seen a bird of Such Small size With so Large & beautifull an Eye. I took it home to James Perrie's Es[qr] and had the pleasure of drawing it While a live and full of Spirit, it often Made off from My fingers by starting Suddenly and unexpectedly, and then would hop round the room as quick as a Carolina or Winter Wren would have done, uttering its tweet tweet tweet all the while, and Snapping every time I took it up. I put it in a Cage for a few Moments but it obstinatly forced the fore part of its head through the Lower part of the Wires and I relieved it by Confining it in My hat for the Night anxious to see More of its Movements—

Joseph unwell With a sick head Hacke—Length of the *Cypress Swamp Fly Catcher Muscicapa Rara,* 5¼ Inches—breadth 7¾ Inches [upp] Whole upper parts handsome ash Color appearing blue at a distance, the front of the Head mixed with Yellow, a Yellow Line Surround the Eye that is very Large Iriss deep brown, pupil Black, between the eye & Bill & under the eye shaded with darker ash Tail Coverts lighter than the back, tail slightly forked of 12 Feathers all plain bronish ash shafts deep brown as well as those of the wings under tail Coverts Long & White—the Throat breast belly & Vent Rich citron Yellow without intermission of Shade in any of these parts. Breast spotted with black [spots] forming small chains fallin to the beginning of the spurious wings—Bill, hooked at the Tip and broad at the base. Legs feet & Claws horn Colors the last Long & Sharp Nostrils very prominent, Tongue much Jagged, Mouth flesh color^{d} & furnish outwardly With Many Long black bristles—it proved a Male. Gizzard fleshy filled with wings of different Insects—Cheeks also ash col^{d} My Drawing a very excellent one—finding this Bird very Weak in the Morning Killed it and put it in Whiskey—

MONDAY 20th August—1821

I spent the Night Nearly in pursuit of the Wood Ibiss and although I Killed One I could never find it this Morning. some Fox or Racoon had No doubt a good Repast of it—I saw 4 Coming, sailing & flapping alternatly; their Necks & Legs strached out a little over the tops of the Trees a few Moments after sun Setting. No Note, they alighted on the Largest Top Branches of the dead trees in a Large Cotton Plantation, drew their Necks & Heads on their shouldiers, Standing perpendicularly. Now & then arranging the feathers of their breast as if to put their Immense bill at Rest on it—I approach^{d} them untill they Were immediatly over My head, but they never minded Me, its growing Dusky I shot at the Largest, it open its Wings and sailed to the hearth without a groan, the others raised and Sailed on Some other trees, the Darkness of the Night then prevented My seing them any More and Also forced Me to Look for the One I had undoubtedly Killed—After a Long Search this Morning begun with the dawn of Day I had to return, fatigued & Much Disapointed—the Planter and Negroes assured Me that for Many Years these Birds some time as Many as 60 & 70 in Numbers, and at other times only a few had resorted to these Dead Trees to roost Nearly said they the round Year, Missing a few Weeks only early in Spring and in Winter. Could Not ascertain the Months—About 2 Weeks since 3 had been Killed the Negroes pronounced their flesh Excellent food—

While sitting waiting for the arrival of these curious Birds saw Several flocks of the White Ibiss and bleu Herons, Moving from the Lake to this rendez-vousing place of Rest to a Large Sand Barr, at the Mouth of Thompson's Creek that empty a few Miles below Bayou Sarah—the first flew in [Waving] single Waving Lines silently—the bleu Herons in acute Angles passing the Word of March from the first to the Last in a Simply *Qua*, these are easily Known by their drawn in Necks and Notes, the others allways Keeping their Necks at its full Length

these passages take place every Evening from about one hour before Sunset untill Dark When the Noises of the One and the Pure Whiteness of the others are the only evidence of [their] Straglers still being going over—

—August 25—

Finished drawing a very fine Specimen of a Rattle Snake—that measured 5 7/12 feet Weight 6 1/4l^{b} had 10 Rattles—

Anxious to give it such a Position as I thought would render it most interesting to Naturalist, I put it in that which that Reptile generally takes when on point to Inflict a Most severe wound—I have examined the Fangs of Many before and their position along the Superior Jaw Bones, but had never seen one Shewing the Whole exposed at the same time having before this thought that the probability was that those Laying Inclosed below the Upper one in Most Specimen Were to Replace these upper ones Which I thought [also] might drop periodically as the Animal Changes its Skin and Rattles—however on Dissection of these from the Ligament by Which they are fixed to the Jaw bones I found them Strongly and I think [with a] permanently [tendency to be] attached & as follows—Two Superior Next the upper Like (I speak of one side of the Jaw only) connected Well together at the bases & running parrallel their Whole Length. They had appartures on the upper & Lower side of their bases to receive the Venom connectivly and the discharging one a short distance from the Sharp points on the inner part of the fangs—the Two next Fangs about one quarter of an Inch below connected and running in the same Manner, but with only one base aperture on the Lower Side of each and the one at the point that issues the venom to the Wound—the 5th rather smaller is also about a quarter of an Inch below, Lonely appertures as in the Secondarius the scales of the belly to the Under part of the Mouth where they finished Numbered 170 and 22 from the vent to the tail—My Drawing I Hope Will give you a good Idea of a Rattle Snake although the Heat of the weather Would not permit me to Spend More than 16 hours at it—My amiable Pupil Miss Eliza Perrie also drew the same Snake; it is With Much pleasure that I now Mention her Name expecting to remember often her sweet disposition and the Happy Days spent near her—

Oct[r] 10th sent 100$ to M[rs] A—

Oc[r] 20th 1821

This Morning about 6 o'clock We Left *M[r] Perrie's* Plantation for New Orleans, Which Place we Reach[d] on Monday the 21st at 2 o'clock but before I alight in that City, I Must Poise Myself and give you a short a/c of the Most Remarkable Incident that have taken Place With us during our Stay at Oakley the Name of James P. Plantation—

Three Months out of the 4 we lived there Were Spent in peaceable tranquility; giving regular Daily Lessons to Miss P. of Drawing, Music, Dancing, Arithmetick, and Some trifling acquirements such as Working Hair &[c] Hunting and Drawing My Cherished Birds of America; Seldom troublesome of Disposition, and not Caring for or Scarcely ever partaking or Mixing with the constant Transient Visitors at the House, *We* Were Called *good Men* and Now & then received a Chearing Look from the Mistress of the House and *sometimes* also one Glance of Approbation of the More Circumspect Miss Eliza—Governor Robertson Visited us and then I formed a Still Stronger Opinion of that Man agreableness and Strength of Mind than I had before & Consider him as a really true Philosopher of the Age—Amongst our other Visitors the Brother of Henry *Clay* M[r] John Clay of this City I found a good agreable Man to all appearance a Rather Singular Caracter Rich W[am] Brand also spent some days at the House and Married in the Neighbourhood—All Kindly Polite to us;

Miss P. had No Particular admirers of her beauties but several very anxious for her fortune amongst Which a Certain M[r] Colt a Young Lawer Who appeared quite Pressing although Very uncivilly Received at First—

M[r] P. a Man of Strong Mind but extremely Weak of Habit and degenerating sometimes into a State of Intoxication, remarkable in its Kind, Never associating With any body on such occasions and Exibiting all the Madman's Actions With under its Paroxism—When Sober; truly a good Man a *Free Mason*, generous and Entertaining—his wife Raised to opulence by Dint of Industry an Extraordinary

Woman—Generous I believe but giving Way for Want of understanding at times to the Whole force of her Violent Passions—found of quizing her husband and Idolatring her Daugher Eliza—

This Daugher Eliza of age 15 Years of a good form of *body*, not Handsome of face, proud of her Wealth and of herself cannot well be too Much fed on Praise—and God Knows how hard I tryed to Please her in Vain—and God Knows also that I have vowed Never to try as much again for any Pupil of Mine—as usual *I* had to do $^2/_3$ of all *her* Work of Course her progresses Were Rapid to the Eyes of every body and truly astonishing to the eyes of some good observers—

a Sister M[rs] Smith I cannot say that I Knew or rather I never did Wish to Know; of Temper Much like her Mother, of Heart Not so good Yet God forgive her the Injuries She did me—

her Husband a good, Honest Man and Citizen Viewed all the faults of her he Wedded With Patient Kindness and felt his reward through his own Correctness of Conduct—I admire him Much—

I saw there a M[rs] Harwood of London England a *good* Little Woman [and] Very Kind to us in Mending our Linen &[c]—her Little Daugher a sweet Child about 5 Years Old, Hated by M[rs] P.—a Certain Miss Throgmorton Was also good deal Disliked, the poor Girl was nearly drove off as We Were by the Ladies—although she had been Invited there to Spend the Summer—

about a Month before We Left Miss P was taken seriously Ill, and as she was the only remaining Child unmaried and the 2[d] of 7, 5 of which had died in the Course of a very few years, Much fears were entertained of the Survival of this One, and No doubt Much too Much Care Was taken of her; Kept in bed Long after She was convalescent and Not permitted to leave her room for a Long time She became, Low of flesh and Crabed of Speech; every thing Must have gone on the Smoothest way to hurt her feelings her Phisician the *Man she Loved* Would not permit her reassuming her Avavocations near Me and told the Mother that it would be highly Improper Miss Eliza Should Draw, Write &[c] untill some Months; but that She Might Eat any thing Pleasing to her fancy—this fancy Was not Confined into small bounds. She Eat so plentifully of evering that could be procured that she had several Relapses of fevers—I saw her during this Illness at appointed hours as if I was an Extraordinary ambassador to some Distant Court—had to Keep the utmost Decorum of Manners and I believe Never Laughed Once With her the Whole 4 Months I was there—

Mrs P. on the 10th of Octr Dismissed me—Not anxious to revisit New Orleans so soon, I begged of her that We Should remain 8 or 10 Days Longer if the familly Would please to Consider us as Visitors, this agreed on, I Continued My Close Application to My Ornithology Writting every day from Morning untill Night, Correcting, arranging from My Scattered Notes All My Ideas and posted up partially all My Land Birds—the great Many Errors I found in the Work of Willson astonished Me I tried to speak of them With Care and as seldom as Possible; Knowing the good Wish of that Man [and] and the Hurry he was in and the Vast Many hear say he depended on—

We perceived however during all this While that a remarkable Coolness had taken place from the Ladies toward us, seldom seing any of them except at table and then With Looks far from Chearing My Spirits that Were during the Whole of My Stay there unfortunately very Low—Mrs Smith took an utter dislike to Me and one day While I was engaged in finish a Portrait of Mrs P. begun by her Daughter Eliza, M^{rs} S addresed the Work and Me in the grossest Words of Insult, and afterwards Never Looked Directly at Me—

She Busted at another time in a ridiculous Laugh at table, When her good Husband Interfered and told her She ought to Make Me some Amends for her Conduct—I Left the table unwilling to hear any More of this—Saturday Came and a Settlement of Money Matter Was Necessary I charged for 10 days of Miss E. Ill time My Bill Was 204$ and Mrs P. in a perfect Rage fit told me that I Cheated her out of 20$—My Coolness sufered all her Vociferations to flow, I simply told her our former mutual Engagements on that score—I figured the Bill and sent it to M^{r} P Who Was then Labouring under one of his unfortunate fits of antoxication—

he Came to see Me, apologized in the Kindest Manner for his Lady's Conduct; Ordered his Son in Law Mr Smith to pay Me, and shewed Me all the Politeness he his possessed of—Mr Smith Congratulated My Firmness of Acting—and All Went on pretty Well that Day—

the Ladies early that Morning Left for St. Francisville Without bidding us any adieu, and expected that on their Return at Night We would be gone; this however was a disapointment for Mr *Perrie* requested We should Stay, representing how easily We could reach the Steam Boat the Next Morning before her time of departure in the Course of this Afternoon M^{rs} P sent for Joseph and presented him with a full suit of fine clothes of her Deceased son—to the acceptation of Which I positively refused to acquiesce, Knowing too Well how far some gifts are talked of—and Not Willing that My Companion should diminish the Self Respect I think Necessary for every Man to Keep towards himself however poor, when able by *Talents*, Health and Industry to Procure his own Necessities—

unfortunately there was Much Company in the evening [and] sometimes after supper We Left our Room were M[r] Pirrie and M[r] Smith had Joined us on leaving the Table to go and bid our farewell to the female Part of the familly—My Entry before the Circle possessed none of that Life and Spirit I formerly Enjoyed on Such Occasion, I would gladly have wished to be excused from the fatiguing Ceremonies yet I Walk[d] in followed by Joseph and [Walking] approaching to M[rs] P. bid her good bye as simply as ever any Honest Quaker Did, touch[d] Slightly M[rs] Smith's Hand as I boughed to her—My Pupil Raised from the Sopha and Expected a Kiss from Me—but None Were to be disposed off, I pressed her Hand and With a general Salute to the Whole Made My Retreat No doubt Much to the great Surprise of Every one Present Who had heard those very Women Speak Constantly before of Me in Highest Terms of Respects, scarcely Deigning to Look at me Now—as Joseph Was following me he received a Voley of farewells from the 3 Ladies of the House put after him Ridiculously to Affect Me, but the Effect Was lost and it Raised a Smile on My Lips [and]—We Joined again the Two good Husbands in our Lodging Chamber—they remained with us untill bed time; Cordially parted With us, retired to repose Without Joining the Company—

Day Light of Sunday Saw us Loading our Trunks and Drawing Table. Vaulted our Sadles and Left this abode of unfortunate Opulence without a single Sigh of regret—

Not so with the sweet Woods around us, to leave them was painfull, for in them We allways enjoyed Peace and the sweetest pleasures of admiring the greatest of the Creator in all his unrivalled Works. I often felt as if anxious to retain the fill of My lungs with the purer air that Circulate through them. Looked With pleasure and sorrow on the few Virgin blooming Magnolias—the 3 Colored Vines and as We descended the Hills of St Francisville bid that farewell to the Country, that under diferent Circumstances We Would have Willingly divided With the Ladies of Oakley—

We Left the Mouth of Bayou Sarah at 10 o'clock in the Morning in the Steam Boat Ramaso with a Medley of Passengers and with a few Stoppages to Land and take occasional Travellers Reached the City on Monday—

the Weather Cool and Rainy, I left the Boat and Walked to My good acquaintance R. Pamar—I had perceived that My Long flowing buckled hair was Looked on with astonishment by the Passengers on board and saw that the effect Was stronger in town—My Large Loose Dress of Whitened Yellow Nankeen and the unfortunate *Cut* of My features Made me decide to be dressed as soon as Possible Like other folks and I had My Chevelure parted from My head the Reception of Mr Pamar's familly was very gratefull to My Spirits. I Was Looked upon as of a Son returned from a Long Painfull Voyage, the Children, the Parents the servants all hung about Me; What Pleasures for the Whole of us—

I dined there, afterwards Visiting the famous hunter Lewis Adam—and the Dimitry familly who also Received Me very Kindly

Rented une Chamber garnie in Rue St Anne N° 29 for 16$ per Month and removed our baggage thereto from the boat—

We Spent Tuesday Wednesday & Thursday, Looking over the City for a Suitable House for My Litle Familly—this appeared a very dificult task and I nearly Concluded to take one we visited in Dauphine Street—

My Clothes being extremely Shabby and forced against My Will to provide some New ones, I bought some clothes and Now Wait very impatiently on the Gentleman Taylor for them that I May go and Procure some Pupils with a better grace—

having renewed our early Morning Visits to the Market to Look at all there—We found it allmost as well suplied of Vegetables, fruits, fish, Meats, flowers, &c as in the Spring—delightfull radishes Letuce &c plenty—Wrote to My Wife yesterday—per Mail it is Now 15 days since I forwarded her a Draft on Mr Gordon Which May probably have reach to day—

I found at Messrs Gordon Grant & C° a Box of Oil Colors and a Letter from Mr Briggs, I was sorry to See both, the first did not Contain What I Wish, and I Cannot say that I felt any pleasure in Reading the Latter—

I answered Mr Briggs's Letter this Evening—the 25th of Octr—

Octr 25th 1821

Raining hard the Whole of the day spent the greater part of it at R. Pamar and his Relation *Lewis Adam the hunter;* Rented a House in Dauphine Street for 17$ Per Month—

Joseph found the time rather Lost to himself Not a Thing to be found to Draw—

26th Wrote a few Lines to James Pirrie Esqr to Inform him that Messrs D. & G.

Flower had not Paid the House of Gordon & C° One hundred Dollars according to Promise—Wrote a few Words to My Lucy and forwarded it by Mail Covering Brigg's Letter;

Walked a good Deal, Visited the familly Dimitry, Spend Much of the Day at Pamar's—in the evening Went some distance down the Levee, the Sky beautifull & serene—Miss Pamar Much Improved in Music and Manners—Many Men formerly *My Friend* passed Me without uttering a Word to me and *I as Willing* to Shun those Rascalls—

fatigued of being *Idle* so powerfull are habits of all Kinds that to spend a Month thus would render me sick of Life—

Hetchberger Visited us Much Pleased at My addition of Drawings since I Left Cincinnati Octr 12th 1820 I have finished 62 Drawings of *Birds & Plants*, 3 quadrupeds, 2 Snakes, 50 Portraits of all sorts and My Father *Don Antonio* have Made out to Live in humble Comfort with Only My Talents and Industry, Without *One Cent* to begin on at My Departure—

I have Now 42 Dollars, health, and as much anxiety to pursue My Plans of Accomplishing My Collection as Ever I had and Hope God Will Grant Me the same Powers to Proceed

My Present Prospects to Procure Birds this Winter are More Ample than ever being now Well Known by the Principal hunters on Lake Borgne, Barataria—Pontchartrain, and the Country of Terre a Boeuf—

Octr 27th SUNDAY

Dressed all new, Hair Cut, my appearance altered beyond My expectations, fully as much as a handsome Bird is when robbed of all its feathering, the Poor thing Looks, Bashfull dejected and is either entirely Neglected or Look^d upon With Contempt; such was my situation Last Week—but When the Bird is Well fed, taken care of, sufered to Enjoy Life and dress himself; he is cherished again, Nai admired—Such my situation this day—Good God that 40 Dollars should thus be *enough* to Make a *Gentleman*—ah My Beloved Country When will thy Sons value more Intrinsectly each Brother's Worth? Never!!

Exibited My Drawings at My good acquaintance Pamar's—received much valuable Intelligence coroborating With My own observations on these things that trully pleases My feelings—Dined there—

Payd a Visit to Mrs Clay and the young Ladies there, with My Portfolio—

unknown, Passed for a *German* until the latter part of My Stay—the Company Much Pleased With My Work—but no pupil as I expected to have—

took a Long Walk to the Canal, talked Much With My Hunter *Gilbert* Who Leaves for Barataria Tomorrow—Weather Beautifull and very warm, good Deal of Game in Market this Morning—

Green Back Swallows Gamboling over the City and the River the Whole day have great Hopes of ascertaining their Winter quarters Not far from this

Octr 29th 1821

Spent this unfructuously in search of Employ^mt Visited several Public Institutions where I cannot say that I Was very politely received; in one or Two Notable ones (Not Willing to Mention Names) I was invited to Walk in and then out in very quick order—Dined at Pamar; Was Visited by John Gwathway of Louisville K^y—Wrote to J^d Rozier to Procure my Drawing of the *Male Grous* or Prairie Hen—Determined on Exibiting some of My Drawings at Public Places for I Well recollect the effect of *Lafontaine* Fable that says that "*a l'oeuvre on connoit* L' *Artizan*"—unknown by most people here, I am like Many others who appear as advanturors look^d on with Care, and Suspicion—but so Moves the World, and no doubt it is *Wright* it should be so—

Visited Rich Mrs *Brand* was then very Politely received, "Must Call again"—Mrs Brand Married a Large fortune, the Honey Moon is not yet *Set* and she Looks well even on her Decline, promising full fullness bye & Bye—

Rec^d a Letter from My Beloved Friend My Lucy unfortunately of old Date, and the one also sent by Mr Ecard dated Nearly 2 Months—

Octr 30th 1821 NEW ORLEANS

Returned to W^am Brand and Procured his Son for a Pupil at 2$ per Leson of one hour, and have some Hopes of having Mrs B. a pupil of *French* and Painting—

Visited another College. Politly Received by the Ladies Who examined My Port Folio with apparent siaisfaction. No Pupils however, a Certain Mr *Torain* having antecedented Me every Where—

Dined at Pamar and Drew My American Hare—to Exibit to the Public—Joseph at Work Preparing Father Antoine Coat the Market Well suplied With game & Vegetables have resumed our Habit of taking a Walk there as soon as the Day Dawns—

The day Warm, Swallows Plenty and quite as gay in their flight as in June—

to find here those Birds in aboundance 3 Months after they have left the Midle States, and to Know that they Winter Within 40 Miles in Multitudes is one of the Gratifications the Most Exquisite I ever Wish^{d} to feel in Ornithological Subjects and that Puts [an] compleat *Dash* over *all* the Nonsense Wrote about their Torpidity during Cold Weather; No Man could ever have enjoyed the Study of Nature in her all Femine Bosomy Wild and err^{d} so Wide—

Octr 31st 1821 NEW ORLEANS

Begun giving Lessons of Drawing to Mrs Brand and Young Master W^{am} Brand this Day at 3 Dollars per Lesson—

Spent some time at Work on Father Antoine and My Drawing of My American Hare—

Received a Visit of Mr Pamar, Mr Dimitry and Dumatras

Weather Warm in the Morning, Much fish Condemned in the Market—also some Game—Excellent regulations—the Wind Shifted to the Northwest and I premidited Cold Weather by the Swallows flying South about noon at Night quite Cool—What Knowledge these Litle Creatures possess and how true they are in their Movements—

November 1st 1821

Weather beautifull—gave Lessons at W^{am} Brands—Dined at My good Pamar's House—

Very unwell at Night with Violent Cholicks and was forced to My bed at seven o'clock a thing I have not done for Very Many Years—Visited M. Basteros, Painter—

at 12 o'clock Disturbed by the Cries of Fire—but as it Was Not in our imediate Neighbourhood did not Suffer *Joseph* to go—

November 2d—FRIDAY—

felt Well at day light and Went to see the Market Much Game, but nothing for Me—

gave My Lessons at Wam Brand's—Much pleased to find his Lady possessed of a *Natural* talent, for Painting Wam Brand extremely Kind and Polite, very anxious to give his Son a good Education—

Recd the Visit of *Brutter* the Painter, the good Man Very sorry to see My *Father* Antonio—fearing an Engraving after it—

I determined to have My Drawing framed although it Cost Me about 30$ having some Hopes that it Would procure Me some Pupils of Note—

SATURDAY 3d November

give My Lessons—John Gwathwey announced me the Death of My *Constant* Enemy Mrs Bakewell My Wife's Mother's in Law—God forgive her faults—Etchberger the Painter spent the evening With us—Mr Hails borowed My Shot Bag at 10 o'clock P.M.—sufered some Mortifications this Morning in a House Were I showed My Birds—Weather Cloudy & Raw—

SUNDAY 4th November N. ORLEANS

Breakfasted at Pamar's saw a School Mistress there who requested that I should Call at her House to shew My Drawings—Did so at 11 o'clock Tolerable reception there, the Lady Drawing Well herself anxious to acquire My Style but Complaining Much of the extraordinary price I asked her—expect there several Pupils but Nothing very Certain—

Dined at Pamar's, Steam Boat Ramapo arrived Without *James Pirrie* Esqr Much Disapointed on a/c of the 100$ that he was to Pay Messrs Gordon Grant & Co on the 20th ultimo—

took a Long Walk Down the River Shore and out to the Swamps—Swallows More Plenty than yesterday, generally Moving Eastwardly to the Lakes—Weather Delightfull, Much such that is felt in May in Kentucky—Many of the trees having a beautifull set of New foliage, Vast Many Plants in Bloom—Particularly the *Elder*—No Lessons at Wam Brand to day—

—MONDAY 5th—

give My Lecons at W. Brand—Drawing at My F. Antoine—Paid Mr Forestal 100$

—TUESDAY 6th—

gave My Lecons at Wam Brand—Swallows plenty—Appearance of *Indian Summer* Took a Long Walk and Much Work Done at My Drawing

—WEDNESDAY 7th—

Gave My Lecon to Mrs Brand & Son—Procured Two Pupils to begin Next Monday—Drew an American Avoset—Weather Beautifull—

Brown Pelican—

Length 4 feet 2½ Inches from the Tip of the Bill to the end of the Toes, which extend about 1½ Inches beyong the Tail—The Bill Measured 12½ Inches the upper Mandible armed with a strong Hooked Point projecting beyond the blade of the upper Mandible ½ Inch—and fitting the Lower one in its Whole Length to the Nail [the] Which runs to the forehead in Two furrows and Contains the Nostrils that are Extremely small, Linear, within ½ an Inch of the forehead and scarce perceivable—the outer edges of this Mandible as Well as that of the Lower and Sharp Cutting edges—3 Process edges are contained in the inner part of the upper Mandible also Sharp edged the whole of this upper Part long stiff & Strong substance—of a Greyish Blue edging in Yellow—the edges of the Lower Mandible are the same Color averaging in Width about ½ an Inch truncated at the end and Capable of distention from their Natural [weigdth] depth of 1Ð of an Inch to 6 Inches and furnished with a tail or pouch of a Bluish Distending Skin begaining at the under point of the Lower Mandible and Loosing itself along the Neck about 9½ Inches below the Junction of the Mandible from Which part to its utmost stratch point with the hand it Measured a foot—the Tongue is a Mere Knot about 12 Inches from the Tip of the Lower Mandible fast to this Bag—the Eye is Brown pretty Large and situated in the Skin that [so] covers the Cheeks and Jaws of the Bill—the upper of the Head and side of the Neck running along the Pouch of a

Mole Cole Color, the hind head ornemented with a Crest of slender feathers of 1½ Inch in Length—the upper Plumage of the Neck, assuming a Silky appearance and Much Worn by resting on the back and Shoulders of the Bird—Shoulders & Back Covered With pointed, Small feathers the former Light ash in their Centers edged With Rufous & Some with brown—the Latter silvery in the Center edge With deep black to the Rump Where the Plumage is Larger yet Pointed ash & Rufous—the Tail Rounded Composed of 18 feathers quills Black shaded Silvery Ash—the Wings Extended Measured 7½ feet—the second Joint 9 Inches Closing on the Body reaching to the begainging of the Neck—and When closed the Tips reach the end of the Tail—the first 9 Primary quills White to their points below and about ½ above; the feathering [dark] brownish black cast secondaries Much the same—Tertials Broad falling over the back part of the body to the root of the Tail; feathers of the Shoulders of a Light Ash some edged With brown, others With Black, quills very slender and Black—Whole under part White and in some Specimen Silvery—Legs Strong and Muscular far behind—4 Toes Webbed in Connection—the Whole of a Bluish greenish Yellow—Claws blunt, Much Hooked the Longest Pectinated Inwardly—the Bird emitting a Strong disagreable fishy Smell Weighed 6½^{lb} femorals Much as the Shoulders—on dissection it was a Male—the Stomack Very Long and slender, fleshy—Containing only about 50 slender Blue Worm all alive about 2½ Inches Long—the Gut Measured 10 feet about the size of a Moderate Swans quill—the Bird Was Killed on Lake Barataria by Mr. Hunter Gilbert—the rump and the root of the Tail Was Covered With a Thin Coating of Oily Yellow fat extremely rancid, and Much air Was Contained between the Whole of the Skin and flesh of the body; the Bones of the Wings & Legs although Extremely hard and dificult to Breake, were very thin [and] Light and perfectly empty—

—THURSDAY 8th—

gave My Lecons at Mrs Brand's—Weather extremely sultry—Anxious to hear from My Wife—

NEW ORLEANS November, FRIDAY 9th 1821

Weather quite Cool a diference in the atmosphere of 22 Degrees from that of yesterday and the Swallows that Were Numerous Last Evening are all gone for the Present—gave My Lecons at Mrs Brand's—

Carried My Port follio to M[rs] Dimitry this morning to Show Miss Euphrosine the Progress of Joseph—Dined and Breakfasted at Good *Pamar's*—

Visited this afternoon Miss *Bornet's* academy of Young Ladies and shew[d] some Drawings—but all to No purpose the Ladies there entirely in favor of M[r] Torane Talents—

M[r] Hawkins during My absence brought an Engraving of Vanderling's head of Adriane for Me to Copy and requested Joseph to tell Me Not to spare My Time on it—M[r] Bartrop also Called While I was out—My feelings Much harrassed about My Beloved Wife from Whom I have Not heard for 2 Weeks—

—Saturday 10th—

Gave My Lecons at M[rs] Brand—Called on M[r] Hawkins who visited me to see father Antoines Drawing—Concluded to have the Engraving he Wished Me to Copy for not exceeding 50$ wishing it could be done as soon as possible—Saw M[r] John Clay—very Polite to Me—

Weather Very beautifull but Cold Drew a female of the *Gadwall Duck* a remarkably fine Specimen sent Me by My Hunter Gilbert—

M[r] Basterop at My Lodgings—Wished that I should Join him in a Painting of a Panorama of this City—but My Birds My Beloved Birds of America feel all my time and nearly My thoughts I do not Wish to See any other *Perspective* than the Last Specimen of them Drawed—No News from My Beloved Lucy nor Children, Very uneasy on their Silence—Mississipi falling fast—

NEW ORLEANS SUNDAY November 11th 1821

Saw John Gwathway early this Morning, who told me that My Wife Intended Leaving Louisville about the first Instant in a Small Steam Boat for this Place—and this News Kept me Nearly Wild all day. Yet No Boat arrived No Wife No Friend yet near—

The Weather Beautifull & Warm Dined at Pamar—Drew a good Deal. Visited Basterop—Joseph hunting all day With Young Dimitry. Killed Nothing—Swallows Very plenty and as gay as could be—Saw some Common Gull, but Not a *Fish Crow* come yet—M[r] Bermudas Visited Me a short time this Evening—The Nearer the Moment that I Expect to see My Beloved Lucy Approaches the greater My Impatience, my disapointment Dayly When evening draws on—

MONDAY 12th 1821

Began Given Lessons to Miss Delafosse and Miss at 2 Dollars per hour for both—

Gave My Lecon also at Mrs Brand—Saw Eliza Throgmorton there—Weather Beautifull but *No Ducks*—Drew a good Deal—Dined at Pamar—had a Conversation Mr John Clay Respecting Mr P—Mr Bermudas brought Me a *Green Winged* Teal as a great Rarity—No News from My Wife yet—

TUESDAY 13th

Gave Lecons at Mrs Brand but Miss Lafosse Wishes only to receive them 3 times per Week—Drew a Wild Goose Not represented by Willson—Weather quite Cool—Very Busy all day (The white fronted)

WEDNESDAY NEW ORLEANS November 14th 1821

Wednesday—gave My Lecons at Mrs Brand's and Miss Delafosse's

Work Constantly the whole day—Drew a female of the *White Nun* or *Smew Merganser*—Weather Rainy & Raw—Dined on Bread & Cheese—Recd a Letter from Mrs A. the purport of Which Lower My Spirits very Considerably—alas were does Comfort Keep herself now; retired certainly on a Desolate Rock unwilling to Cast even a Look on our Wretched Species—

Mr E. Fiske formerly our Agent in this City presented Me this morning with a bill of Fellows & Rugles—I Spoke to him on that subject in Terms that astonished him, but My determination is bent and I Shall Philosophise Now on all things—Little Expectations of seing My familly before the Latter Part of Winter—

THURSDAY 15th

Gave Lesson at Mrs Brand's—

Drew Closely all day finishing 3 Drawings of Birds and Continued after sun set by Candle light until 10 at Vanderlyn's Head—

Weather fair but Cool—Very Low of Spirits Wished Myself off this Miserable Stage—

16th
FRIDAY

Gave Lesson to Mrs Brand—but Miss Delafosse Was unwell and Postponed taking untill tomorow—Sent a Bill to Dr Heermann—it was accepted and Promised Joseph Payment for Next Week

17th
SATURDAY

Gave Lessons at Mrs Brand's & Miss Delafosse also—her Mother Knew My father—Drew Much to day and late this Evening—[Gilbert the Hunter Returnd]

NOVEMBER SUNDAY 18th 1821

Drawing all day; Dined at Mr Pamar Recd a Visit of Philip Guesnore—also one from the famous Hunter Louis Adams Who however had No Knowledge of the Small Merganzer I had drawn

—19th—

Monday Gave Lessons at Mrs Brand, and Miss Dellfosse—these Latters have conetracted to have one every day Much to My satisfaction—Needing this acumulation of Income very Much—

Drew a Black Bellied Darter Male a Superb Specimen—

Was Visited by Mr Hawkins, an agreable Man Possessing Taste, Information & Judgment—

—20th—

Tuesday Gave Lessons at Mrs Brand & Miss Dellfosse Drew Much and finished both My Black Bellied Darter and the Vanderlyn's Head—Shabbily used by Dr Heermann—Who refused Paying My Well earned Bill—Visited by My good acquaintance Pamar—Basterop the Painter, Much Talk With My good Hunter Gilbert Who Procured Me a Superb Specimen of the *Great Sand Hill Crane*—

Sufering Much from Sore Eyes and Violent Headache the Whole Day—

—WEDNESDAY 21st—

Gave Lessons at Mrs Brand & Miss Delfose—Drawing all Day at My Whooping Crane—Weather Extremely Sultry—

November THURSDAY 22d 1821

Gave Lessons at Mrs Brand—My fair Pupill Miss Delfosse engaged otherwise—Recd 40$ on a/c from Mr Hawkins who appeared to be Much Pleased With the Drawing I give him of *Arianne*—Weather Summer Heat—Swallows Plenty—Recd a Letter from My Wife. My Spirits yet very Low—Drew Much to day—received 100$ from Mr Forestal as Mr Gordon had Not Paid any Money to My Wife at Louisville—

23d

Friday—Gave My Lessons to Mrs Brand and Miss Dellfosse—Rainy and Warm—drawing all day—bought a Portfollio from Vigny for 8$—

24th

Gave My Lessons at Mrs Brand and Miss Dellfosse and also at Pamar to his Daugher Who exibited the brightest Genius I believe I ever Met With—Miss Dellfosse beautifull and extremely agreable—

Wrote to My Wife and Wam Bakewell and forwarded to cash a Check on the U. S. Bank of Philadelphia Received by Mr Bermudas for me and for Which I Paid ten per cent—of 100$ Weather extremely changed Cold and Windy—

25th
SUNDAY

Weather Very Cold and raw—Gave Two Lessons of Drawing to Euphemie Pamar—

26

Gave My Lessons to Mrs Brand. Miss Dellfosse and Euphemie Pamar—Weather rather Mild—Swallows plenty—Ship Fulton arrived 120 Passengers on Board

November TUESDAY 27th 1821

Gave My Lessons at Both Places—Visited the Maire's Lady Mrs Rosignol Who had evinced a desire to see some of My Pencil Productions—Expect her Daughter for a Pupil Weather Charming—Drew 2 Ducks. Called by Willson the *Tufted Duck* Male & female the Clerk of the Steam Boat Ramapo Called on us Mr *Laurent* an agreable Man

28th

Wednesday Gave My Lessons at Both Houses Weather fine Rec^{d} a Visit from Mr Brand & his Lady—Basterop Not Much Pleased at this; Drew a good deal

—29th—

Thursday—Gave Lessons to Mrs Brand's only Rec^{d} a Letter from My Wife of an older date than the former one—Weather beautifull—Painted Joseph Likeness—

—30th—

Friday—Gave My Lessons to Mrs Brand's and at Miss Dellfosse—

NEW ORLEANS Decr 1st 1820 (1821)

[Friday] Saturday Gave My Lessons to both houses and Miss Pamar—Rec^{d} a Letter from My Beloved Wife—Expect her in a few Days U. S. Steam Boat arrived—weather uncommonly pleasant Rec^{d} this Evening a Non Descript Hawk—

—SUNDAY 2[d]—

Gave 2 Lessons to Euphemie Pamar and One to Miss Dellfosse *only*—Weather beautifull finished My Drawings of the *Crested Hawk* Which proved a femalle with Many Minute Eggs how Rare this bird is I Need Not Say being the Only specimen I Ever have Met with—although I once before found some Tail feathers of another Killed by a Squatter on the Ohio—Which Tail feathers having Kept—Corespond exactly With those of the present bird—

Regret Much that I Cannot Save the Skin but the Weather being Warm and My Drawing having taken Nearly Two Days it was not possible to skin it—

—Monday 3[d]—

Gave Lesson to E. Pamar at M[rs] Brand and Miss Delfosse—saw M[r] Wheeler who arrived this day—how little I Expected to have Ever Met with him—Weather quite Warm—

—Tuesday 4—

Gave Lessons to M[rs] Brand at Miss Delfosse also Rec[d] 40$ from the former and paid the rent of the house in Dauphine Street Weather very Warm drew an American Bitern—

NEW ORLEANS Decemb[r] 5th 1821

½ quire of Paper to Miss Josephine

Wednesday—Gave Lessons at M[rs] Brands and Miss Delfosse—Weather Cool and Rainy—Visited Basterop and Was Introduced to M[r] Sell another of the Trade—another day of Disapointment My Lucy Not arrived—Thursday Begun taking & Giving Lesson to M[r] Lombar Gave My Lessons at M[rs] Brands and today Miss Delfosse—Mr. Pirrie Arrived to day

—7th—

Friday Gave Lessons at Mr Brand's & Miss Delfosse—Mr Lombar Give Me Lesson on the Violin & I to his Son of Drawing in Exchange—Rainy Cold all day

—8th—

Saturday—Gave Lessons at Mrs Brand's and gave her 1 dozn Black Chalks Making 1½ 6 Pencil handles at Miss Delfosse and Miss Pamar—at Night Music & Drawing With young Lombard had the Pleasure of seing Mr *Harwood* from London and Recd by him My Bitch *Belle*—Weather Rain & Cloudy

—9th—

Sunday Weather Extremely Disagreable the Steam Boat Hero arrived from Louisville but No Regular Information about My familly by it—Gave 2 Lessons at Mrs Pamar where I spent the greater part of the Day—Young Lombar Drawing at my House the Whole Day—

—10th—

Gave Lessons 2 to Miss Pamar, Mrs Brand's house and Miss Delfosse—Received a Visit from Mr Selle & Mr. Jany Painters—

NEW ORLEANS Dec 11th TUESDAY 1821

I have but little time to spare at present to write the Many Incidents connected with the Life I am forced to follow for My Maintainance and of Course hundreds of them are passed and forgotten although I am well assured that a Rearsal would at future period amuse My thoughts. *One* however so curious appeared Me this day that I Cannot let it escape—May you My Dear Sons reap some benefit from the details

I am a Teacher of Drawing and have some Pupils My Style of giving Lessons and the high rate I charge for My Tuition have proved Me the Ill will of Every other artist in the City who Knows me or has heard of My Maxims—I Called on a *Bastard* of *Appollon* this Day to see his *Labours* I was unknown, tolerably well received and had the pleasure of seing the *Animal in* action—I also heard his *Barkings* and saw his Eyes gladening at the sight produced on the canvass before them—a Third unfortunate Dauber came in who it appeared Was an old acquaintance that Criticised at once with ease on all that Was around us—as Every day arrivals by Sea and Land bring New hands to the bellows the Names of Many Were Called forth and Mine amongst them—I Kept Myself and Waited & the following Picture was given Me Without any Varnishing I assure You—"That Man Came No one knows from Whence—he goes throu the streets Like the Devil I am told that he has as Many Pupils as he Wishes for and Makes a Wonderfull quantity of What he Calls Portraits and Assures the good folks who employ him that in a few Months by his Method any One May become able Painter—and yet from What I am told the Man Never Drew [but] has bought a set of handsome Drawings of Beasts Birds Flowers, &c Which he Shews and Says are his own—all this a Lye and take in While I Who Was Naturally Intended to Paint Teach &c am Without a Pupil or Portrait."—

here I Took My Hat told the Gentleman where I resided and that I Would be happy to See him giving him the [Two firsts] Initials of My Name only for a Guide—from this Eloquent Member of the *Sans Culottes of the Trade,* I Moved pretty briskly to Mr Basterop's Room were in a few Minutes I had the Satisfaction of Seing Messrs Jelle & Janin—all artists and agreable Men. Not Well setled about the a/c of Myself I had so lately heard but thinking how strangely the good Man Will feel when *he* Calls on Me—if ever he does—

I Gave My Lessons first to Miss Pamar—Mrs Brand at Miss Dellfosse and then according to Promise Went to a Pensionnat (were My Young Friend Miss Pamar receives the *Larger* portion of her Education) to give her regular Lessons of Drawing—When I Entered the Room, I saw a Degree of Coolness in the appearance of the *Lady* of the Institution that along With My unfortunate or foolish Natural awkwardness in similar Cases rendered My stay extremely disagreable; My Pupil who is generally Lively and full of Confidence in her actions Was at this time so astonishingly astray from her Work that not a line was properly Copied—I perceived the sarcastic Looks of the Diferent Teachers who Were present going the round and Was highly relieved When the Clock struck My Departure—a few Expressions uttered on My Entering, Joined to a few that reachd My Ears (that burned all this while) as I was Making My Escape Made Me take the resolution Never again to trespass on that Threshold or any other Without first Knowing Well how it May agree with the Will of the first Caracters attached to them [It seems that there are good and bad (—) for everybody]

My Lovely Miss Pirrie of Oackley Passed by Me this Morning, but did not remember how beautifull I had rendered her face once by Painting it at her Request with Pastelles; She Knew not the Man Who with the utmost patience and in fact attention *Waitted* on her Motions to please her—but thanks to My humble talents I can run the gantlet throu this World without her help

WEDNESDAY 12th

Gave My Lessons at Mrs Brand's—Miss Delfosse absent to day—gave Lesson also to E. Pamar having taken all in Consideration I put aside the Shadow and hang to the bone—and with a *White Lye* arranged the Matter quite Well With Mrs Pamar to Whom I had said that I should Not return there any more—

December THURSDAY 13th 1821

Gave Lessons at Mrs Brand and E. Pamar Miss Dellfosse finding the Weather Too Cold—So Anxious Am I during the Whole of My present days to see My familly that My head is scarce at right With My Movements and yet I Must feel My sad Disapointments and retire to rest without the comfort of her so much Wanted Company

I saw to day a Work on Natural History with Colored Plates rather better than usual—

14

FRIDAY Gave Lessons to all this afternoon—but Miss Dellfosse appeared dejected and Work but Indiferently—

It is Now 26 days since the Last Letter I have from My Wife is dated. Three Steam Boats have arrived since from Louisville and No News of her departure have reached Me—My anxiety renders every Moment painfull and Irksome—

I Met quite unexpectedly My Pupil Miss Pirrie at Mrs Brands, the interview was Short more friendly than I expected and We parted as if We Might see each other again With some Pleasure at some future Period—

—15th—

SATURDAY Gave My Lessons to All this afternoon Weather Cool & Clear; feeling much relieved from My Anxiety about My Familly having heard that the Steam Boat the Rocket by Which they are to Come had not Left on the 28th ultimo and that Probably they would Not arrive for 4 or 5 Days yet—

Mr Jany Visited us this Evening and stayed Very Late—

December 16th 1821

SUNDAY—Gave 2 Lessons to E. Pamar & 2 to Miss Delfosse only—Weather fine but Cool begun Drawing a Young *Swan* Sent me by My Hunter Gilbert from Barataria—

Young Lombar at Work all Day in My Room Drawing—

17th

Monday Gave My Lessons all round—I Drew at My *Swan* Mr Jarviss paid me a Visit and I returned it imeditly—Gave him 3 pieces of Canvas—Weather very Dark & Rainy

—18th—

TUESDAY—Gave a Lesson to Miss Pamar and One to Mrs Brand—My Wife & My Two sons arrived at 12 o'clock all in good health—I took them to Mr Pamar Where We all dined and then Moved to our Lodgings in Dauphine Street after 14 Months absence the Meeting of all that renders Life agreeable to Me, was gratefully wellcomed and I thanked My Maker for this Mark of Mercy

—19th—

WEDNESDAY I only gave Lesson to Mrs Brand having much to do arranging

My familly—Examined My Drawings & found them not so good as I Expected them to be; When compared With those Drawn since Last Winter—Bonaparte's Service Was performed this Day here—

—20th—

THURSDAY—Gave My Lessons all round—Weather Extremely Warm—Recd a *Nondescript Rail*

21st

FRIDAY Gave Lessons at Mrs Brand's & Miss Delfosse but so Wet and Damp that I declined going to see E. Pamar—Drew a Streaked Rail

NEW ORLEANS December 22d 1821

SATURDAY—Gave My Lessons all round—Weather very Disagreable—Recd 20$ on a/c from R. Pamar—Young Lombar Resumed his Lessons this Evening having Missed Coming Whilst I was arranging My familly at home—

—23d—

SUNDAY Gave 2 Lessons to E. Pamar and 1 to Miss Delfosse—Weather Extremely Cold—having seen Ice this Morning Nearly one Inch Thick

—24—

MONDAY Gave a Lesson to E. Pamar Young Brand and 2 to Miss Delfosse—

Weather very Cold—Mr Rozier Came to Pay us a Visit—it is Eleven Years since he [and], I, and My Familly, Were all together—

25

TUESDAY—Gave 2 Lessons to Miss Delfosse but not any were else—

Snowed from Day Light untill Twelve o'clock and afterwards froze hard—

26

WEDNESDAY—Gave 2 Lessons to Miss Delfosse and 1 to Miss Pamar—M[r] Gordon Visited us this Morning and M[r] Colas the Miniature Painter this Evening to see My Birds—

27

THURSDAY—Gave 2 Lessons to E. Pamar 2 to Miss Delfosse and One at M[rs] Brand—Weather beautifull—Paid our Rent this Morning—

—28—

FRIDAY Gave My Lessons all round 2 to Miss D.

—29th—

SATURDAY—Gave My Lessons all round—

Dec[r] SUNDAY—30th 1821

Gave 1 Lessons to E. Pamar and 1 to Miss Delfosse—M[r] Pamar Dined with us—I Drew this Day a Ferruginous Thrush and am to Draw 99 Birds in that Number of Days for Which I am to Pay One Dollar for each to *Robert* the Hunter—Who is to furnish Me With One hundred Specimen of Diferent Kinds Should he Not fullil the Contracts, he is to have only 50[cts] for each furnished

—31—

MONDAY—Gave 2 Lessons to Miss Pamar 1 to M[rs] Brand and 1 at Miss Delfosses & Drew an *ampellis Americana*

[The following material appears at the back of the journal and is written in from the back cover as a first page. It has no apparent connection with the regularly kept journal.]

Copy of a Letter Written to the Honorable Henry Clay, Speaker of the House of Representatives, Lexington, Kentucky—

SIR

After having Spent the greater part of Fifteen Years in Procuring and Drawing the Birds of the United States with a view of Publishing them; I find Myself possessed of a Large Number of such Specimen as usually resort to the Midle States Only, having a desire to complete the Collection before I present it to My Country in perfect Order, I intend to Explore the Territories Southwest of the Mississipi.

I Shall leave this place about the midle of Sepr for the purpose of Visiting the Red River, Arkansas and the Countries adjacent, and Well aware of the good Reception that a few lines from one on Whom our Country looks up with respectfull Admiration, would procure me; I have taken the liberty of requesting such Introductory Aid, as you, May deem Necessary to a Naturalist, While at the Frontier forts and Agencies of the United States

I Remain
Respectfully
Your Ob^t^ Servt
J. J. A.

Cincinnati Ohio Augt 12th 1820

Descriptions of the Water Birds of the United States, with their Generic [arrangement] characters according to the arrangement of Latham as Described by Alexander Willson—the Species discovered by Myself are Marked with My Initials

SPOON BILL

Bill Long, thin, the tip dilated, Orbicular, flat, nostrils Small, placed near the base of the Bill, tongue Short, pointed; feet four toed, Semipalmate

Roseate Spoonbill, Platalea ajaja. 7.123

La Spatule couleur de Rose de Brisson—

I sent Willson from Natchez, Measured 2 feet 6/12 and 4 feet in extent, Bill 6½ Inches Long from the Corner of the Mouth 7 from its upper lap, 2 Inches its greatest Width ¾ Inches Narowest place—½ Black covered with Scaly protuberances like the edge of oister shells—waitish stained with red—Nostrils oblong in the Midle of the upper Mandible—a deep groove runs along the Mandible about ¼ Inch from the edge—Crown & Chin bare covered with a greenish Skin; that below the Lower Mandible dilatable as in Pelicans—orange round the eye, Irides blood red; cheeks & hind head bare black Skin; Neck Long covered with Short White feathers, tipt on the Neck with Crimson; breast White its Sides burnt brown Color, a Long tuft of hair Like plumage proceeds from the breast pale rose Color back White Slightly tinged with brownish, Wings pale rose Color, Shafts of same Lake; Shoulders of the Wings covered with Long hairy plumage deep & splendid Carmine; upper & Lower tail Coverts Same; belly rosy; rump paler; tail equal at the end 12 bright brownish orange feathers, Shalf redish; Legs and Naked part of the thigh dark dirty red; feet ½ Webbed—toes Very long particularly hind one, inside of the Wing richer than Outside

---

List of the Watter Birds of North America taken from Turton's Linne

AMERICAN AVOSET, head, Neck & Breast rufous—Bill Black, Legs pale blue—

BLUE CRANE, Head, & Neck dark purple—3 Long narrow pendant feathers 6 Inches beyond Eye—Length—23 Inches extent 3 feet

SNOWY HERON—2 1/12 feet—extent 3 2/12—orange yellow round the Eye—Irids vivid Orange—Whole plumage White—(Head Largely Crested—4 Inches) breast D° upper back feathers recumbent and Loose—

NIGHT HERON, Bill Blk 4¼ Inch Skin about eye Blue, Irids Red, Cap[d] deep Blue—3 White feathers Issuing—Back deep blue—Vent & belly White, Legs & feet Light Buff—Length 2 4/12—extent 4 feet—

GREAT WHITE HERON Bill yellow, Legs & feet claws Blk; Whole body White, Back feathers falling far over tail—Length 3 feet 6 to end of tail, 7 or 8 Inches Long to extremities of back feathers—

STORMY PETREL—

GREAT TERN, Bill & feet red, Black Cap[d] tail very forked, upper parts, Light Bluish Ash—Belly white

LESSER TERN, Bill & Leg Yellow, Blk Cap[d] tail very forked—

SHORT TAIL TERN, Bill & Legs Black, Cap & Cheek Blk, Tail Shorter than Wings—

BLACK SKIMMER—

SPOTTED SAND PIPER; Bartram, Sand Snipe, ring plover, Sanderling plover, Golden P. Kildeer P—

RED BREASTED SNIPE—Long Legged Avoset Bill Blk Red purpleish red, Eye red, whole upper part. Deep Olive, Solitary Sand piper, Yellow Shank Snipe—Tell tail D°

[HERONS]

TURN STONE, Bill Black Leg deep Orange—Breast Side of Neck and Spot under the Eye Black; Much of Brick Color on the Wings—belly White—4 Toes

ASH COLORED SANDPIPER Legs dull Yellow—4 Toes

PURE—Bill Black, Legs & feet D° 4 Toes

BLACK BELLIED PLOVER—4 Toes the Hind one small and very high

RED BREASTED SANDPIPER—Bill Short Vent White, Bell breast and under Neck Deep Rufous—

ESQUIMAUX CORLEW—Legs Greenish Blue—4 Toes

RED BACKED SNIPE—Bill Much Curved Blk belied Bend

SEMI PALMATED SNIPE—have this

MARBLED GODWIT—Bill Long rather inclined upwards—

LOUISIANNA HERON, Legs Yellow Bill blue—upper head Purpled, Crest White, Back Feathers very Long Light Buff—Tail, Wing & Back Deep Blue—

PIN OISTER CATCHER—

HOOPING CRANE—Black Tips—Bill yellow

LONG BILL CURLEW, Legs Bluish—

YELLOW CROWNDED HERON Bill dark Blue, Legs Yellow White Crested Very Long, throat and Head black with an Oval White Spot—

GREAT HERON, thigh Deep Rufous, under Crest Long Blk, upper White—Breast, & Back fathers Long & Loose

AMERICAN BITTER, Dirty Yellow Oakre, Zig Zaged with dark Brown—a Black Triangular Line runs from the Mouth to the Back of Neck

LEAST BITTERN—

WOOD IBIS GROS BEAK

SCARLET IBIS

FLAMINGO

WHITE IBIS

BLACK SURF DUCK Singular Bill—3 White Marks about the head—Legs Red—

BUFFEL HEADED DUCK—the Spirit—

CANADA GOOSE

TUFTED DUCK Bill Blue—Head Neck and breast Blk a Rufous bend on the Lower part of Neck

GOLDEN EYE DUCK

SHOVELLER DUCK

GOOSANDER, Head Changeable Green

PIN TAIL DUCK Sprigg Tail

BLUE WING[D] TEAL

SNOW GOOSE

PIED DUCK Legs & feet Light Ockre, Ring round the Neck Connected with Back, Back Tail Belly and Line over the Eye Black, primaries also Black, the Remainder White

RED BREASTED MERGANSER—have this

AMERICAN WIDGEON, Bald Pate

BLUE BILL—head Neck & Breast Dark Brown—Back Canvassed—Rump Tail & Tail Coverts Blk

HOODED OR CRESTED MERGANSER

LONG TAIL DUCK—Old Wife

SUMMER DUCK

GREEN WINGED TEAL

CANVAS BACK DUCK

RED HEADED DUCK—Head and neck half way—Bright Rufous—Lower part of Neck & Breast Blk. Back Canvassed—

MALARD

CADWALL DUCK

EIDER DUCK Male & femelle

the SMEW, the BLACK SPECTACLED GOOSANDER

RUDDY DUCK, Blue Bill, Legs Rosy—upper head Black, Neck & whole upper part Brick Color—Cheeks, Chin, & side of Head White—femelle Dark olive all upper Part—sides of Head Dusty Yellow—Breast transversally Barred with Brick Colored Lines—Belly & Vent dirty Yellow—

BRANT

SCOTTER DUCK Wholy Black, upper Mandible Yellow—Lower & Nails Blk—Protuberances Red—Legs Red

VELVET DUCK—have it Large Blk Sea Duck

HARLEQUIN DUCK a Narrow regular White ring around the Lower part of Neck—

A BLACK DUCK A DUSTY DUCK

MARSH TERN Black Cap[d] Wings, Back & Tail Light Ash or Blue—the first Very Long the Latter Slightly forked—Legs & palmate feet Lead—

SOOTY TERN—Bill Black, front and whole under part White, upper Blk tail Much forked, the Tips White edged inwardly with Blk

The 9th Volume I believe Contains—the Loon = Purple Galinulle = Coot = Darter = Black Headed Gull = Great Footed Hawk, &e =

AT HOME IN THE FELICIANAS

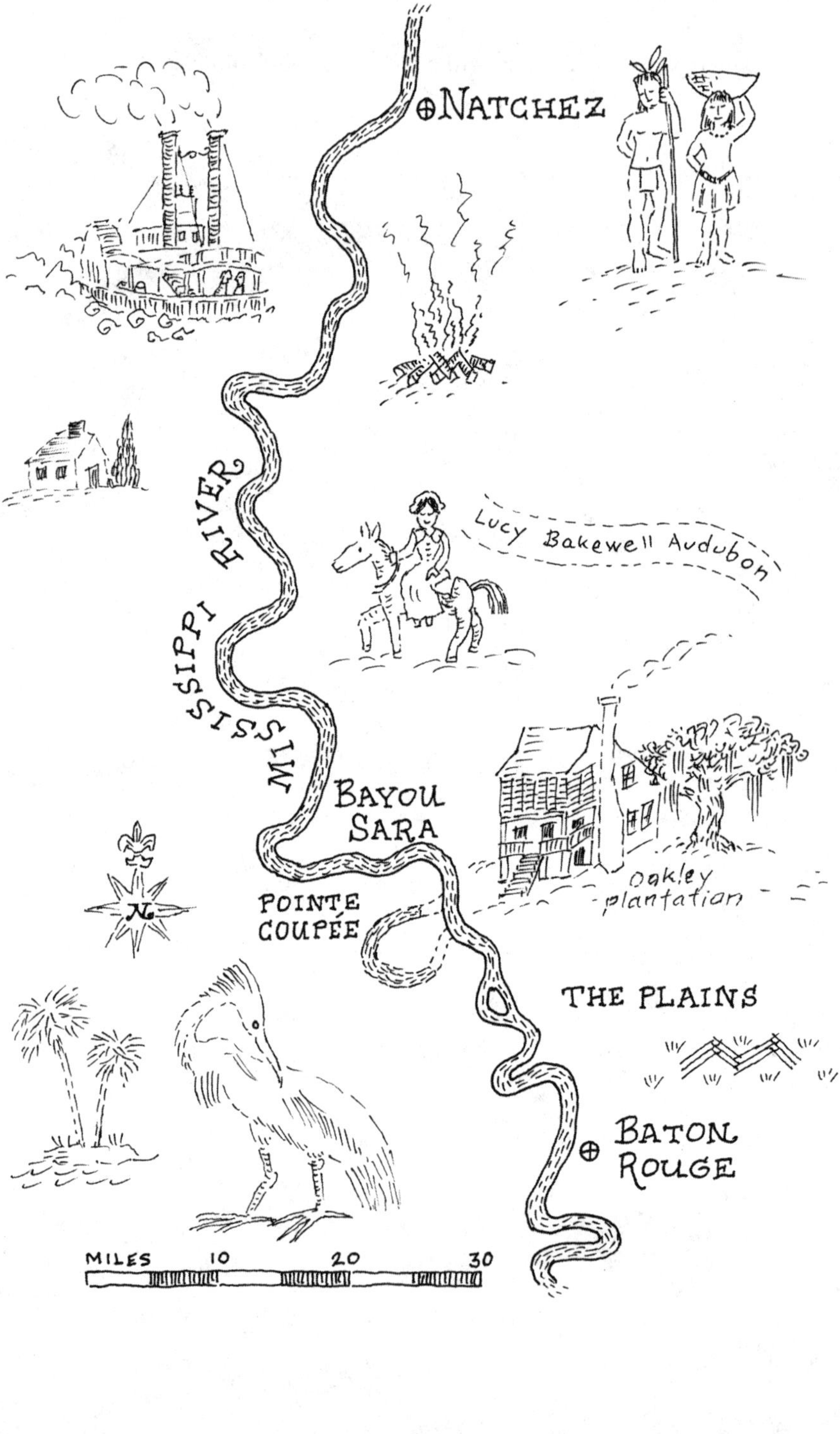

LOUISIANA EPISODES

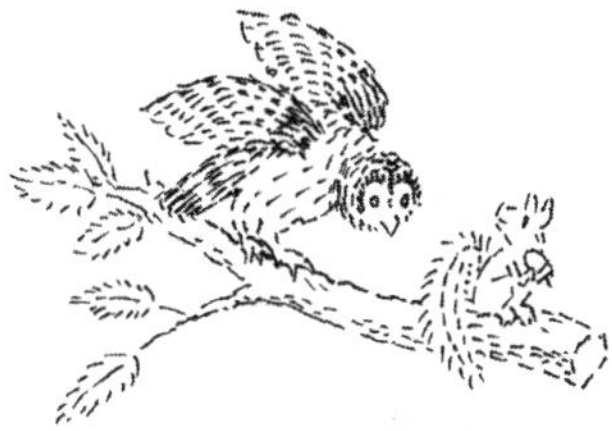

Early in the evening on May 17, 1826, Audubon left New Orleans on the ship *Delos*, bound for Liverpool with a full load of cotton and two other passengers. Already seasick, he watched the Mississippi banks move by slowly in the dusk as the steamboat *Hercules* tugged them out to the Gulf to set sail. On his way to England. determined to find an engraver and launch *Birds of America*, he had no idea how long he would be gone. The voyage turned out to last three years, three long years away from Lucy and his two sons. By the time he returned, Audubon had seen his efforts crowned with glory, popular and professional. His birds had been exhibited, written about, and praised to the stars by some of the leading men of science in England. He had enrolled his first subscribers, had convinced one good engraver, then another, to take on the project, and he had begun printing the spectacular, oversized book serially in sections of five folio plates each. But when he stepped onboard the *Delos* he did so with much the same dread and doubt that had clouded his departure down the Ohio and the Mississippi in 1820. The difference this time was the impact of lower Louisiana on his art, and the high cards he now held to make good his new wager.

By May of 1826 Audubon had spent well over four years in Louisiana. On the *Delos*, when the weather allowed, he slept on deck with a "hard-pressed bale of cotton" as a bed, but he also carried in his luggage a Louisiana parting gift of incalculable value. The tin-lined wooden box that served as a traveling case for his drawings contained more than half the watercolors that would make up the total number of 435 plates when the *Birds of America* was completed in 1838. Most of these had been painted in Louisiana. Audubon had worked ceaselessly, increasing his knowledge of Louisiana bird lore and perfecting his art. He had spent most of his time outdoors. He was already an accomplished woodsman when he arrived, having developed his skills for twelve years on the Kentucky frontier. In Louisiana those skills deepened into a rare expertise. No man alive knew the Louisiana swamps and woodlands better than Audubon. And no man alive had ever painted the natural life of Louisiana birds with more brilliance and authority. He had made himself over as an artist, and he had taken full possession of the Louisiana birthright he had desired in his youth and often had shamelessly claimed. The power of the willing exile and convert comes from being able to choose an identity that is imposed by blind fate on the often indifferent native-born.

Bored on the *Delos,* with plenty of time on his hands, Audubon drew sketches of the crew, helped with the fishing, read the books on board at least twice, wrote copiously in his journal, and dreamed about Louisiana. The memories were not always magnolias and birdsong. There had been staggering hardships and bitter deceptions. There had been the time spent, and lost, in New Orleans, and much indecisive drifting back and forth between the Felicianas and Natchez sleeping in rented rooms, with no real base of operations. When Lucy opened her school at Beech Woods early in 1823, Audubon's status as an occasional tolerated guest depended entirely on the mercurial mood of Jane Percy. She exercised her powers as mistress of the plantation with little patience for the wayward artist with a mind of his own. The sense of losing ground, even as the stacks of masterpieces were mounting, led Audubon to leave Lucy again, in October 1823, this time to search for professional support and a publisher in Philadelphia. Over the years, the study of natural sciences in the United States had flourished in Philadelphia, thanks largely to the seeds planted and nurtured by the Bartrams, father and son. On the way to Philadelphia, Audubon left his elder son Victor with his Berthoud in-laws in Kentucky to train as an apprentice under Nicholas Berthoud. Audubon remained in Louisville over the winter, painting signs and steamboat murals to pay his way.

Philadelphia turned out to be a disaster. Arriving in April 1824 after five months in Louisville, Audubon expected at least a fair showing of his birds. Philadelphia was the home of the American Philosophical Society, founded by one of Audubon's heroes, Benjamin Franklin, and the Academy of Natural Sciences, whose members included William Bartram (who had died the year before) and the shining lights of the new generation: Titian Peale and Thomas Say. Unfortunately, the vice president of the Academy was George Ord, a self-taught ornithologist who had taken on the task of completing Alexander Wilson's *American Ornithology* when Wilson, himself an early member of the Academy, had died in Philadelphia in 1813. Ord considered himself the keeper of Wilson's flame and fame, and jealously attacked Audubon's paintings and credentials. Without Ord's approval, Audubon lost all chance of membership in the Academy or success in Philadelphia. Ord's venom did not prevent other artists and naturalists in the city from individually taking sides with Audubon. Three of them became lasting allies: the well-known portrait painter Thomas Sully, the physician and zoologist Richard Harlan, and the budding ornithologist Charles Lucien Bonaparte, nephew of Napoleon, who was writing his own *American Ornithology* to include birds that Wilson had missed. Audubon also met Edward Harris, a generous gentleman farmer from New Jersey. Harris helped him financially and became a staunch lifelong friend.

Audubon seems to have absorbed the shock of the Philadelphia rejection with unusual aplomb. He may have realized that Ord's hostility was a perverse form of praise. At any rate, the rupture with Ord soon proved a blessing in disguise. It turned Audubon's thoughts with a vengeance toward England as the place to go to bring his birds into print and the public eye. England had long been fascinated with the exotic flora and fauna of the American frontier, and England abounded in skilled engravers. Closer to home, the Philadelphia misadventure brought Audubon back to Louisiana with a renewed strength of purpose, but not right away. Someone following in his footsteps after the four months in Philadelphia would have been puzzled about the pathway of his return. Instead of going south, he went north, first to New York to test his chances in another urban center of universities and scientific research. He was well received by Samuel Mitchill, a highly respected doctor, botanist, and genial eccentric (he once proposed that the United States be renamed Fredonia). Audubon had first met Mitchill during the winter of 1806–1807 when he was living in New York trying to learn a trade under the supervision of Lucy's uncle, Benjamin Bakewell, owner of a large import/export business. The polymathic Mitchill had hired Audubon to help him mount specimens for his collection of fish, birds, and mammals, and allowed him the benefit of his wide-ranging conversation and his considerable library. Mitchill had since become the president of the Lyceum of Natural History, which he had founded in 1817 as a deliberately more democratic open-door version of the American Philosophical Society. Presenting his portfolio of birds to his former mentor almost twenty years after their first meeting, Audubon and his work were duly praised; he was made a member of the Lyceum, and he wrote a brief account of the cliff swallow, accepted for publication that same year (1824) in the Lyceum's annals.

The reception in New York was gratifying, but not enough to ignite any new hope for the portfolio. Once more Audubon turned north, zigzagging into colder climes, somewhat like a solitary Romantic hero fleeing the complications of society. For Audubon this was not the case. He was simply biding his time until he was ready to return to Lucy empty-handed. Audubon had a genius for biding his time while getting things done. He headed to Albany to meet Governor DeWitt Clinton, a good friend of Samuel Mitchill, but the governor was out of town. He had always wanted to see Niagara Falls, so he next went to Buffalo, a worn-out blanket for a mattress strapped to his back. He was so impressed by the majestic spectacle of the Falls that he decided a drawing could never do it justice. When he later wrote one of his episodes about the experience (*Niagara*), he advised his readers to see it for themselves: "go to Niagara, reader, for all the pictures you may see, all the descriptions you may read of these mighty Falls, can only produce in your mind the faint glimmer of a glowworm compared with the overpowering glory of the meridian sun." Leaving Niagara behind, he continued on to Erie, Pennsylvania, via Lake Erie by ship. He then made his way to Pittsburgh, where he stayed six weeks giving lessons and painting portraits. After a difficult trip on the Ohio, battered by the elements, carried along in turn by skiff, keelboat, and then steamboat, he arrived in Louisville, made sure his son was doing well, and booked passage on another steamboat, finally heading home to Lucy and Louisiana. On November 24, 1824, he landed in Bayou Sara. He had been away for fourteen months.

It is tempting to look at Audubon's return as a mocking repetition of his momentous plunge into lower Louisiana in 1820. The separation with Lucy before she joined him in New Orleans then had lasted fourteen months as well. His pockets once more were almost empty. The difference was that this time he was coming back to a Louisiana he now knew and loved, and he was coming back to Lucy, who had settled in and had given the Audubon family a semipermanent Louisiana home. There were other differences too. Audubon had not found an engraver or publisher in Philadelphia, thanks to the hard-nosed Ord, but his portfolio had been seen and praised by his peers. Philadelphia now knew who he was. And Audubon himself now knew his birds were bound to prevail. All he had to do was to continue to bide his time and keep working.

The work was made all that easier by the improved circumstances that welcomed him back. Jane Percy, wholly won over by Lucy's success with the school, put aside her misgivings about Audubon and invited him to move into Lucy's cottage. For the next year and a half he and Lucy, and their younger son John, lived together in a nest of nearly ideal conditions. For Audubon, Beech Woods joined Oakley as the Louisiana that would become fixed in his memory as a celestial map of everything he held dear. In the lushness of the Felicianas he and Lucy could once more indulge, side by side, their mutual love of the natural world. They swam together in the evening; they took rides in the woods (Lucy was as good with horses as Audubon); they read their favorite books out loud to each other; they talked about the day's work; they slept in the same bed. Audubon resumed his search for Louisiana birds, and continued to produce masterpiece after masterpiece: the wild turkey, the barred owl, the bobwhites attacked by the red-shouldered hawk, the pileated woodpecker, and many others.

Audubon was also fortunate in his friends. Somewhat miraculously, his former young clerk in the general store in Henderson, Nat Pope, had in the meantime studied medicine, had become a doctor, and had married and moved to St. Francisville. He and Audubon renewed their friendship and made the Felicianas their personal stamping ground for hunting expeditions. A local Creole planter, Augustin Bourgeat, the most skilled alligator hunter in the area, often joined them. Under Audubon's expert guidance, both Pope and Bourgeat refined their knowledge of birds, learning more than enough to help spot and track down the ones he needed next: the more keen eyes the better.

When Audubon arrives in England the following year he often looks back to the recent months in Beech Woods and notes in his journal his depression at having left all those carefree outings behind. The spells of homesickness sting the most when he compares the polite cultural rituals of the English gentry, the formal dining for example, with the relaxed makeshift pleasures of the Louisiana woods: "The sumptuous dinners of this country are quite too much for me. This is not with friend Bourgeat on the Flat Lake roasting the orange fleshed Ibis, and a few Sun Perch. Neither is it on the heated Banks of Thompson Creek on the 4th of July swallowing quickly the roasted eggs of a large soft shell Turtle."

When Audubon stepped off the *Delos* on July 21, 1826, and walked into Liverpool after two months at sea, it did not take him long to make the waves of Louisiana nostalgia work to his advantage. The first two days in the big port did not unroll auspiciously. There was a steady drizzle. He choked on the thick smoke of the city's coal fires. The midsummer English cold made his teeth chatter. When he woke up the first morning at the hotel he heard a lark sing, but the lark was in a cage. He learned he had to pay duty on his watercolors. Within a week, however, the tide had turned. He met the influential Rathbone family, who were quickly seduced by both his birds and his unaffected charm. Through the Rathbones he met William Roscoe, founder of the Liverpool Botanic Garden. The Rathbones and Roscoe introduced Audubon to everyone who counted in Liverpool's scientific community and social register. He was suddenly overwhelmed with invitations and dined out almost every night. He exhibited his watercolors in the Liverpool Royal Institution. He signed up subscribers to the book to be, though he had not yet found a publisher. Within two months his birds had been shown and applauded in Liverpool and Manchester, and he was on his way to Edinburgh.

From his first days in Liverpool Audubon had announced himself as a Louisiana citizen (with its implication of native origins). Just before leaving New Orleans he had secured a paper from the governor of Louisiana, Henry Johnson, stamped with the seal of the state to serve as passport. This appealed to the superstitious soul in Audubon given his early history as an exile in need of a biography that would not raise too many eyebrows. When he had sailed for America from France in 1803, his father had handed him a passport with Louisiana marked as his place of birth. More to Audubon's purpose in his new venture were the birds he had painted in the Felicianas. They confirmed his kinship with Louisiana and were always the ones he insisted on showing first. The display would begin with the dazzling wild turkey he had killed on a hunting trip with Jane Percy's son Robert near Beech Woods in February 1825. He had spent twenty-three hours during three days painting it, and his skill with colors and composition in the great bird looking back through the cane stalks was obvious in the whole, and in every square inch from the retracted conical snood to each orange dusted tail feather tip.

The triumph of the Feliciana turkey followed Audubon from Liverpool to Manchester, and finally to Edinburgh where the breakthrough towards the making of *Birds of America* was not long in coming. He arrived in Edinburgh on October 25, 1826, and within a week had met with Robert Jameson, one of the leaders of the city's elite scientific circle. Jameson was the founder of the Wernerian Natural History Society, and editor of the *Edinburgh New Philosophical Journal.* Through Jameson's influence, Audubon's progress proceeded at whirlwind pace. Just before meeting Jameson, Audubon had shown the engraver William Lizars several of his drawings, and within two weeks Lizars had agreed to engrave and publish the first plates of *Birds of America*. Not surprisingly, Lizars had been stunned by the wild turkey, and by the mockingbird, the whooping crane, the covey of bobwhites attacked by the hawk, and, most of all, the brace of peregrine falcons ("with bloody rage at their beaks' ends, and cruel delight in the glance of their daring eyes," as Audubon later wrote): all Louisiana compositions.

By this time the turkey had become a talisman for Audubon. Early in November he received in the mail the gift of a gold seal (impressed with his own image of the turkey) from the elder Madame Hannah Rathbone, matriarch of the Liverpool Rathbones. Audubon affectionately baptized it by sealing a letter to Lucy with hot red wax. By then he had decided to print up calling cards stamped with the turkey to be handed out as bait for potential subscribers. On November 19 he finished a large oil painting he called *Wild Turkey Cock, Hen, and Young* to be given to the Edinburgh Royal Institute. Audubon had been deeply touched by the Institute's invitation to hold an exhibition, and by the gracious reception of the members, in total contrast to the dismissive glares of Ord and Company in Philadelphia.

On November 20 Audubon wrote a description of the turkey painting and read it to John Wilson of *Blackwood's Edinburgh Magazine*, one of England's foremost literary journals. Audubon by now had decided on publishing, in a separate volume, or volumes, a series of bird biographies (rivaling those in Alexander Wilson's *American Ornithology*) to provide background on the paintings, and began gathering his notes in what he called his letterpress. He had already written a lengthy rough account of the wild turkey in 1824 for Charles Bonaparte, who included sections of it in the first volume (1825) of his *American Ornithology: The Natural History of Birds Inhabiting the United States, Not Given by Wilson*. Bonaparte expressed his gratitude with a handsome (though to Audubon, insufficient) tribute: "We have particular satisfaction in acknowledging the kindness of Mr. John James Audubon, from whom we have received a copious narrative, containing a considerable portion of the valuable notes collected by him, on this bird, during twenty years that he has been engaged in studying Ornithology, in the only book free from error and contradiction, the great book of nature."

Thus, with the Louisiana wild turkey cock, whose favorite food, according to Audubon, was the pecan nut, the idea of the *Ornithological Biography* was born. John Wilson kept the Edinburgh turkey text to suggest revisions, and from then on made sure Audubon's name was featured in *Blackwood's* columns. Two months later, under his pseudonym Christopher North, Wilson published a review of the

Royal Institute exhibition, saluting Audubon with the highest praise he had yet received in a major journal: "Birds painted to the life. Almost the whole American Ornithology, true to nature, as if the creatures were in their native haunts in the forests, or on the sea-shores. Not stiff and staring like stuffed specimens—but in every imaginable characteristic attitude, perched, wading, or a-wing—not a feather, smooth or ruffled, out of its place—every song, chirp, chatter, or cry, made audible by the power of genius." These were strong words by the fiery champion of Wordsworth, Coleridge, and Shelley, among others.

With his portfolio the center of attention in one of the great world capitals of natural science, and the *Birds of America* now in the process of becoming a palpable reality, Audubon was pressed by his new friends in Edinburgh to write his own history of the birds, and other creatures of the southern wilds if he so desired. He accepted reluctantly at first, but soon warmed to the challenge. As he realized, he had little formal scientific training, and even less experience writing for publication in English, or in French for that matter. On the other hand, he had carefully studied the classic ornithological literature, and knew Wilson's nine volumes by heart. And he trusted his own firsthand observations in the field above those of anyone else, alive or dead.

There were two problems facing Audubon when he began to write his essays (all based on the Louisiana years) in Edinburgh. The first concerned the nature of his audience and their expectations. The papers were to be read (before publication) at meetings of the Wernerian Society, each member an eminent scholar in his field. Scientific accuracy was necessarily a prime consideration. On the other hand, these same scholars enjoyed probing Audubon to expand on the wilder side of his life beyond the edges of civilization. Earlier on, in Liverpool, he had been encouraged to entertain dinner parties with his mimicry of frontier sounds: "I was asked to imitate the wild turkey call, and I did so to the surprise of the whole circle, hooted like the barred owl, and cooed like the doves—I am glad that I was not desired to bray!"

As the last remark suggests, Audubon felt his powers of live performance might be overplayed and would distract attention from his art. Still, he was all too willing to oblige. He knew this was part of the game; a dinner guest should be able to return the favor of an invitation. Even in Edinburgh, thronging with high-minded universities and high-collared research institutes, one of the dignitaries had suddenly stood up at a banquet and imitated a buzzing bee chasing him from his seat. Audubon was amused, no doubt secretly pleased that he could bark like an alligator if called upon to do so.

Audubon was a naturally gifted teller of tales, and he loved to please when the company was primed to listen with willing ears. To work its will, and to hold its audience, a well-made story is bound by its own sovereign decrees. Historical or scientific fact must be twisted and tangled at times for narrative truth and coherence to shine forth. When Audubon was asked to prepare his papers for the Wernerians, he was not always sure of how best to reconcile science and storytelling, and the confusion shows up in these early essays. But he was learning, and would not make the same mistake with the *Ornithological Biography* four years later.

The second problem was time. Along with his Royal Institute exhibition, his painting in oil (he was learning the new medium by trial and error), his supervision of the engravings, his correspondence, his courting and then hobnobbing with the Edinburgh illuminati, and his almost daily dinner invitations, he was now put to the task of writing scientific papers with deadlines. On December 6 he had eagerly proposed an account of the turkey buzzard to Professor Jameson. Jameson accepted, and later the same day Audubon learned that he was expected to turn it on the 7th. He rushed home to "scribble" (as he called it) and managed to finish just on time. In Edinburgh he rarely went to sleep before the early hours in the morning.

During the next two months Audubon wrote four more substantial papers: on the carrion crow, the alligator, the wild pigeon, and the rattlesnake. All of them were worthy efforts, filled with informed detail and personal observation. They were all read at Wernerian meetings before being published by the *Edinburgh New Philosophical Journal* or the *Edinburgh Journal of Science.* Audubon gave himself over to these commitments with customary gusto and self-absorption. Bent over his desk writing about wild pigeons, he claimed he could hear the "noise of their wings." When we consider that Audubon was writing from memory, with no documents other than his journal notes to rely on, his accomplishment was close to miraculous. Unfortunately, in the account of the rattlesnake, some of the content as he hurried to finish was as miraculous as the performance. Trapped by a promise, and improvising as if he were sitting by a campfire, he threw together a patchwork of fact and fable, stitching in a few of the tall tales he had heard up and down the Mississippi. One common myth Audubon repeated concerned the rattlesnake's ability to hypnotize its victims. Audubon may have felt authorized in making his claim by having read the same theory in Bartram's *Travels.*

At any rate, the rattlesnake paper relit the feud with George Ord. Ord joined forces with his friend Charles Waterton, an English naturalist and travel writer who had met Audubon in Philadelphia in 1824, and had followed Ord's lead in disparaging Audubon's birds. Audubon never forgot. Waterton's popular *Wanderings in South America* was published in 1825. In Edinburgh, when the subject was raised, Audubon was quick to scoff at Waterton's boasting of his exploits, such as riding an alligator. Charles Darwin, a young medical student in Edinburgh at the time, heard Audubon read several of his papers at the Wernerian Society and remembered him taking direct aim at Waterton's book. Waterton learned from friends in Edinburgh of Aububon's jesting at his expense, and he and Ord launched a public campaign against Audubon's credibility, citing both the rattlesnake paper and the one on the turkey buzzard. Audubon should have let the matter die in the bud. Instead, he defended himself, soliciting testimony from friends, and the controversy burned on for years.

One of its main effects was felt when Audubon's painting of the Louisiana mockingbirds under attack by a fierce rattlesnake coiled in a swirl of yellow jasmine was engraved for the *Birds of America* (plate 21). Ord and Waterton duly took up their poison pens, and once more Audubon was forced to waste precious time countering their criticism. A better tactic would have been to let the painting speak for itself, as it did then and does today. Rattlesnakes may or may not climb trees. Actually they can, though they rarely do so. A rat snake, much the same size and color, certainly climbs to kill, and does so often. What does it matter? What counts is the life of the composition, tensely focused on the male mockingbird's beak poised to pierce the bulging eye of the snake. The truth of the painting lies in its dramatic power to compel, like the grip of a story we can never forget.

Of Audubon's writings on Louisiana that were meant to be published, his various papers for the Wernerian Society in late 1826 and early 1827 were the first to see print. Of these, the account of the alligator seems to have made the strongest impression. As was the custom, Audubon presented it as a letter to a member, or members, of the Wernerians. In this case, he addressed it to William Jardine and Prideaux John Selby, both prominent figures among the Edinburgh naturalists. Jardine was vice president of the Wernerian Society and an authority on Alexander Wilson. He happened to be, as well, the brother-in-law of Audubon's first engraver, William Lizars. He and Lizars later joined forces to publish the forty-volume series, the *Naturalist's Library,* a chief source in the mid-nineteenth century for British interest in the natural sciences. Selby was an ornithologist and artist who was in the process of finishing his *Illustrations of British Ornithology* and had just embarked on a major collaboration with Jardine, *Illustrations of Ornithology.* Despite their lofty standing in Edinburgh, both Jardine and Selby were honest enough to recognize Audubon's superiority as an artist. They both took drawing lessons from him and became two of his most faithful friends in Edinburgh.

It is easy to understand the appeal of the alligator essay. Audubon needed to go back only a year to draw on his hunting expeditions with his Creole friend Augustin Bourgeat in the Felicianas. The memories were fresh in his mind, bringing the events closer to the present and to his natural speaking voice. All he needed to do was dip into the well of his recent experience and tell the story in language that was immediate, direct, and physical. There was no last-minute panic to cast around for secondhand accounts (as there would be with the rushed rattlesnake paper a month later). He knew what the audience expected, given the topic. The Wernerians did not want nature at peace with itself, Shelley skylarks or Keats nightingales, or moonbeams dancing on the wild hyacinths. They wanted to be informed, as men of science, but they also wanted to be shocked out of their seats by the unspeakable terrors of the southern swamp.

Audubon delivered on all fronts, with a master's confidence in the wealth of details at his fingertips and the tip of his tongue. He began generally, downplaying the reputation of the alligator as an aggressive or even very dangerous animal, and then plunged suddenly into a vision of particular Louisiana swamps, seething with the fury of the survival of the fittest, all tooth and claw and beak, feeding, fighting, killing, and being killed. In a few sentences he listed a dozen creatures, all engaged in a Hobbesian (or Darwinian) struggle, including an old grey squirrel's castration of a younger rival. The castration may or may not have been part imagination, understandable in the murky gloom of cypress trees and Spanish moss. The rest was verifiable naturalistic fact. Here was the dark underside of all God's creation, doomed to destroy in order to live.

Once the scene was set, Audubon shifted the focus to himself, portraying himself blithely wading waist-deep in muddy water among the half-hidden prehistoric snouts, watched by a thousand eyes. Since he was the center of his tale, and since he was now standing in front of the Wernerians, alive and well, he could easily afford not to parade his courage, matter-of-factly implying that the daredevil risks he had faced were simply part of the job. He had come through fearsome dangers, obvious to everyone in the room, with a maximum of self-restraint, motivated wholly by the lure of the hunt, scientific curiosity, and the love of knowledge (dear to the assembled naturalists listening to a fellow searcher). Heroics were the realm of the trumpet-blowing self-promoters, like Waterton.

The reading of the alligator paper took place on January 13, 1827. After it was published two months later in the *Edinburgh New Philosophical Journal*, it did not take much time for Audubon's exploits and observations to make their way out of the exclusive halls of the Wernerians and strike a chord in the public imagination and the popular press, first in Edinburgh and then all over England. One widely read British weekly, the *Mirror of Literature, Amusement, and Instruction*, was particularly struck by Audubon's story of finding stone-like objects in alligator stomachs. The *Mirror* introduced him as "Mr Audubon of Louisiana" and cited him at length in an issue early in 1830, not long before he began publishing his other Louisiana essays. Audubon's theory that the stones were related to the alligator's digestion had concluded his paper with a bold stroke that must have had the Wernerians leaning forward with all ears when they first heard it read. Audubon the ornithologist, thinking of gizzards and grit, was not the first naturalist to remark on the stones in alligators. His inspired speculation on their function was not new either, but remarkable in having been made without any knowledge of the rare documents on the subject. The stones are now called gastroliths and are thought (by some) to provide ballast allowing the alligator to float with only the head above water. There is no clear consensus, however. Audubon was well ahead of his time, both in his perceptions in the field and in his theorizing.

Audubon's later essays, or episodes, on Louisiana were written for the *Ornithological Biography*, begun in 1831 in Edinburgh. After three years in England, Audubon returned to the United States in May 1829, having been seasick, as usual, during the entire voyage. He carried with him the first fifty prints of *Birds of America*, bound in a massive volume. In England, the base of operations for the engraving of *Birds of America* had suddenly shifted, during the summer of 1827, from Edinburgh to London. The workers of the Edinburgh engraver, William Lizars, had gone on strike after having printed only ten plates. Audubon was forced to search for a replacement, and fortunately found the Robert Havells in London, father and son, far superior to Lizars. The Havells (notably the son) took on the *Birds of America* and continued until it was completed in 1838, acclaimed ever since as one of the finest works of aquatint in the history of printmaking. Leaving his birds in good hands when he left England, Audubon's immediate future was fixed: he would travel, search, and paint birds all over the United States until the great work was finished, signing up subscribers along the way.

Though he arrived back in the United States in May 1829, he did not return to Lucy, and to Louisiana, until the middle of November. His excuse was the need to search and paint. The truth of the matter was less professional. After three years of separation, he and Lucy were hesitant to confront, face to face, a resumption of their life together. The slowness and unreliability of the transatlantic mail had not helped. There had been a relentless ebb and flow of delays, crossed signals, misunderstandings, recriminations, conjecture in solitude: all of which raised doubts and fears. In 1827 Lucy had moved her school from the Beech Woods plantation after a disagreement with Jane Percy, and had settled a few miles away at another plantation, Beech Grove, owned by William Garret Johnson. There, from May to November, she waited for Audubon to make a decision. More letters, none conclusive, were exchanged. Finally, as was always the case with Audubon, and with Lucy, the desire for reconciliation prevailed, leading to yet another passionate reunion. During the night of November 17, Audubon stepped off the steamboat at Bayou Sara fearing the worse, with an epidemic of yellow fever raging up and down the river. He borrowed a horse, and later wrote in his journal:

> It was so dark that I soon lost my way, but I cared not, I was about to rejoin my wife, I was in the woods, the woods of Louisiana, my heart was bursting with joy! The first glimpse of dawn set me on my road, at six o'clock I was at Mr. Johnson's house; a servant took the horse, I went at once to my wife's apartment; her door was ajar, already she was dressed and sitting by her piano, on which a young lady was playing. I pronounced her name gently, she saw me, and the next moment I held her in my arms. Her emotion was so great I feared I had acted rashly, but tears relieved our hearts, once more we were together.

The next day he wrote a letter to his good friend in Philadelphia, Richard Harlan. Reunited with Lucy, back in Louisiana (*Magnificent Louisiana!*), his mood was exultant. He was convinced that all he had to do was to continue what he had begun, keep *Birds of America* on track with new birds and paintings, return to England as soon as possible to begin the *Ornithological Biography*, do everything in his power to sell subscriptions, and maintain a steady rhythm of engraving. Both his sons were safely employed with Lucy's relatives in Kentucky, and Lucy herself was delighted with the prospect of helping her husband, by his side, in her native country. In his biography of Audubon, *Audubon the Naturalist* (1917, 1938), Frances Hobart Herrick transcribed the letter in full:

Rich[d] Harlan Esq[r]. M.D. &c &c &c
Philadelphia Pens[a]

St Francisville Louisiana Novemb[r] 18th 1829—

My Dear Friend.—

You will see by the data of this the rapidity with which I have crossed *two thirds* of the United States. I had the happiness of pressing my beloved wife to my breast Yesterday morning; saw my two sons at Louisville and all is well.—from Philadelphia to Pittsburgh I found the Roads, the Coaches, horses Drivers and Inns all much improved and yet needing a great deal to make the traveller *quite comfortable*—The slownesse of the stages is yet a great bore to a man in a hurry—I remained part of a day at Pittsburgh where of course I paid my respects to the Museum! I was glad to see the germ of one—it is conducted by a very young man named Lambdin—I made an arrangement with him [place of seal—paper gone] &[c] &[c] &[c] at Cincinnati I also visited the Museum [paper gone] it scarsely improves since my last view of it, except indeed by wax figures and such other shows as are best suitable to make money and the least so to improve the mind.—I could not see D [illegible] my time was very limited.—The Ohio was in good order for Navigation and I reached Louisville distant from you about 1,000 Miles in one week.=as you spoke of travelling westwardly I give you here an a/c of the Fare.—to Pittsburgh all included 21$.—to Louisville 12$.—and 25$ more to Bayou Sarah where I Landed. 30$ is the price from Louisville to N. Orleans.=our Steam Boats are commodious and go well—but my Dear Friend the most extraordinary change has taken place in appearance as I have proceeded.—The foliage had nearly left the Trees in Pennsylvania, the Swallows had long since disappeared severe frost indeed had rendered Nature gloomy and uninteresting—Judge of the contrast: I am now surrounded by *Green Trees* and Swallows gambole around the house as in Pennsylvania during June & July=The mock bird is heard to sing and during a

Walk with my Wife yesterday I collected some 20 or 30 Insects=that is not all, a friend of mine here *says* that he has discovered 2 or 3 *New Birds!!!*—new Birds are new birds our days, and I shall endeavor to shew you the Facts Simile when again I shall have the pleasure of shaking your hand—

although so lately arrived, I have established the fact that Mrs A. and myself will be on our way towards "Old England" by the 15th of Jan.$^{y.}$ we will ascend the Mississipi and after resting ourselves at Louisville with our sons and other relatives about one month and then proceed with the Rapidity of the Wild Pigeon should God grant our wishes!—

have you seen or heard any thing of Ward?—have you the little sketch of Dear?= we had a passenger on Board the Huntress named *Potts* from your City who knows you well a lively young Gentleman; has a Brother (a Clergyman) established and married at Natchez.—

I will begin Drawing next week having much scratching with the Pen to perform this one, and I am also desirous to make [paper gone] Large Shipment of aborigines both animal and vegetal as soon as possible.—Turkeys, Aligators, Oppossums, Paroekett, and plants, as Bignonias &^{c} &^{c} &$^{c.}$ will be removed to the Zoological Gardens of London, from the Natural ones of this Magnificent Louisiana!—meantimes I will not forget my *Friends in Phil*$^{a.}$ No I would rather forgive all, to all my Ennemies there.—assure Dr Hammersley that Ivory Billed and Peleated Woodpeckers will be skinned, and who knows but I may find something more for him.—I will give free leave to Dr. Pickkering to chuse amongst the Insects and who knows but I may find something new for him. Remember me most kindly to both, nay not in the common manner of saying "Mr Audubon begs to be remembered" no not [at] all. This way M^{r} A remembers *you* and *you* and *I* will remember *you* and *you* and *I* always!!—

May I also beg to be remembered in humble words to a fine pair of Eyes; divided, not by the Allegany Mountains; but by a nose evidently imported from far *East,* to a placid forehead, to a mouth speaking happiness to—[dash nearly across page.]

Should you see Friend Sully remember me to him also—and should you see George Ord Esqr *Fellow* of all the Societies *Imaginable* present him my most humble—[dash line across the page]

Should you see that good woman where I boarded at Camd'en tell her that I am well and thankful to her for her attentions to me.—

I cannot hope the pleasure of an answer from you here but you may do so, and I say pray do so, directed to the care of N. Berthoud Esq[r] Louisville Kentucky.—by the bye my sons are taller than me, the eldest one so much altered that I did not know him at first sight, and yet I have *Eyes*—

God bless You, Your Friend
John J. Audubon.

[The following is written across the first page:]

I reopened my letter to say that I have Just *now* killed a Large *New Falcon yes positively a new Species of Hawk almost black about* 25 Inches Long and 4 feet broad tail square Eyye yellowish White, Legs and Feet bare short & strong.—I will skin it!!!—

Remember me to Lehmman—

What I have said about the Hawk to You must be *Lawful* to *Academicians* and you will please announce *Falco Harlanii* by

John J. Audubon
F.L.S.L.

It took two months for Audubon and Lucy to set their affairs in order at Beech Grove, and set sail for England. On April 1, after visiting their sons and relatives in Kentucky, they left on the oddly but reassuringly named ship, the *Pacific,* to cross the Atlantic. They were not aware of it at the time, but they were leaving behind Louisiana for good. Audubon made one more trip along the Gulf coast in 1837, but for the remainder of his life Louisiana reigned alone in his memory, and in Lucy's, as the charmed place of struggles overcome, a world of natural beauty and promise that had regenerated his ambitions, had provided a measure of stability and respect thanks to Lucy's school, had reknitted the family in a common purpose, and had generally worked its wizardry to make the *Birds of America* a reality.

In England, Lucy visited with her sister Anne Gordon in Liverpool while Audubon was in London to discuss the progress of the engravings with the Havells.

After another three months of traveling together, rounding up new subscribers and confirming the old, Audubon and Lucy arrived in Edinburgh where he had decided to begin work on the *Ornithological Biography.* Edinburgh was his favorite city in Great Britain, now sanctified as the birthplace of the *Birds of America* as a physical book. Surrounded by friends, particularly the members of the Wernerian Society, Audubon knew he could rely on the help and encouragement of some of the leading naturalists in England. There were also the familiar resources of the University library and Natural History Museum. Audubon and Lucy settled in his old lodgings with Mrs Dickie at Number 6, George Street, in October 1830. He hired a young naturalist, William MacGillivray, to help him prepare the text of the bird biographies. MacGillivray was a member of the Wernerians, a former assistant to the founder, Robert Jameson, and until recently the supervisor of the Natural History Museum. He was the perfect choice for the job. MacGillivray would be responsible for the scientific descriptions at the end of each biography, and would help Audubon smooth out the "asperities" in the main body of the text.

Along with the individual bird biographies, aimed above all at the scientific community, Audubon decided to include a number of personal essays interspersed as *episodes* scattered at random in each volume. These would give him free rein to write down his tales of life on the frontier without being held to account by purely factual constraints. He referred to the episodes collectively as *delineations of American scenery and character.* As the *Ornithological Biography* progressed, from volume one in 1831 to the final volume five in 1839 (a year after the fourth and last volume of *Birds of America* was published), he wrote sixty episodes, all included in the first three volumes. They drew on the whole history of his travels and experiences, reaching years back to the early days in Kentucky and ranging up to the present. Every expedition he undertook over the next few years to make sure *Birds of America* would be definitive, from South Carolina to Florida, and north to Maine and Newfoundland, brought new vistas and escapades to be sifted, seasoned, and baked into another good story.

In writing the episodes, digging deep in the journals and notes he had kept for over twenty years, Audubon shifted back and forth between the frontier sketch, the informal essay, and straightforward autobiography. Though he somehow combined these various genres into original hybrids of his own invention, he was writing within clearly established traditions. He was of course steeped in the literature of natural history, and studied the texts of Bartram and Wilson like schoolbooks. Travel memoirs of all kinds, perennially popular, were staple features of early nineteenth-century magazines, newspapers, almanacs, even practical river guides such as the one by Zadok Cramer that Audubon read on the keelboat down the Ohio and Mississippi in 1820. Audubon seems to have read, and absorbed, everything that fell into his hands. His journals are filled with references to poems and novels, and to the writers he admired: Byron, Scott, Edgeworth, Thomson, Smollett, Sterne, La Fontaine. One of these was Washington Irving, widely celebrated for such stories as *Rip Van Winkle* and *The Legend of Sleepy Hollow*, and considered by the time Audubon began publishing his own efforts the founder of the new short story and the living father of an independent American literature. Audubon took the trouble of searching him out (the way he had done to meet Walter Scott in Edinburgh) on a visit to Washington in 1833. He immediately received Irving's help in securing a government boat for his expedition to Florida. Irving's literary skills were beyond Audubon's reach or ambition, but the dramatic unified effect of the short story form definitely made an impact on his imagination.

Closer to Audubon's experience on the frontier, and his taste for the far-fetched, was the fashion in eastern newspapers for *eyewitness* accounts of the remote regions of the Southwest and the untamed personalities they hosted. This blossomed into the first full-scale literary movement in the South, the short stories and tales of the so-called Southwestern Humor writers, beginning with Augustus Baldwin Longstreet's *Georgia Scenes* in 1831 and continuing with the books of Johnson Jones Hooper and others, on through the early sketches of Mark Twain. The Civil War ended humor in the South for a long time, and the movement died. Audubon himself never mentioned the early southwestern humorists, but he breathed the same air, remarkably anticipating some of the most original popular writing of his time.

The Louisiana episodes in the *Ornithological Biography* were distributed unevenly throughout the first three volumes, four in the first (1831), two in the second (1834), and five in the third (1835). They were not arranged chronologically, though the first to appear, *The Prairie*, the haunting tale that inspired Robert Penn Warren, looks back to his earliest trip to Louisiana during the winter of 1810–1811 (related in *Journey up the Mississippi*) when he traveled to and from Ste. Geneviève. All the others evoke his six years in Louisiana between 1820 and 1826, with a concentration on his descent of the Mississippi River in 1820. *A Long Calm at Sea* and

Still Becalmed conclude the Louisiana series with the departure from New Orleans to England in 1826 onboard the ship *Delos*, drifting away past the barrier islands on the Gulf coast waiting for wind.

In writing the episodes Audubon could rely on a richness of experience still stirring in his mind, and the extensive notes and diaries he could always *ransack* (to use one of his favorite words) when memory sputtered. Embryos of the published essays can be found all through his journals of 1820 and 1826. One day in Liverpool in October 1826, the same day he drew the thumbnail sketch of the wild turkey the elder Hannah Rathbone needed for her gift of the signet seal, one of the Rathbone servants had been sent to the house where Audubon had been dining and playing the flute to offer him a ride in the family gig and the use of a greatcoat against the cold. Audubon politely declined, and wrote in his journal: "How little he knew how often I have laid to rest, wet, hungry, Harassed & full of Sorrow with millions of Mosquitoes buzzing me awake, lessening [listening] to the chuck wills widow, the Horned owl, and the roaring Bullfrogs=Impatiently waiting for the return of day to enable me to hunt the Forests and feed my eyes on their beautiful Inhabitants." These reflections led him on to write a short rough draft on the hospitality he had received from *squatter* families in the southern backwoods, a theme he returned to and expanded several times in the episodes.

The first of the episodes for the *Ornithological Biography* to be presented to the public was *A Flood*, in a paper read to the Wernerians the evening of February 19, 1831. By then he had completed the first volume of the *Biography*, five hundred tightly printed pages written in less than four months. Since Louisiana birds overwhelmingly dominated the engravings of *Birds of America* by then well engaged, it is fitting that *A Flood* led the way. Some of the episodes show Audubon submitting to Lucy's more conventional tastes in literary style and form. *A Flood* seems to be almost pure Audubon, dressed for the occasion, but speaking in his own voice. As a gesture to the Wernerians in the audience, he reverts to the design of the papers he had read to them four years earlier. He begins with generalities and enough solid documentary detail to establish his authority and affirm his academic credentials. Then he suddenly appears, alone in his canoe, floating over the swollen bayous and the Mississippi bottomlands, listening to "the mournful bleating of the hemmed-in deer" and the "dismal scream of an eagle or raven." This is the same tactic of the alligator essay. The room in Edinburgh was filled with scientists, eager to learn but yearning for the vicarious thrills of the solitary explorer face to face with the fierceness of primitive nature in all its stark brutality. Audubon once more knew how to please. He was one of them, a devoted member of the Wernerians, yet he did not quite belong to their world. For the audience, at least in a corner of their minds, Audubon's swamps were theaters of the fabulous, with the alligator playing the role of the dragon. For Audubon it was home.

The one Louisiana episode never published in Audubon's lifetime, *The Fair Incognito,* is chronologically the first he composed. He inserted it in a letter to Lucy that he copied in the *Louisiana Journal.* Oddly, though he often despaired of becoming an author, the literary intentions of *The Fair Incognito* are obvious. They show up particularly in the unusual way the narrative is urged forward by means of dialogue instead of description, a method he used when he imagined talking to Lucy in some of his journals. This is a device that he probably adopted from Sterne. What is curious is his desire to convert a private self-involved confession into a kind of public art, a play in prose or a dramatic short story. I think it safe to say that, read by itself as a completely independent piece of work, without anyone else, especially Lucy, having a hand in it, it stands out as the most original of all his writings. For that reason alone, more than its simple position in the timeline of Audubon's life in Louisiana, it deserves its place as the first of the Louisiana episodes.

SOURCES AND TRANSCRIPTION

All the Louisiana episodes except *Observations on the Natural History of the Alligator* and *The Fair Incognito* are taken from Maria Audubon's two-volume *Audubon and his Journals,* first published in 1897. Her edition of the episodes (with notes by Elliot Couës), reprinted from the *Ornithological Biography,* was their first complete appearance in the United States. Audubon had not included them in the American octavo edition of *Birds of America* (1840–1844), which did, however, contain the bird biographies. Some of the episodes were broken up and used in the first full-scale biography of Audubon, Mrs. Horace St. John's *Life of Audubon, the Naturalist of the New World: His Adventures and Discoveries* (1856). A selection of the episodes provided by Lucy Audubon appeared in Robert Buchanan's British biography, *The Life and Adventures of John James Audubon, the Naturalist* (1868), and in Lucy's subsequent reedited edition, *The Life of John James Audubon* (1870). Many of the episodes were translated into French by Eugène Bazin in his *Scènes de la Nature dans les Etats-Unis et le Nord de l'Amérique* (1857, 1868). Most modern collections of the episodes use the Maria Audubon edition. *Observations on the Natural History of the Alligator* is taken directly from the *Edinburgh New Philosophical Journal,* vol. 2 (October–April). *The Fair Incognito* (along with the full letter to Lucy that contains it) is held in the

John James Audubon Papers (B/Au 25) in the manuscript collection of the American Philosophical Society in Philadelphia. I would like to thank the Society and its staff for kindly allowing me to examine, photocopy, and transcribe the text. Audubon's account of his homecoming to the Felicianas in 1829 is taken from Maria Audubon's introduction to *Audubon and his Journals*. The three quotations from the 1826 journal are taken from Daniel Patterson's edition *John James Audubon's Journal of 1826* (2011).

The Fair Incognito

I was accosted on the day of x 1821, at the corner of Street & Street the former I take allmost daily not to be seen so much on Levée Street lugging my Port Folio, the astonishment of many—by a femelle of a fine form but whose face thickly covered by a Veil that I could not then distinguish it and who addressed Me quickly in about the following words in French—"Pray sir are you the one sent by the French Academy to draw the Birds of America" I answered that I drew them for my pleasure, "You are he that draws Likenesses in Black Chalk so remarkably strong," I answered again that I took Likenesses in that Style, "then call in 30 Minutes at N° in Street and walk upstairs I will wait for you"—I Bowed, "do not follow me now" I bowed again and as I went from her the course I had at first, took my pencil and put down the Street and Number; I soon reached a Book store where I waited some time having a feeling of astonishment undescribable, recollecting however how far I had to Walk, I Started and suppose the Stranger Employed a Carriadge—

I arrived, and as I walk[d] upstairs I Saw her apparently waiting "I am glad you have come, walk in quickly." My feeling became so agitated, that I trembled like a Leaf—this she perceived, shut the door with a double lock and throwing her veil back Shewed me one of the most beautiful face, I ever saw "have You been or are you Married" yes madam "Long" 12 years "is your Wife in this City" no Madam "Your name Audubon" yes Madam "Set down and be easy" and with the smile of an Angel "I will not hurt you" I felt such a blush and such Deathness through me I could not answer. She raised and handed Me a Glass of cordial, so strange was all this to me that I drank it for I needed it, but awkwardly gave her the glass to take back—

She sat again imediatly opposit me, and looking Me steadfastly ask[d] me if I thought I could draw her face—indeed I fear not answer[d] I "I am sure You can if you will but before I say more what is Your price"? Generally 25$ she smiled again most sweetly "will you Keep my name if you discover it and My residence a secret?" If you require it "I do" You must promess that to me, keep it for ever sacred although I care not about any thing else" I promised to keep her name and her place of residence to Myself "have you ever Drawn a full figure" Yes "*Naked*" had I been shot with a 48 pounder through the Heart my articulating Powers could not have been more suddenly stopped. "well why do you not answer" I answered Yes; She raised walk[d] the room a few times and sitting again Said, "I want you to draw my Likeness and the whole of my form naked but as I think you cannot work now, leave your Port Folio and return in one hour be silent."

She had Judged of my feelings precisely, I took my hat She open[d] the door and I felt like a Bird that makes his escape from a strong Cage, filled with sweet Meats. had I met a Stranger on the stairs No doubt I would have been suspected for a Thief. I walk[d] away fast looking behind me—

My thoughts rolled on her conduct, She look[d] as if perfect Mistress of herself and yet look[d] *I then thought* too young not supposing her more than Sixteen (a mistake however) and apparently not at all afraid to disclose to my eyes her sacred beauties—I tried to prepare myself for the occasion, the time passed and I arrived again at the foot of the Stairs.

She again was waiting for me and beckoned to me to move quick, she shut the door as the first time, then coming to me "Well how do you feel now, still trembling a litle, what a Man you are—come, come, I am anxious to see the outline you will make, take time and be sure do not embellish any parts With your briliant Imagination, have your paper Suficiently large, I have some beautifull and good Chalks, the drawing will be compleated in this room and You will please do it on this" raising She gave me a Large Sheet of Elephant paper out of an *Armoire.*

The die was cast, I felt at once Easy, ready and pleased—

I told her I was waitting for her convenience, She repeated the urgency of Secrecy which I again promessed—

The *Couch* in the room was supperbly decorated She drew the Curtains and I heard her undress. "I must be nearly in the Position you will see me unless your taste Should think proper to alter it by Speaking" very well was my answer Although I felt yet very Strange and never will forget the moment "please to draw the Curtains and arrange the light to Suit yourself." when drawing hirelings in company with 20 more I never cared but for a good outline, but Shut up with a beautifull young woman as much a Stranger to me as I was to her, I could not well reconcile all the feelings that were necessary to draw well, without mingling with them some of a very diferent nature—

Yet I drew the Curtain and saw this Beauty. "will I do so" I Eyed her but drop[d] my black Lead pencil "I am glad you are so timid, but tell me will I do so" perceiving at once that the Position, the light and all had been carefully Studied before, I told her I fear[d] She Look[d] only too well for my talent—She Smiled and I begun—

I drew 55 Minutes by my watch, when She desired I should close the curtains—She dressed in an Instant and came imediatly to look at what was done—

"Is it like me? will it be like me? I hope it will be a likeness. I am a litle Chill, can you work any more without me today." I told her I could correct my Sketch "Well be contented and work as much as you can, I wish it was done, it is a folly but all our Sex is more or Less so." She remarked very appropriety an error and made me correct it—She pulled a Bell, a Woman came in with a Waiter covered with Cakes & Wine, left it and passed through a door I had not perceived before—She insisted on My resting a while, made me drank, asked me a thousand questions about my familly, residence, Birds, Way of travelling of living &[c] &[c] and certainly is a well informed femelle, using the best expressions and in all her actions possessing the manners necessary to Insure Respect & wonder—

I worked nearly Two hours more and Casted all the outlines of the Drawing, it pleased her apparently very much I soon found She had received good Lessons, I begged to know her name "not to day and if you are not Carefull and silent you never will see me again" I assured her I would "I have thought well of you from hear say and hope not to be mistaken."

I felt now very diferent thoughts from those I had while She was undressing in her curtains and ask[d] when I must return again "every day at the same hour until done but never again with your port folio I will manage this once through your drawings of Birds." for Ten days at the exception of one Sunday, that She went out of the City, I had the pleasure of this beautifull woman's Company about one hour naked, and Two talking on diferent subjects, She admired my work more every day at least was pleased to say So, and on the 5th sitting she worked herself in a style much superior to mine—

The second day She desired to Know What I would ask her for my work, I told her I would be satisfied with whatever she would please give me "I take you at your word it will be *un souvenir* one who hunts so much needs a good Gun or Two, this afternoon see if there is one in the City & give this on a/c if you wish to please me to the last" She handed me five Dollars "I must see it and if I do not like it you are not to have it." I thank[d] her and told her it proprobly would be a high price piece "well that probably is Necessary to insure the good quality, do what I bid." I took the note, when gone, felt very undetermined, Yet I hunted through the stores, found a good one & gave the Note on a/c with my name, telling the Shopmen that I did not well Knew when I would Call for it.—She was much gratified I had found One, when I told her I was asked One hundred and Twenty Dollars "No more well we will say nothing about it until I see how I am pleased with your part." I worked from day to day. drawing besides a 25 Dollars Likeness every day, to be sure a litle at the expense of my Eyes at night, but how could I complain, how many artists would have been delighted of such lessons.

I finished my Drawing, or rather she did for when I returned every day I allways found the work much advanced, she touch[d] it she said not because she was fatigued of my company daily but because She felt happy in mingling her talents with mine in a piece She had had in contemplation to have done, even before She left the country She came from. I supposed from Italy or France but never could assertain—

She often took my pencil to compose a devise to have engraved on the Gun Barrels, and ask[d] me to have one of my own, but this I declined and left to her taste and will. She at last decided on one which she said I must absolutely have done for her sake and ordered me to have it finished in a few days.

She had a beautifull frame the last morning I went, on which she ask[d] my opinion, this of Course I gave as I thought her desires inclined. She put her Name at the foot of the Drawing as if *her own* and mine in a dark Shadded part of Drapery; when I had Closed it and put it in a true light She gazed at it for some moments and assured Me her wish was at last Gratified, and taking me by one hand gave me a delightfull Kiss—"had you acted otherwise than you have, you would have received a very diferent recompense, go, take this (125$) be happy think of me sometimes as you rest on your Gun, keep for ever my name a secret" I begged to kiss her hand, She held it out freely—We parted probably for ever—

it is well that I should say that She had had heard of me in a circle a few days after I had taken the B.C. [British Consul] Likeness and of my collection of Birds that she then understood, I was from France purposely sent by the Academy of Paris, but soon was assured of the contrary by my way of living; and that she ask[d] me to try my veracity—that she employed a servant to watch my ways and that for several nights this servant had remained very late to see if I absented from the Boat and that in fact She knew every Step I had taken since the day She had resolved on employing me. She ask[d] Me if I had not seen a Mulatto Man standing near the Gates of Madame A. x dry. two days Successively; this I recollected but never supposed that any one watch[d] my Steps—She praised the few drawings of Birds I shewed her the first day and assured me that She had no doubt I would be well recompensed for Such a Collection—

She never ask[d] me to go see her when we parted, I have tried several times in vain, the Servants allways saying Madame is absent. I have felt a great desire to see the Drawing since to Judge of it as I allways can do best after some time.

here my Journal takes to another subject and I Leave to thee to conclude what thou may of this extraordinary femelle—

Since, I have wished often I could have Shewn that Drawing to M[r] Venderlyn.—

The Lady was kind, the Gun is good and here is the Inscriptions on it—Ne refuse pas ce don d'une amie qui t'est reconnaisante puisse t'il t'égaler en bonté—and under the Raming Rod—Property of Laforest Audubon February 22[d] 1821.—her name *I* Engraved on it where I do not believe it will ever be found—

Observations on the Natural History of the Alligator

(In a Letter to Sir William Jardine, Baronet, and Prideaux John Selby, Esq. By John J. Audubon, Esq. Member of the Wernerian Natural History Society, &c.)

My Dear Sirs,

One of the most remarkable objects connected with the Natural History of the United States, that attracts the traveller's eye, as he ascends through the mouths of the mighty sea-like river Mississippi, is the Alligator. There, along the muddy shores, and on the large floating logs, these animals are seen either lying basking and asleep, stretched to their full length, or crossing to and fro the stream in search of food, with only the head out of water. It is here neither wild nor shy, neither is it the very dangerous animal represented by travellers. But, to give you details that probably may not be uninteresting to you, I shall take you to their more private haunts, and relate what I have experienced and seen respecting them and their habits.

In Louisiana, all our lagoons, bayous, creeks, ponds, lakes, and rivers, are well stocked with them,—they are found wherever there is a sufficient quantity of water to hide them, or to furnish them with food, and they continue thus, in great numbers, as high as the mouth of the Arkansas River, extending east to North Carolina, and as far west as I have penetrated. On the Red River, before it was navigated by steam-vessels, they were so extremely abundant, that, to see hundreds at a sight along the shores, or on the immense rafts of floating or stranded timber, was quite a common occurrence, the smaller on the backs of the larger, groaning and uttering their bellowing noise, like thousands of irritated bulls about to meet in fight, but all so careless of man, that unless shot at, or positively disturbed, they remained motionless, suffering boats or canoes to pass within a few yards of them, without noticing them in the least. The shores are yet trampled by them in such a manner, that their large tracks are seen as plentiful as those of sheep in a fold. It was on that river particularly that thousands of the largest size were killed, when the mania of having either shoes, boots, or saddle-seats, made of their hides, lasted. It had become an article of trade, and many of the squatters and strolling Indians followed for a time no other business. The discovery that the skins are not sufficiently firm and close-grained, to prevent water or dampness long, put a stop to their general destruction, which had already become very apparent. The leather prepared from these skins was handsome and very pliant, exhibiting all the regular lozenges of the scales, and able to receive the highest degree of polish and finishing.

The usual motion of the alligator, when on land, is slow and sluggish; it is a kind of laboured crawling, performed by moving alternately each leg, in the manner of a quadruped when walking, scarce able to keep up their weighty bodies from dragging on the earth, and leaving the track of their long tail on the mud, as if that of the keel of a small vessel. Thus they emerge from the water, and go about the shores and the woods, or the fields, in search of food, or of a different place of abode, or one of safety to deposit their eggs. If, at such times, when at all distant from the water, an enemy is perceived by them, they droop and lie flat, with their nose on the ground, watching the intruder's movements with their eyes, which are able to move considerably round, without affecting the position of the head. Should a man then approach them, they do not attempt either to make away or attack, but merely raise their body from the ground for an instant, swelling themselves, and issuing a dull blowing sound, not unlike that of a blacksmith's bellows. Not the least danger need be apprehended; then you either kill them with ease or leave them. But, to give you a better idea of the slowness of their movements and progress of travels on land, when arrived at a large size, say 12 to 15 feet, believe me when I tell you, that, having found one in the morning 50 yards from a lake going to another in sight, I have left him unmolested, hunted through the surrounding swamps all the day, and met the same alligator within 500 yards of the spot, when returning to my camp at dusk. On this account they usually travel during the night, they being then less likely to be disturbed, and having a better chance to surprise a litter of pigs, or of land-tortoises, for prey.

The power of the alligator is in his great strength; and the chief means of his attack or defence is his large tail, so well contrived by nature to supply his wants, or guard him from danger, that it reaches, when curved into half a circle, his enormous mouth. Woe be to him who goes within the reach of this tremendous thrashing instrument for no matter how strong or muscular; if human, he must suffer greatly, if he escapes with life. The monster, as he strikes with this, forces all objects within the circle towards his jaws, which, as the tail makes a motion, are open to their full stretch, thrown a little sidewise, to receive the object, and, like battering-rams, to bruise it shockingly in a moment.

The alligator, when after prey in the water, or at its edge, swims so slowly towards it, as not to ruffle the water. It approaches the object sidewise, body and head all concealed, till sure of his stroke; then, with a tremendous blow, as quick as thought, the object is secured, as I described before.

When alligators are fishing, the flapping of their tails about the water may be heard at half a mile; but, to describe this in a more graphic way, suffer me to take you along with me, in one of my hunting excursions, accompanied by friends and negroes. In the immediate neighbourhood of Bayou Sarah, on the Mississippi, are extensive shallow lakes and morasses, that are yearly overflowed by the dreadful floods of that river, and supplied with myriads of fishes of many kinds, amongst which trouts are most abundant, white-perch, cat-fish, and alligator-gars, or devil-fish. Thither, in the early part of autumn, when the heat of a southern sun has exhaled much of the water, the squatter, the planter, the hunter, all go in search of sport. The lakes are then about 2 feet deep, having a fine sandy bottom; frequently much grass grows in them, bearing crops of seeds, for which multitudes of water-fowls resort to those places. The edges of these lakes are deep swamps, muddy for some distance, overgrown with heavy large timber, principally cypress, hung with Spanish beard, and tangled with different vines, creeping plants and cane, so as to render them almost dark during the day, and very difficult to the hunter's progress. Here and there in the lakes are small islands, with clusters of the same trees, on which flocks of snake-birds, wood-ducks, and different species of herons, build their nests. Fishing-lines, guns, and rifles, some salt and some water, are all the hunters take. Two negroes precede them,—the woods are crossed,—the scampering deer is seen,—the racoon and the opossum cross before you,—the black, the grey, and the fox squirrel, are heard barking,—here on a tree close at hand, is seen an old male pursuing intensely a younger one; he seizes it, they fight desperately, but the older attains his end, *vincit, castratque juniorem*. (Now, my dear Sirs, if this is not mental power illustrated, what shall we call it)? As you proceed farther on, the *hunk hunk* of the lesser ibis is heard from different parts, as they rise from the puddles that supply them with cray-fishes. At last the opening of the lake is seen; it has now become necessary to drag one's-self along through the deep mud, making the best of the way, with the head bent, through the small brushy growth, caring about nought but the lock of your gun. The long narrow Indian canoe kept to hunt those lakes, and taken into them during the fresh, is soon launched, and the party seated in the bottom is paddled or poled in search of water-game. There, at a sight, hundreds of alligators are seen dispersed over all the lake, their head, and all the upper part of the body, floating like a log, and, in many instances, so resembling one, that it requires to be accustomed to see them to know the distinction. Millions of the large wood-ibis are seen wading through the water, mudding it up, and striking deadly blows with their bills on the fish within. Here are a hoard of blue herons,—the sand-hill-crane rises with his hoarse note,—the snake-birds are perched here and there on the dead timber of the trees,—the cormorants are fishing,—buzzards and carion-crows exhibit a mourning train, patiently waiting for the water to dry and leave food for them,—and far in the horizon the eagle overtakes a devoted wood-duck, singled from the clouded flocks that have been bred there. It is then that you see and hear the alligator at his work,—each lake has a spot deeper than the rest, rendered so by those animals who work at it, and always situate at the lower end of the lake near the connecting bayous, that, as drainers, pass through all those lakes, and discharge sometimes many miles below where the water had made its entrance above, thereby ensuring to themselves water as long as any will remain. This is called by the hunters the Alligator's Hole. You see them there lying close together. The fish that are already dying by thousands, through the insufferable heat and stench of the water, and the wounds of the different winged enemies constantly in pursuit of them, resort to the Alligator's Hole to receive refreshment, with a hope of finding security also, and follow down the little currents flowing through the connecting sluices: but, no ! for, as the water recedes in the lake, they are here confined. The alligators thrash them and devour them whenever they feel hungry, while the ibis destroys all that make towards the shore. By looking for a little on this spot, you plainly see the tails of the alligators moving to and fro, splashing, and now and then, when missing a fish, throwing it up in the air. The hunter, anxious to prove the value of his rifle, marks one of the eyes of the largest alligator, and, as the hair-trigger is touched, the alligator dies.

Should the ball strike one inch astray from the eye the animal flounces, rolls over and over, beating furiously with his tail all about him, frightening all his companions, who sink immediately, whilst the fishes, like blades of burnished metal, leap in all directions out of the water, so terrified are they at this uproar.* Another and another receives the shot in the eyes, and expires; yet those that do not feel the fatal bullet, pay no attention to the death of their companions till the hunter approaches very close, when they hide themselves for a few moments, by sinking backward.

So truly gentle are the alligators at this season, that I have waded through such lakes in company of my friend Augustin Bourgeat, Esq. to whom I owe much information, merely holding a stick in one hand to drive them off, had they attempted to attack me. When first I saw this way of travelling through the lakes, waist-deep, sometimes with hundreds of these animals about me, I acknowledge to you that I felt great uneasiness, and thought it fool-hardiness to do so; but my friend, who is a most experienced hunter in that country, removed my fears by leading the way, and, after a few days, I thought nothing of it. If you go towards the head of the alligator, there is no danger, and you may safely strike it with a club, four feet long, until you drive it away, merely watching the operations of the point of the tail, that, at each blow you give, thrashes to the right and left most furiously.

The drivers of cattle from the Appelousas, and those of mules from Mexico, on reaching a lagoon or creek, send several of their party into the water, armed merely each with a club, for the purpose of driving away the alligators from the cattle; and you may then see men, mules, and those monsters, all swimming together, the men striking the alligators, that would otherwise attack the cattle, of which they are very fond, and those latter hurrying towards the opposite shores, to escape those powerful enemies. They will swim swiftly after a dog, or a deer, or a horse, before attempting the destruction of man, of which I have always remarked they were afraid, if the man feared not them.

Although I have told you how easily an alligator may be killed with a single rifle ball, if well aimed, that is to say, if it strike either in the eye or very immedi-

*This so alarms the remaining alligators, that, regularly, in the course of the following night, every one, large and small, removes to another hole, going to it by water, and probably for a week not one will be seen there.

ately above it, yet they are quite as difficult if not shot properly; and, to give you an idea of this, I shall mention two striking facts.

My good friend, Richard Harlan, M. D. of Philadelphia, having intimated a wish to have the heart of one of those animals to study its comparative anatomy, I one afternoon went out about half a mile from the plantation, and seeing an alligator that I thought I could put whole into a hogshead of spirits, I shot it immediately on the skull bone. It tumbled over from the log on which it had been basking into the water, and, with the assistance of two negroes, I had it out in a few minutes, apparently dead. A strong rope was fastened round its neck, and, in this condition, I had it dragged home across logs, thrown over fences, and handled without the least fear. Some young ladies there, anxious to see the inside of its mouth, requested that the mouth should be propped open with a stick put in vertically; this was attempted, but at this instant the first stunning effect of the wound was over, and the animal thrashed and snapped its jaws furiously, although it did not advance a foot. The rope being still around the neck, I had it thrown over a strong branch of a tree in the yard, and hauled the poor creature up, swinging free from all about it, and left it twisting itself, and scratching with its fore-feet to disengage the rope. It remained in this condition until the next morning, when finding it still alive, though very weak, the hogshead of spirits was put under it, and the alligator fairly lowered into it with a surge. It twisted about a little, but the cooper secured the cask, and it was shipped to Philadelphia, where it arrived in course.

Again, being in company with Augustin Bourgeat, Esq. we met an extraordinary large alligator in the woods whilst hunting; and, for the sake of destruction I may say, we alighted from our horses and approached it with full intention to kill it. The alligator was put between us, each of us provided with a long stick to irritate it, and, by making it turn its head partly on one side, afford us the means of shooting it immediately behind the fore-leg and through the heart. We both discharged five heavy loads of duck-shot into its body, and almost all into the same hole, without any other effect than that of exciting regular strokes of the tail, and snapping of the jaws, at each discharge, and the flow of a great quantity of blood out of the wound, and mouth and nostrils of the animal; but it was still full of life and vigour, and to have touched it with the hand would have been madness; but as we were anxious to measure it, and to knock off some of its larger teeth, to make powder chargers, it was shot with a single ball just over the eye, when it bounded a few inches off the ground, and was dead when it reached it again. Its length was seventeen feet; it was apparently centuries old; many of its teeth measured three inches. The shots taken were without a few feet only of the circle that we knew the tail could form, and our shots went *en masse.*

As the lakes become dry, and even the deeper connecting bayous empty themselves into the rivers, the alligators congregate into the deepest hole in vast numbers; and, to this day, in such places, are shot for the sake of their oil, now used for greasing the machinery of steam-engines and cotton-mills, though formerly, when indigo was made in Louisiana, the oil was used to assuage the overflowing of the boiling juice, by throwing a ladleful into the kettle whenever this was about to take place. The alligators are caught frequently in nets by fishermen: they then come without struggling to the shore, and are killed by blows on the head given with axes.

When autumn has heightened the colouring of the foliage of our woods, and the air feels more rarified during the nights and earlier part of the day, the alligators leave the lakes to seek for winter quarters, by burrowing under the roots of trees, or covering themselves simply with earth along their edges. They become then very languid and inactive, and, at this period, to sit or ride on one, would not be more difficult than for a child to mount his wooden rocking-horse. The Negroes who now kill them, put all danger aside by separating, at one blow with an axe, the tail from the body. They are afterwards cut up in large pieces, and boiled whole in a good quantity of water, from the surface of which the fat is collected with large ladles. One single man kills oftentimes a dozen or more of large alligators in the evening, prepares his fire in the woods, where he has erected a camp for the purpose, and by morning has the oil rendered.

I have frequently been very much amused when fishing in a bayou, where alligators were numerous, by throwing a blown bladder on the water towards the nearest to me. The alligator makes for it at once, flaps it towards its mouth, or attempts seizing it at once, but all in vain. The light bladder slides off; in a few minutes many alligators are trying to seize this, and their evolutions are quite interesting. They then put one in mind of a crowd of boys running after a football. A black bottle is sometimes thrown also, tightly corked; but the alligator seizes this easily, and you hear the glass give way under its teeth as if ground in a coarse mill. They are easily caught by Negroes, who most expertly throw a rope over their heads when swimming close to shore, and haul them out instantly.

But, my dear sirs, you must not conclude that alligators are always thus easily conquered: there is a season when they are dreadfully dangerous; it is during spring, during the love season. The waters have again submerged the low countries; fish are difficult of access; the greater portion of the game has left for the northern latitudes; the quadrupeds have retired to the high lands; and the heat of passion, joined to the difficulty of procuring food, render these animals now ferocious and very considerably more active. The males have dreadful fights together, both in the water and on the land. Their strength and weight adding much to their present courage, exhibit them like colossuses wrestling. At this time no man swims or wades among them; they are usually left alone at this season.

About the first days of June the female prepares a nest; a place is chosen forty or fifty yards from the water, in thick bramble or cane, and she gathers leaves, sticks, and rubbish of all kinds, to form a bed to deposite her eggs; she carries the materials in her mouth, as a hog does straw. As soon as a proper nest is finished, she lays about ten eggs, then covers them with more rubbish and mud, and goes on depositing in different layers until fifty, or sixty, or more eggs are laid. The whole is then covered up, matted and tangled with long grasses, in such a manner that it is very difficult to break it up. These eggs are the size of that of a goose, more elongated, and, instead of being contained in a shell, are in a bladder, or thin transparent parchment-like substance, yielding to the pressure of the fingers, yet resuming its shape at once, like the eggs of snakes and tortoises. They are not eaten even by hogs. The female now keeps watch near the spot, and is very wary and ferocious, "going to the water from time to time only for food. Her nest is easily discovered, as she always goes and returns the same way, and forms quite a path by the dragging of her heavy body. The heat of the nest, from its forming a mass of putrescent manure, cause the hatching of the eggs, not that of the sun, as is usually believed.

Some European writers say, that at this juncture the vultures feed on the eggs, and thereby put a stop to the increase of those animals. In the United States, I assure you, it is not so, nor can it be so, were the vultures ever so anxiously inclined; for, as I have told you before, the nest is so hard, and matted, and plastered together, that a man needs his superior strength, with a strong sharp stick, to demolish it.

The little alligators, as soon as hatched (and they all break shell within a few hours from the first to last), force themselves through, and issue forth all beautiful, lively, and as brisk as lizards. The female leads them to the lake, but more frequently into small detached bayous for security's sake; for now the males, if they can get at them, devour them by hundreds, and the wood ibis and the sand-hill cranes also feast on them.

I believe that the growth of alligators takes place very slowly, and that an alligator of twelve feet long, for instance, will most probably be fifty or more years old. My reasons for believing this to be fact is founded on many experiments, but I shall relate to you one made by my friend Bourgeat. That gentleman, anxious to send some young alligators as a present to an acquaintance in New York, had a bag of young ones, quite small, brought to his house. They were put out on the floor, to shew the ladies how beautiful they were when young. One accidentally made its way out into a servant's room, and lodged itself snug from notice into an old shoe. The alligator was not missed, but, upwards of twelve months after this, it was discovered about the house, full of life, and, apparently, scarcely grown bigger; one of his brothers, that had been kept in a tub and fed plentifully, had grown only a few inches during the same period.

Few animals emit a stronger odour than the alligator; and, when it has arrived at great size, you may easily discover one in the woods in passing fifty or sixty yards from it. This smell is highly musky, and so strong, that, when near, it becomes insufferable; but this I never experienced when the animal is in the water, although I have, whilst fishing, been so very close to them, as to throw the cork of my fishing line on their heads, to tease them. In those that I have killed, and, I assure you, I have killed a great many, if opened, to see the contents of the stomach, or take fresh fish out of them, I regularly have found round masses of a hard substance, resembling petrified wood. These masses appeared to be useful to the animal in the process of digestion, like those found in the craws of some species of birds. I have broken some of them with a hammer, and found them brittle, and as hard as stones, which they resemble outwardly also very much. And, as neither our lakes nor rivers, in the portion of the country I have hunted them in, afford even a pebble as large as a common egg I have not been able to conceive how they are procured by the animals, if positively stones, or by what power wood can become stone in their stomachs.

The Prairie

On my return from the Upper Mississippi I found myself obliged to cross one of the wide prairies which, in that portion of the United States, vary the appearance of the country. The weather was fine; all around me was as fresh and blooming as if it had just issued from the bosom of Nature. My knapsack, my gun, and my dog were all I had for baggage and company. But, although well moccasined, I moved slowly along, attracted by the brilliancy of the flowers, and the gambols of the fawns around their dams, to all appearance as thoughtless of danger as I felt myself.

My march was of long duration; I saw the sun sinking below the horizon long before I could perceive any appearance of woodland, and nothing in the shape of man had I met with that day. The track which I followed was only an old Indian trace, and as darkness overshadowed the prairie I felt some desire to reach at least a copse, in which I might lie down to rest. The Night Hawks were skimming over and around me, attracted by the buzzing wings of the beetles which form their food, and the distant howling of wolves gave me some hope that I should soon arrive at the skirts of some woodlands.

I did so, and almost at the same instant, a firelight attracting my eye, I moved towards it, full of confidence that it proceeded from the camp of some wandering Indians. I was mistaken: I discovered by its glare that it was from the hearth of a small log cabin, and that a tall figure passed and repassed between it and me, as if busily engaged in household arrangements.

I reached the spot, and presenting myself at the door, asked the tall figure, which proved to be a woman, if I might take shelter under her roof for the night. Her voice was gruff, and her attire negligently thrown about her. She answered in the affirmative. I walked in, took a wooden stool, and quietly seated myself by the fire. The next object that attracted my notice was a finely formed young Indian, resting his head between his hands, with his elbows on his knees. A long bow rested against the log wall near him, while a quantity of arrows and two or three Raccoon skins lay at his feet. He moved not; he apparently breathed not. Accustomed to the habits of Indians, and knowing that they pay little attention to the approach of civilized strangers (a circumstance which in some countries is considered as evincing the apathy of their character), I addressed him in French, a language not infrequently partially known to the people in that neighborhood. He raised his head, pointed to one of his eyes with his finger, and gave me a significant glance with the other. His face was covered with blood. The fact was that an hour before this, as he was in the act of discharging an arrow at a Raccoon in the top of a tree, the arrow had split upon the cord, and sprung back with such violence into his right eye as to destroy it forever.

Feeling hungry, I inquired what sort of fare I might expect. Such a thing as a bed was not to be seen, but many large untanned Bear and Buffalo hides lay piled in a corner. I drew a fine time-piece from my breast, and told the woman that it was late, and that I was fatigued. She had espied my watch, the richness of which seemed to operate upon her feelings with electric quickness. She told me there was plenty of venison and jerked buffalo meat, and that on removing the ashes I should find a cake. But my watch had struck her fancy, and her curiosity had to be gratified by an immediate sight of it. I took off the gold chain that secured it, from around my neck, and presented it to her; she was all ecstasy, spoke of its beauty, asked me its value, and put the chain round her brawny neck, saying how happy the possession of such a watch would make her. Thoughtless, and as I fancied myself in so retired a spot secure, I paid little attention to her talk or her movements. I helped my dog to a good supper of venison, and was not long in satisfying the demands of my own appetite.

The Indian rose from his seat, as if in extreme suffering. He passed and repassed me several times, and once pinched me on the side so violently that the pain nearly brought forth an exclamation of anger. I looked at him. His eye met mine, but his look was so forbidding that it struck a chill into the more nervous part of my system. He again seated himself, drew his butcher-knife from its greasy scabbard, examined its edge, as I would do that of a razor suspected dull, replaced it, and again taking his tomahawk from his back, filled the pipe of it with tobacco, and sent me expressive glances, whenever our hostess chanced to have her back towards us.

Never until that moment had my senses been awakened to the danger which I now suspected to be about me. I returned glance for glance to my companion, and rested well assured that, whatever enemies I might have, he was not of their number.

I asked the woman for my watch, wound it up, and under pretence of wishing to see how the weather might probably be on the morrow, took up my gun, and walked out of the cabin. I slipped a ball into each barrel, scraped the edges of my flints, renewed the primings, and returning to the hut gave a favorable report of my observations. I took a few Bear skins, made a pallet of them, and calling my faithful dog to my side, lay down, with my gun close to my body, and in a few minutes was, to all appearance, fast asleep.

A short time had elapsed when some voices were heard, and from the corner of my eye I saw two athletic youths making their entrance, bearing a dead stag on a pole. They disposed of their burden, and asking for whiskey, helped themselves freely to it. Observing me and the wounded Indian, they asked who I was, and why the devil that rascal (meaning the Indian, who, they knew, understood not a word of English) was in the house. The mother—for so she proved to be—bade them speak less loudly, made mention of my watch, and took them to a corner, where a conversation took place, the purport of which it required little shrewdness in me to guess. I tapped my dog gently. He moved his tail, and with indescribable pleasure I saw his fine eyes alternately fixed on me, and raised towards the trio in the corner. I felt that he perceived danger in my situation. The Indian exchanged a last glance with me.

The lads had eaten and drunk themselves into such a condition that I already looked upon them as *hors de combat;* and the frequent visits of the whiskey bottle to the ugly mouth of their dam, I hoped would soon reduce her to a like state. Judge of my astonishment, reader, when I saw this incarnate fiend take a large carving-knife, and go to the grindstone to whet its edge; I saw her pour the water on the turning machine, and watched her working away with the dangerous instrument, until the cold sweat covered every part of my body, in despite of my determination to defend myself to the last. Her task finished, she walked to her reeling sons, and said: "There, that'll soon settle him! Boys, kill yon ——, and then for the watch."

I turned, cocked my gun-locks silently, touched my faithful companion, and lay ready to start up and shoot the first who might attempt my life. The moment was fast approaching, and that night might have been my last in this world, had not Providence made preparations for my rescue. All was ready. The infernal hag was advancing slowly, probably contemplating the best way of despatching me, whilst her sons should be engaged with the Indian. I was several times on the eve of rising and shooting her on the spot; but she was not to be punished thus. The door was suddenly opened, and there entered two stout travellers, each with a long rifle on his shoulder. I bounced up on my feet, and making them most heartily welcome, told them how well it was for me that they should have arrived at that moment. The tale was told in a minute. The drunken sons were secured, and the woman, in spite of her defence and vociferations, shared the same fate. The Indian fairly danced with joy, and gave us to understand that, as he could not sleep for pain, he would watch over us. You may suppose we slept much less than we talked. The two strangers gave me an account of their once having been themselves in a somewhat similar situation. Day came, fair and rosy, and with it the punishment of our captives.

They were now quite sobered. Their feet were unbound, but their arms were still securely tied. We marched them into the woods off the road, and having used them as Regulators were wont to use such delinquents, we set fire to the cabin, gave all the skins and implements to the young Indian warrior, and proceeded, well pleased, towards the settlements.

During upwards of twenty-five years, when my wanderings extended to all parts of our country, this was the only time at which my life was in danger from my fellow-creatures. Indeed, so little risk do travellers run in the United States that no one born there ever dreams of any to be encountered on the road; and I can only account for this occurrence by supposing that the inhabitants of the cabin were not Americans.

Will you believe, good-natured reader, that not many miles from the place where this adventure happened, and where fifteen years ago, no habitation belonging to civilized man was expected, and very few ever seen, large roads are now laid out, cultivation has converted the woods into fertile fields, taverns have been erected, and much of what we Americans call comfort is to be met with? So fast does improvement proceed in our abundant and free country.

Improvements in the Navigation of the Mississippi

I have so frequently spoken of the Mississippi that an account of the progress of navigation on that extraordinary stream may be interesting even to the student of nature. I shall commence with the year 1808, at which time a great portion of the western country, and the banks of the Mississippi River, from above the city of Natchez particularly, were little more than a waste, or to use words better suited to my feelings, remained in their natural state. To ascend the great stream against a powerful current, rendered still stronger wherever islands occurred, together with the thousands of sand-banks, as liable to changes and shiftings as the alluvial shores themselves, which at every deep curve or bend were seen giving way, as if crushed down by the weight of the great forests that everywhere reached to the very edge of the water, and falling and sinking in the muddy stream by acres at a time, was an adventure of no small difficulty and risk, and which was rendered more so by the innumerable logs, called *sawyers* and *planters*, that everywhere raised their heads above the water, as if bidding defiance to all intruders. Few white inhabitants had yet marched towards its shores, and these few were of a class little able to assist the navigator. Here and there a solitary encampment of native Indians might be seen, but its inmates were as likely to prove foes as friends, having from their birth been made keenly sensible of the encroachments of the white men upon their lands.

Such was then the nature of the Mississippi and its shores. That river was navigated, principally in the direction of the current, in small canoes, pirogues, keelboats, some flatboats, and a few barges. The canoes and pirogues, being generally laden with furs from the different heads of streams that feed the great river, were of little worth after reaching the market of New Orleans, and seldom reascended, the owners making their way home through the woods, amidst innumerable difficulties. The flatboats were demolished and used as fire-wood. The keel-boats and barges were employed in conveying produce of different kinds besides furs, such as lead, flour, pork, and other articles. These returned laden with sugar, coffee, and dry goods suited for the markets of St. Geneviève and St. Louis on the upper Mississippi, or branched off and ascended the Ohio to the foot of the Falls near Louisville in Kentucky. But, reader, follow their movements, and judge for yourself of the fatigues, troubles, and risks of the men employed in that navigation. A keel-boat was generally manned by ten hands, principally Canadian French, and a patroon or master. These boats seldom carried more than from twenty to thirty tons. The barges frequently had forty or fifty men, with a patroon, and carried fifty or sixty tons. Both these kinds of vessels were provided with a mast, a square sail, and coils of cordage known by the name of *cordelles*. Each boat or barge carried its own provisions. We shall suppose one of these boats under way, and, having passed Natchez, entering upon what were the difficulties of their ascent. Wherever a point projected, so as to render the course or bend below it of some magnitude, there was an eddy, the returning current of which was sometimes as strong as that of the middle of the great stream. The bargemen therefore rowed up pretty close under the bank, and had merely to keep watch in the bow, lest the boat should run against a planter or sawyer. But the boat has reached the point, and there the current is to all appearance of double strength, and right against it. The men, who have all rested a few minutes, are ordered to take their stations, and lay hold of their oars, for the river must be crossed, it being seldom possible to double such a point, and proceed along the same shore. The boat is crossing, its head slanting to the current, which is, however, too strong for the rowers, and when the other side of the river has been reached, it has drifted perhaps a quarter of a mile. The men are by this time exhausted, and, as we shall suppose it to be twelve o'clock, fasten the boat to the shore or to a tree. A small glass of whiskey is given to each, when they cook and eat their dinner, and after repairing their fatigue by an hour's repose, recommence their labors. The boat is again seen slowly advancing against the stream. It has reached the lower end of a large sand-bar, along the edge of which it is propelled by means of long poles, if the bottom be hard. Two men called bowsmen remain at the prow, to assist, in concert with the steersman, in managing the boat, and keeping its head right against the current. The rest place themselves on the land side of the footway of the vessel, put one end of their poles on the ground, the other against their shoulders, and push with all their might. As each of the men reaches the stern, he crosses to the other side, runs along it, and comes again to the landward side of the bow, when he recommences operations. The barge in the meantime is ascending at a rate not exceeding one mile in the hour.

The bar is at length passed, and as the shore in sight is straight on both sides of the river, and the current uniformly strong, the poles are laid aside, and the men being equally divided, those on the river side take to their oars, whilst those on the land side lay hold of the branches of willows, or other trees, and thus slowly propel the boat. Here and there however, the trunk of a fallen tree, partly lying on the bank, and partly projecting beyond it, impedes their progress, and requires to be doubled. This is performed by striking it with the iron points of the poles and gaff-hooks. The sun is now quite low, and the barge is again secured in the best harbor within reach. The navigators cook their supper, and betake themselves to their blankets or Bear skins to rest, or perhaps light a large fire on the shore, under the smoke of which they repose, in order to avoid the persecutions of the myriads of mosquitoes which are found along the river during the whole summer. Perhaps, from dawn to sunset, the boat may have advanced fifteen miles. If so, it has done well. The next day, the wind proves favorable, the sail is set, the boat takes all advantages, and meeting with no accident, has ascended thirty miles, perhaps double that distance. The next day comes with a very different aspect. The wind is right ahead, the shores are without trees of any kind, and the canes on the bank are so thick and stout that not even the cordelles can be used. This occasions a halt. The time is not altogether lost, as most of the men, being provided with rifles, betake themselves to the woods, and search for the Deer, the Bears, or the Turkeys that are generally abundant there. Three days may pass before the wind changes, and the advantages gained on the previous fine day are forgotten. Again the boat proceeds, but in passing over a shallow place, runs on a log, swings with the current, but hangs fast, with her lee side almost under water. Now for the poles! All hands are on deck, bustling and pushing. At length, towards sunset, the boat is once more afloat, and is again taken to the shore, where the wearied crew pass another night.

I shall not continue this account of difficulties, it having already become painful in the extreme. I could tell you of the crew abandoning the boat and cargo, and of numberless accidents and perils; but be it enough to say that advancing in this tardy manner, the boat that left New Orleans on the first of March often did not reach the Falls of the Ohio until the month of July,—nay, sometimes not until

October; and after all this immense trouble, it brought only a few bags of coffee, and at most one hundred hogsheads of sugar. Such was the state of things in 1808. The number of barges at that period did not amount to more than twenty-five or thirty, and the largest probably did not exceed one hundred tons burden. To make the best of this fatiguing navigation, I may conclude by saying that a barge which came up in three months had done wonders, for, I believe, few voyages were performed in that time.

If I am not mistaken, the first steamboat that went down out of the Ohio to New Orleans was named the "Orleans," and, if I remember right, was commanded by Captain Ogden. This voyage, I believe, was performed in the spring of 1810. It was, as you may suppose, looked upon as the *ne plus ultra* of enterprise. Soon after, another vessel came from Pittsburgh, and before many years elapsed, to see a vessel so propelled had become a common occurrence. In 1826, after a lapse of time that proved sufficient to double the population of the United States of America, the navigation of the Mississippi had so improved, both in respect to facility and quickness, that I know no better way of giving you an idea of it than by presenting you with an extract from a letter written by my eldest son, which was taken from the books of N. Berthoud, Esq., with whom he at that time resided.

"You ask me in your last letter for a list of the arrivals and departures here. I give you an abstract from our list of 1826, showing the number of boats which plied each year, their tonnage, the trips they performed, and the quantity of goods landed here from New Orleans and intermediate places:—

	Boats.	Tons.	Trips.	Tons.
1823, from Jan. 1 to Dec. 31,	42	7860	98	19,453
1824, " " " " Nov. 25,	36	6393	118	20,291
1825, " " " " Aug. 15,	42	7484	140	24,102
1826, " " " " Dec. 31,	51	9386	182	28,914

The amount for the present year will be much greater than any of the above. The number of flatboats and keel-boats is beyond calculation. The number of steamboats above the Falls I cannot say much about, except that one or two arrive at and leave Louisville every day. Their passage from Cincinnati is commonly fourteen or sixteen hours. The "Tecumseh," a boat which runs between this place and New Orleans, which is of 210 tons, arrived here on the 10th inst. in nine days, seven hours, from port to port; and the "Philadelphia," of 300 tons, made the passage in nine days, nine and a half hours, the computed distance being 1650 miles. These are the quickest trips made. There are now in operation on the waters west of the Alleghany Mountains 140 or 150 boats. We had last spring (1826) a very high freshet, which came four and a half feet deep in the counting-room. The rise was 57 feet 3 inches perpendicular."

All the steamboats of which this is an account did not perform voyages to New Orleans only, but to all points on the Mississippi, and other rivers which fall into it. I am certain that since the above date the number has increased, but to what extent I cannot at present say.

When steamboats first plied between Shippingport and New Orleans, the cabin passage was a hundred dollars, and a hundred and fifty dollars on the upward voyage. In 1829, I went down to Natchez from Shippingport for twenty-five dollars, and ascended from New Orleans on board the "Philadelphia," in the beginning of January, 1830, for sixty dollars, having taken two state-rooms for my wife and myself. On that voyage we met with a trifling accident, which protracted it to fourteen days, the computed distance being, as mentioned above, 1650 miles, although the real distance is probably less. I do not remember to have spent a day without meeting with a steamboat, and some days we met several. I might here be tempted to give you a description of one of these steamers of the western waters, but the picture having been often drawn by abler hands, I shall desist.

A Flood

Many of our larger streams, such as the Mississippi, the Ohio, the Illinois, the Arkansas, and the Red River, exhibit at certain seasons the most extensive overflowings of their waters, to which the name of *floods* is more appropriate than the term *freshets,* usually applied to the sudden risings of smaller streams. If we consider the vast extent of country through which an inland navigation is afforded by the never-failing supply of water furnished by these wonderful rivers, we cannot suppose them exceeded in magnitude by any other in the known world. It will easily be imagined what a wonderful spectacle must present itself to the eye of the traveller who for the first time views the enormous mass of waters, collected from the vast central regions of our continent, booming along, turbid and swollen to overflowing, in the broad channels of the Mississippi and Ohio, the latter of which has a course of more than a thousand miles, and the former of several thousands.

To give you some idea of a *Booming Flood* of these gigantic streams, it is necessary to state the causes which give rise to it. These are, the sudden melting of the snows on the mountains, and heavy rains continued for several weeks. When it happens that, during a severe winter, the Alleghany Mountains have been covered with snow to the depth of several feet, and the accumulated mass has remained unmelted for a length of time, the materials of a flood are thus prepared. It now and then happens that the winter is hurried off by a sudden increase of temperature, when the accumulated snows melt away simultaneously over the whole country, and the southeasterly wind, which then usually blows, brings along with it a continued fall of heavy rain, which, mingling with the dissolving snow, deluges the alluvial portions of the western country, filling up the rivulets, ravines, creeks, and small rivers. These delivering their waters to the great streams, cause the latter not merely to rise to a surprising height, but to overflow their banks, wherever the land is low. On such occasions the Ohio itself presents a splendid, and at the same time, an appalling spectacle; but when its waters mingle with those of the Mississippi, then, kind reader, is the time to view an American flood in all its astonishing magnificence.

At the foot of the Falls of the Ohio, the water has been known to rise upwards of sixty feet above its lowest level. The river, at this point, has already run a course of nearly seven hundred miles from its origin at Pittsburgh in Pennsylvania, during which it has received the waters of its numberless tributaries, and overflowing all the bottom lands or valleys, has swept along the fences and dwellings which have been unable to resist its violence. I could relate hundreds of incidents which might prove to you the dreadful effects of such an inundation, and which have been witnessed by thousands besides myself. I have known, for example, of a cow swimming through a window, elevated at least seven feet from the ground, and sixty-two feet above low-water mark. The house was then surrounded by water from the Ohio, which runs in front of it, while the neighboring country was overflowed; yet, the family did not remove from it, but remained in its upper portion, having previously taken off the sashes of the lower windows, and opened the doors. But let us return to the Mississippi.

There the overflow is astonishing, for no sooner has the water reached the upper part of the banks than it rushes out and overspreads the whole of the neighboring swamps, presenting an ocean overgrown with stupendous forest-trees. So sudden is the calamity that every individual, whether man or beast, has to exert his utmost ingenuity to enable him to escape from the dreaded element. The Indian quickly removes to the hills of the interior, the cattle and game swim to the different strips of land that remain uncovered in the midst of the flood, or attempt to force their way through the waters until they perish from fatigue. Along the banks of the river, the inhabitants have rafts ready made, on which they remove themselves, their cattle, and their provisions, and which they then fasten with ropes or grape-vines to the larger trees, while they contemplate the melancholy spectacle presented by the current, as it carries off their houses and wood-yards piece by piece. Some who have nothing to lose, and are usually known by the name of *squatters,* take this opportunity of traversing the woods in canoes, for the purpose of procuring game, and particularly the skins of animals, such as the Deer and Bear, which may be converted into money. They resort to the low ridges surrounded by the waters, and destroy thousands of Deer, merely for their skins, leaving the flesh to putrefy.

The river itself, rolling its swollen waters along, presents a spectacle of the most imposing nature. Although no large vessel, unless propelled by steam, can now make its way against the current, it is seen covered by boats, laden with produce, which, running out from all the smaller streams, float silently towards the city of New Orleans, their owners meanwhile not very well assured of finding a landing-place even there. The water is covered with yellow foam and pumice, the latter having floated from the Rocky Mountains of the Northwest. The eddies are larger and more powerful than ever. Here and there tracts of forest are observed undermined, the trees gradually giving way, and falling into the stream. Cattle, horses, Bears, and Deer are seen at times attempting to swim across the impetuous mass of foaming and boiling water; whilst here and there a Vulture or an Eagle is observed perched on a bloated carcass, tearing it up in pieces, as regardless of the flood as on former occasions it would have been of the numerous sawyers and planters with which the surface of the river is covered when the water is low. Even the steamer is frequently distressed. The numberless trees and logs that float along break its paddles, and retard its progress. Besides, it is on such occasions difficult to procure fuel to maintain its fires; and it is only at very distant intervals that a wood-yard can be found which the water has not carried off.

Following the river in your canoe, you reach those parts of the shores that are protected against the overflowings of the waters, and are called *levees*. There you find the whole population of the district at work repairing and augmenting those artificial barriers, which are several feet above the level of the fields. Every person appears to dread the opening of a *crevasse*, by which the waters may rush into his fields. In spite of all exertions, however, the crevasse opens, the water bursts impetuously over the plantations, and lays waste the crops which so lately were blooming in all the luxuriance of spring. It opens up a new channel, which, for aught I know to the contrary, may carry its waters even to the Mexican Gulf.

I have floated on the Mississippi and Ohio when thus swollen, and have in different places visited the submersed lands of the interior, propelling a light canoe by the aid of a paddle. In this manner I have traversed immense portions of the country overflowed by the waters of these rivers, and particularly when floating over the Mississippi bottom-lands I have been struck with awe at the sight. Little or no current is met with, unless when the canoe passes over the bed of a bayou. All is silent and melancholy, unless when the mournful bleating of the hemmed-in Deer reaches your ear, or the dismal scream of an Eagle or a Raven is heard, as the foul bird rises, disturbed by your approach, from the carcass on which it was allaying its craving appetite. Bears, Cougars, Lynxes, and all other quadrupeds that can ascend the trees are observed crouched among their top branches. Hungry in the midst of abundance, although they see floating around them the animals on which they usually prey, they dare not venture to swim to them. Fatigued by the exertions which they have made to reach the dry land, they will there stand the hunter's fire, as if to die by a ball were better than to perish amid the waste of waters. On occasions like this, all these animals are shot by hundreds.

Opposite the city of Natchez, which stands on a bluff bank of considerable elevation, the extent of inundated land is immense, the greater portion of the tract lying between the Mississippi and the Red River, which is more than thirty miles in breadth, being under water. The mail-bag has often been carried through the immersed forests, in a canoe, for even a greater distance, in order to be forwarded to Natchitochez.

But now, kind reader, observe this great flood gradually subsiding, and again see the mighty changes which it has effected. The waters have now been carried into the distant ocean. The earth is everywhere covered by a deep deposit of muddy loam, which in drying splits into deep and narrow chasms, presenting a reticulated appearance, and from which, as the weather becomes warmer, disagreeable, and at times noxious, exhalations arise, and fill the lower stratum of the atmosphere as with a dense fog. The banks of the river have almost everywhere been broken down in a greater or less degree. Large streams are now found to exist, where none were formerly to be seen, having forced their way in direct lines from the upper parts of the bends. These are by the navigator called short-cuts. Some of them have proved large enough to produce a change in the navigation of the Mississippi. If I mistake not, one of these, known by the name of the *Grand Cut-off*, and only a few miles in length, has diverted the river from its natural course, and has shortened it by fifty miles. The upper parts of the islands present a bulwark consisting of an enormous mass of floated trees of all kinds, which have lodged there. Large sand-banks have been completely removed by the impetuous whirls of the waters, and have been deposited in other places. Some appear quite new to the eye of the navigator, who has to mark their situation and bearings in his log-book. The trees on the margins of the banks have in many parts given way. They are seen bending over the stream, like the grounded arms of an overwhelmed army of giants. Everywhere are heard the lamentations of the farmer and planter, whilst their servants and themselves are busily employed in repairing the damages occasioned by the floods. At one crevasse an old ship or two, dismantled for the purpose, are sunk, to obstruct the passage opened by the still rushing waters, while new earth is brought to fill up the chasms. The squatter is seen shouldering his rifle, and making his way through the morass, in search of his lost stock, to drive the survivors home, and save the skins of the drowned. New fences have everywhere to be formed; even new houses must be erected, to save which from a like disaster, the settler places them on an elevated platform supported by pillars made by the trunks of trees. The land must be ploughed anew, and if the season is not too far advanced, a crop of corn and potatoes may yet be raised. But the rich prospects of the planter are blasted. The traveller is impeded in his journey, the creeks and smaller streams having broken up their banks in a degree proportionate to their size. A bank of sand, which seems firm and secure, suddenly gives way beneath the traveller's horse, and the next moment the animal has sunk in the quicksand, either to the chest in front, or over the crupper behind, leaving its master in a situation not to be envied.

Unlike the mountain torrents and small rivers of other parts of the world, the Mississippi rises but slowly during these floods, continuing for several weeks to increase at the rate of about an inch a day. When at its height, it undergoes little fluctuation for some days, and after this, subsides as slowly as it rose. The usual duration of a flood is from four to six weeks, although, on some occasions, it is protracted to two months.

Every one knows how largely the idea of floods and cataclysms enters into the speculations of the geologist. If the streamlets of the European continent afford illustrations of the formation of strata, how much more must the Mississippi, with its ever-shifting sand-banks, its crumbling shores, its enormous masses of drift timber, the source of future beds of coal, its extensive and varied alluvial deposits, and its mighty mass of waters rolling sullenly along, like the flood of eternity.

The Squatters of the Mississippi

Although every European traveller who has glided down the Mississippi, at the rate of ten miles an hour, has told his tale of the squatters, yet none has given any other account of them, than that they are "a sallow, sickly looking sort of miserable beings," living in swamps, and subsisting on pig-nuts, Indian-corn, and Bear's-flesh. It is obvious, however, that none but a person acquainted with their history, manners, and condition, can give any real information respecting them.

The individuals who become squatters, choose that sort of life of their own free will. They mostly remove from other parts of the United States, after finding that land has become too high in price, and they are persons who, having a family of strong and hardy children, are anxious to enable them to provide for themselves. They have heard from good authorities that the country extending along the great streams of the West, is of all parts of the Union, the richest in its soil, the growth of its timber, and the abundance of its game; that, besides, the Mississippi is the great road to and from all the markets in the world; and that every vessel borne by its waters affords to settlers some chance of selling their commodities, or of exchanging them for others. To these recommendations is added another, of even greater weight with persons of the above denomination, namely, the prospect of being able to settle on land, and perhaps to hold it for a number of years, without purchase, rent or tax of any kind. How many thousands of individuals in all parts of the globe would gladly try their fortune with such prospects, I leave to you, reader, to determine.

As I am not disposed too highly to color the picture which I am about to submit to your inspection, instead of pitching on individuals who have removed from our eastern boundaries, and of whom certainly there are a good number, I shall introduce to you the members of a family from Virginia, first giving you an idea of their condition in that country, previous to their migration to the west. The land which they and their ancestors have possessed for a hundred years, having been constantly forced to produce crops of one kind or another, is now completely worn out. It exhibits only a superficial layer of red clay, cut up by deep ravines, through which much of the soil has been conveyed to some more fortunate neighbor, residing in a yet rich and beautiful valley. Their strenuous efforts to render it productive have failed. They dispose of everything too cumbrous or expensive for them to remove, retaining only a few horses, a servant or two, and such implements of husbandry and other articles as may be necessary on their journey, or useful when they arrive at the spot of their choice.

I think I see them at this moment harnessing their horses, and attaching them to their wagons, which are already filled with bedding, provisions, and the younger children, while on their outside are fastened spinning-wheels and looms, and a bucket filled with tar and tallow swings between the hind wheels. Several axes are secured to the bolster, and the feeding-trough of the horses contains pots, kettles, and pans. The servant, now become a driver, rides the near saddled horse, the wife is mounted on another, the worthy husband shoulders his gun, and his sons, clad in plain substantial homespun, drive the cattle ahead, and lead the procession, followed by the hounds and other dogs. Their day's journey is short, and not agreeable; the cattle, stubborn or wild, frequently leave the road for the woods, giving the travellers much trouble; the harness of the horses here and there gives way, and needs immediate repair; a basket, which has accidentally dropped, must be gone after, for nothing that they have can be spared; the roads are bad, and now and then all hands are called to push on the wagon, or prevent it from upsetting. Yet by sunset they have proceeded perhaps twenty miles. Rather fatigued, all assemble round the fire, which has been lighted, supper is prepared, and a camp being erected, there they pass the night.

Days and weeks, nay months, of unremitting toil, pass before they gain the end of their journey. They have crossed both the Carolinas, Georgia, and Alabama. They have been travelling from the beginning of May to that of September, and with heavy hearts they traverse the State of Mississippi. But now, arrived on the banks of the broad stream, they gaze in amazement on the dark deep woods around them. Boats of various kinds they see gliding downwards with the current, while others slowly ascend against it. A few inquiries are made at the nearest dwelling, and assisted by the inhabitants with their boats, and canoes, they at once cross the Mississippi, and select their place of habitation.

The exhalations arising from the swamps and morasses around them have a powerful effect on these new settlers, but all are intent on preparing for the winter. A small patch of ground is cleared by the axe and the fire, a temporary cabin is erected, to each of the cattle is attached a jingling bell before it is let loose into the neighboring cane-brake, and the horses remain about the house, where they find sufficient food at that season. The first trading-boat that stops at their landing, enables them to provide themselves with some flour, fish-hooks, and ammunition, as well as other commodities. The looms are mounted, the spinning-wheels soon furnish some yarn, and in a few weeks the family throw off their ragged clothes, and array themselves in suits adapted to the climate. The father and sons meanwhile have sown turnips and other vegetables; and from some Kentucky flatboat, a supply of live poultry has been procured.

October tinges the leaves of the forest, the morning dews are heavy, the days hot, the nights chill, and the unacclimated family in a few days are attacked with ague. The lingering disease almost prostrates their whole faculties, and one seeing them at such a period might well call them sallow and sickly. Fortunately the unhealthy season soon passes over, and the hoar-frosts make their appearance. Gradually each individual recovers strength. The largest ash-trees are felled; their trunks are cut, split, and corded in front of the building; a large fire is lighted at night on the edge of the water, and soon a steamer calls to purchase the wood, and thus add to their comforts during the winter.

The first fruit of their industry imparts new courage to them; their exertions multiply, and when spring returns, the place has a cheerful look. Venison, Bear's-flesh, Wild Turkeys, Ducks and Geese, with now and then some fish, have served to keep up their strength, and now their enlarged field is planted with corn, potatoes, and pumpkins. Their stock of cattle, too, has augmented; the steamer, which now stops there as if by preference, buys a calf or a pig, together with the whole of their wood. Their store of provisions is renewed, and brighter rays of hope enliven their spirits.

Who is he of the settlers on the Mississippi that cannot realize some profit? Truly none who is industrious. When the autumnal months return, all are better prepared to encounter the ague which then prevails. Substantial food, suitable clothing, and abundant firing, repel its attacks; and before another twelvemonth has elapsed the family is naturalized. The sons have by this time discovered a swamp covered with excellent timber, and as they have seen many great rafts of saw logs, bound for the mills of New Orleans, floating past their dwelling, they resolve to try the success of a little enterprise. Their industry and prudence have already enhanced their credit. A few cross-saws are purchased, and some broad-wheeled "carry-logs" are made by themselves. Log after log, is hauled to the bank of the river, and in a short time their first raft is made on the shore, and loaded with cord-wood. When the next freshet sets it afloat, it is secured by long grape-vines or cables, until the proper time being arrived, the husband and sons embark on it, and float down the mighty stream.

After encountering many difficulties, they arrive in safety at New Orleans, where they dispose of their stock, the money obtained for which may be said to be all profit, supply themselves with such articles as may add to their convenience or comfort, and with light hearts procure a passage on the upper deck of a steamer, at a very cheap rate, on account of the benefit of their labor in taking in wood or otherwise.

And now the vessel approaches their home. See the joyous mother and daughters as they stand on the bank! A store of vegetables lies around them, a large tub of fresh milk is at their feet, and in their hands are plates, filled with rolls of butter. As the steamer stops, three broad straw hats are waved from the upper deck, and soon husband and wife, brothers and sisters, are in each other's embrace. The boat carries off the provisions for which value has been left, and as the captain issues his orders for putting on the steam, the happy family enter their humble dwelling. The husband gives his bag of dollars to the wife, while the sons present some token of affection to the sisters. Surely, at such a moment, the squatters are richly repaid for all their labors.

Every successive year has increased their savings. They now possess a large stock of horses, cows, and hogs, with abundance of provisions, and domestic comfort of every kind. The daughters have been married to the sons of neighboring squatters, and have gained sisters to themselves by the marriage of their brothers. The government secures to the family the lands on which, twenty years before, they settled in poverty and sickness. Larger buildings are erected on piles, secure from the inundations; where a single cabin once stood, a neat village is now to be seen; warehouses, stores, and workshops increase the importance of the place. The squatters live respected, and in due time die regretted by all who knew them.

Thus are the vast frontiers of our country peopled, and thus does cultivation, year after year, extend over the western wilds. Time will no doubt be, when the great valley of the Mississippi, still covered with primeval forests interspersed with swamps, will smile with corn-fields and orchards, while crowded cities will rise at intervals along its banks, and enlightened nations will rejoice in the bounties of Providence.

The Runaway

Never shall I forget the impression made on my mind by the *rencontre* which forms the subject of this article, and I even doubt if the relation of it will not excite in that of my reader emotions of varied character.

Late in the afternoon of one of those sultry days which render the atmosphere of the Louisiana swamps pregnant with baneful effluvia, I directed my course towards my distant home, laden with a pack, consisting of five or six Wood Ibises, and a heavy gun, the weight of which, even in those days, when my natural powers were unimpaired, prevented me from moving with much speed. Reaching the banks of a miry bayou, only a few yards in breadth, but of which I could not ascertain the depth, on account of the muddiness of its waters, I thought it might be dangerous to wade through it with my burden, for which reason, throwing to the opposite side each of my heavy birds in succession, together with my gun, powder-flask, and shot-bag, and drawing my hunting-knife from its scabbard, to defend myself, if need should be, against Alligators, I entered the water, followed by my faithful dog. As I advanced carefully, and slowly, "Plato" swam around me, enjoying the refreshing influence of the liquid element that cooled his fatigued and heated frame. The water deepened, as did the mire of its bed; but with a stroke or two I gained the shore.

Scarcely had I stood erect on the opposite bank, when my dog ran to me, exhibiting marks of terror; his eyes seeming ready to burst from their sockets, and his mouth grinning with the expression of hatred, while his feelings found vent in a stifled growl. Thinking that all this was produced by the scent of a Wolf or Bear, I stooped to take up my gun, when a stentorian voice commanded me to "stand still, or die!" Such a *qui vive* in these woods was as unexpected as it was rare. I instantly raised and cocked my gun; and although I did not yet perceive the individual who had thus issued so peremptory a mandate, I felt determined to combat with him for the free passage of the grounds. Presently a tall, firmly built negro emerged from the bushy underwood, where until that moment he must have been crouched, and in a louder voice repeated his injunction. Had I pressed a trigger, his life would have instantly terminated; but observing that the gun which he aimed at my breast, was a wretched, rusty piece, from which fire could not readily be produced, I felt little fear, and therefore did not judge it necessary to proceed at once to extremities. I laid my gun at my side, tapped my dog quietly, and asked the man what he wanted.

My forbearance, and the stranger's long habit of submission, produced the most powerful effect on his mind. "Master," said he, "I am a runaway; I might perhaps shoot you down; but God forbids it, for I feel just now as if I saw him ready to pass his judgment against me for such a foul deed, and I ask mercy at your hands. For God's sake, do not kill me, master!" "And why," answered I, "have you left your quarters, where certainly you must have fared better than in these unwholesome swamps?" "Master, my story is a short, but a sorrowful one. My camp is close by, and, as I know you cannot reach home this night, if you will follow me there, depend upon *my honor* you shall be safe until the morning, when I will carry your birds, if you choose, to the great road."

The large, intelligent eyes of the negro, the complacency of his manners, and the tones of his voice, I thought invited me to venture; and as I felt that I was at least his equal, while moreover, I had my dog to second me, I answered that I would *follow him*. He observed the emphasis laid on the words, the meaning of which he seemed to understand so thoroughly that, turning to me, he said, "There, master, take my butcher's knife, while I throw away the flint and priming from my gun!" Reader, I felt confounded: this was too much for me: I refused the knife, and told him to keep his piece ready, in case we might accidentally meet a Cougar or a Bear.

Generosity exists everywhere. The greatest monarch acknowledges its impulse, and all around him, from the lowliest menial to the proud nobles that encircle his throne, at times experience that overpowering sentiment. I offered to shake hands with the runaway. "Master," said he, "I beg you thanks," and with this he gave me a squeeze that alike impressed me with the goodness of his heart and his great physical strength. From that moment we proceeded through the woods together. My dog smelt at him several times, but as he heard me speak in my usual tone of voice, he soon left us and rambled around as long as my whistle was unused. As we proceeded, I observed that he was guiding me towards the setting of the sun, and quite contrary to my homeward course. I remarked this to him, when he with the greatest simplicity replied, "Merely for our security."

After trudging along for some distance, and crossing several bayous, at all of which he threw his gun and knife to the opposite bank, and stood still until I had got over, we came to the borders of an immense cane-brake, from which I had, on former occasions, driven and killed several Deer. We entered, as I had frequently done before, now erect, then on "all fours." He regularly led the way, divided here and there the tangled stalks, and, whenever we reached a fallen tree, assisted me in getting over it, with all possible care. I saw that he was a perfect Indian in his knowledge of the woods, for he kept a direct course as precisely as any "Red-skin" I ever travelled with. All of a sudden he emitted a loud shriek, not unlike that of an Owl, which so surprised me, that I once more instantly levelled my gun. "No harm, master, I only give notice to my wife and children I am coming." A tremulous answer of the same nature gently echoed through the tree tops. The runaway's lips separated with an expression of gentleness and delight, when his beautiful set of ivory teeth seemed to smile through the dusk of evening that was thickening around us. "Master," said he, "my wife, though black, is as beautiful to me as the President's wife is to him; she is my queen, and I look on our young ones as so many princes; but you shall see them all, for here they are, thank God."

There, in the heart of the cane-brake, I found a regular camp. A small fire was lighted, and on its embers lay gridling some large slices of venison. A lad nine or ten years old was blowing the ashes from some fine sweet potatoes. Various articles of household furniture were carefully disposed around, and a large pallet of Bear and Deer skins, seemed to be the resting-place of the whole family. The wife raised not her eyes towards mine, and the little ones, three in number, retired into a corner, like so many discomfited Raccoons; but the Runaway, bold, and apparently happy, spoke to them in such cheering words, that at once one and all seemed to regard me as one sent by Providence to relieve them from all their troubles. My clothes were hung up by them to dry, and the negro asked if he might clean and grease my gun, which I permitted him to do, while the wife threw a large piece of Deer's flesh to my dog, which the children were already caressing.

Only think of my situation, reader! Here I was, ten miles at least from home, and four or five from the nearest plantation, in the camp of runaway slaves, and quite at their mercy. My eyes involuntarily followed their motions, but as I thought I perceived in them a strong desire to make me their confidant and friend,

I gradually relinquished all suspicions. The venison and potatoes looked quite tempting, and by this time I was in a condition to relish much less savory fare; so, on being humbly asked to divide the viands before us, I partook of as hearty a meal as I had ever done in my life.

Supper over, the fire was completely extinguished, and a small lighted pine-knot placed in a hollowed calabash. Seeing that both the husband and the wife were desirous of communicating something to me, I at once and fearlessly desired them to unburden their minds, when the Runaway told me a tale of which the following is the substance.

About eighteen months before, a planter, residing not very far off, having met with some losses was obliged to expose his slaves at a public sale. The value of his negroes was well known, and on the appointed day the auctioneer laid them out in small lots, or offered them singly, in the manner which he judged most advantageous to their owner. The Runaway, who was well known as being the most valuable next to his wife, was put up by himself for sale, and brought an immoderate price. For his wife, who came next, and alone, eight hundred dollars were bidden and paid down. Then the children were exposed, and, on account of their breed, brought high prices. The rest of the slaves went off at rates corresponding to their qualifications.

The Runaway chanced to be bought by the overseer of the plantation; the wife was bought by an individual residing about a hundred miles off, and the children went to different places along the river. The heart of the husband and father failed him under this dire calamity. For a while he pined in sorrow under his new master; but having marked down in his memory the names of the different persons who had purchased each dear portion of his family, he feigned illness, if indeed, he whose affections had been so grievously blasted could be said to feign it, refrained from food for several days, and was little regarded by the overseer, who felt himself disappointed in what he had considered a bargain.

On a stormy night, when the elements raged with all the fury of a hurricane, the poor negro made his escape, and being well acquainted with all the neighboring swamps, at once made directly for the cane-brake in the centre of which I found his camp. A few nights afterwards he gained the abode of his wife, and the very next after their meeting, he led her away. The children, one after another, he succeeded in stealing, until at last the whole of the objects of his love were under his care.

To provide for five individuals was no easy task in those wilds, which after the first notice was given of the wonderful disappearance of this extraordinary family, were daily ransacked by armed planters. Necessity, it is said, will bring the Wolf from the forest. The Runaway seems to have well understood the maxim, for under the cover of night he approached his first master's plantation, where he had ever been treated with the greatest kindness. The house-servants knew him too well not to aid him to the best of their power, and at the approach of each morning he returned to his camp with an ample supply of provisions. One day, while in search of wild fruits, he found a Bear dead before the muzzle of a gun that had been set for the purpose. Both articles he carried to his home. His friends at the plantation managed to supply him with some ammunition, and on damp and cloudy days he first ventured to hunt around his camp. Possessed of courage and activity, he gradually became more careless, and rambled farther in search of game. It was on one of his excursions that I met him, and he assured me the noise which I made in passing the bayou had caused him to lose the chance of killing a fine Deer, "although," said he, "my old musket misses fire sadly too often."

The Runaways, after disclosing their secret to me, both rose from their seat, with eyes full of tears. "Good master, for God's sake, do something for us and our children," they sobbed forth with one accord. Their little ones lay sound asleep in the fearlessness of their innocence. Who could have heard such a tale without emotion? I promised them my most cordial assistance. They both sat up that night to watch my repose, and I slept close to their urchins, as if on a bed of the softest down.

Day broke so fair, so pure, and so gladdening that I told them such heavenly appearances were ominous of good, and that I scarcely doubted of obtaining their full pardon. I desired them to take their children with them, and promised to accompany them to the plantation of their first master. They gladly obeyed. My Ibises were hung round their camp, and, as a memento of my having been there, I notched several trees; after which I bade adieu, perhaps for the last time, to that cane-brake. We soon reached the plantation, the owner of which, with whom I was well acquainted, received me with all the generous kindness of a Louisiana planter. Ere an hour had elapsed, the Runaway and his family were looked upon as his own. He afterwards repurchased them from their owners, and treated them with his former kindness; so that they were rendered as happy as slaves generally are in that country, and continued to cherish that attachment to each other which had led to their adventures. Since this event happened, it has, I have been informed, become illegal to separate slave families without their consent.

Natchez in 1820

One clear, frosty morning in December I approached in my flatboat the city of Natchez. The shores were crowded with boats of various kinds, laden with the produce of the Western country; and there was a bustle about them such as you might see at a general fair, each person being intent on securing the advantage of a good market. Yet the scene was far from being altogether pleasing, for I was yet "under the hill"; but on removing from the Lower Town I beheld the cliffs on which the city, properly so called, has been built. Vultures unnumbered flew close along the ground on expanded pinions, searching for food; large pines and superb magnolias here and there raised their evergreen tops towards the skies; while on the opposite shores of the Mississippi vast alluvial beds stretched along, and the view terminated with the dense forest. Steamers moved rapidly on the broad waters of the great stream; the sunbeams fell with a peculiarly pleasant effect on the distant objects; and as I watched the motions of the White-headed Eagle while pursuing the Fishing Hawk, I thought of the wonderful ways of that Power to whom I too owe my existence.

Before reaching the land I had observed that several saw-mills were placed on ditches or narrow canals, along which the water rushed from the inner swamps towards the river, and by which the timber is conveyed to the shore; and, on inquiring afterwards, I found that one of those temporary establishments had produced a net profit of upwards of six thousand dollars in a single season.

There is much romantic scenery about Natchez. The Lower Town forms a most remarkable contrast with the Upper; for in the former the houses were not regularly built, being generally dwellings formed of the abandoned flatboats, placed in rows, as if with the view of forming a long street. The inhabitants formed a medley which it is beyond my power to describe; hundreds of laden carts and other vehicles jogged along the declivity between the two towns; but when, by a very rude causeway, I gained the summit, I was relieved by the sight of an avenue of those beautiful trees called here the Pride of China. In the Upper Town I found the streets all laid off at right angles to each other, and tolerably well lined with buildings constructed with painted bricks or boards.

The agricultural richness of the surrounding country was shown by the heaps of cotton bales and other produce that encumbered the streets. The churches, however, did not please me; but as if to make up for this, I found myself unexpectedly accosted by my relative, Mr. Berthoud, who presented me with letters from my wife and sons. These circumstances put me in high spirits, and we proceeded towards the best hotel in the place, that of Mr. Garnier. The house, which was built on the Spanish plan, and of great size, was surrounded by large verandas overlooking a fine garden, and stood at a considerable distance from any other. At this period the city of Natchez had a population not exceeding three thousand individuals. I have not visited it often since, but I have no doubt that, like all the other towns in the western district of our country, it has greatly increased. It possessed a bank, and the mail arrived there thrice in the week from all parts of the Union.

The first circumstance that strikes a stranger is the mildness of the temperature. Several vegetables as pleasing to the eye as agreeable to the palate, and which are seldom seen in our Eastern markets before May, were here already in perfection. The Pewee Fly-catcher had chosen the neighborhood of the city for its winter quarters, and our deservedly famed Mocking-bird sang and danced gratis to every passer by. I was surprised to see the immense number of Vultures that strode along the streets or slumbered on the roofs. The country for many miles inland is gently undulated. Cotton is produced abundantly, and wealth and happiness have taken up their abode under most of the planters' roofs, beneath which the weaned traveller or the poor wanderer in search of a resting-place is sure to meet with comfort and relief. Game is abundant, and the free Indians were wont in those days to furnish the markets with ample supplies of vension and Wild Turkey. The Mississippi, which bathes the foot of the hill some hundred feet below the town, supplies the inhabitants with fish of various kinds. The greatest deficiency is that of water, which for common purposes is dragged on sledges or wheels from the river, while that used for drinking is collected in tanks from the roofs, and becomes very scarce during protracted droughts. Until of late years the orange-tree bore fruit in the open air; but, owing to the great change that has taken place in the temperature, severe though transient frosts occasionally occur, which now prevent this plant from coming to perfection in the open air.

The remains of an old Spanish fort are still to be seen at a short distance from the city. If I am correctly informed, about two years previous to this visit of mine a large portion of the hill near it gave way, sank about a hundred feet, and carried many of the houses of the Lower Town into the river. This, it would appear, was occasioned by the quicksand running springs that flow beneath the strata of mixed pebbles and clay of which the hill is composed. The part that has subsided presents the appearance of a basin or bowl, and is used as a depot for the refuse of the town, on which the Vultures feed when they can get nothing better. There it was that I saw a White-headed Eagle chase one of those filthy birds, knock it down, and feast on the entrails of a horse which the Carrion Crow had partly swallowed.

I did not meet at Natchez many individuals fond of ornithological pursuits, but the hospitality with which I was received was such as I am not likely to forget. Mr. Garnier subsequently proved an excellent friend to me, as you may find elsewhere recorded. Of another individual, whose kindness to me is indelibly impressed on my heart, I would say a few words, although he was such a man as Fenelon alone could describe. Charles Carré was of French origin, the son of a nobleman of the old regime. His acquirements and the benevolence of his disposition were such that when I first met him I could not help looking upon him as another Mentor. Although his few remaining locks were gray, his countenance still expressed the gayety and buoyant feelings of youth. He had the best religious principles; for his heart and his purse were ever open to the poor. Under his guidance it was that I visited the whole neighborhood of Natchez; for he was acquainted with all its history, from the period at which it had first come under the power of the Spaniards to that of their expulsion from the country, its possession by the French, and subsequently by ourselves. He was also well versed in the Indian languages, spoke French with the greatest purity, and was a religious poet. Many a pleasant hour have I spent in his company; but alas! he has gone the way of all the earth!

The Lost Portfolio

While I was at Natchez, on the 31st of December, 1820, my kind friend, Nicholas Berthoud, Esq., proposed to me to accompany him in his keel-boat to New Orleans. At one o'clock the steam-boat "Columbus" hauled off from the landing and took our bark in tow. The steamer was soon ploughing along at full speed, and little else engaged our minds than the thought of our soon arriving at the emporium of the commerce of the Mississippi. Towards evening, however, several inquiries were made respecting particular portions of the luggage, among which ought to have been one of my portfolios, containing a number of drawings made by me while gliding down the Ohio and Mississippi from Cincinnati to Natchez, and of which some were to me peculiarly valuable, being of birds previously unfigured, and perhaps undescribed. The portfolio was nowhere to be found, and I recollected that I had brought it under my arm to the margin of the stream, and there left it to the care of one of my friend's servants, who, in the hurry of our departure, had neglected to take it on board. Besides the drawings of birds, there was in this collection a sketch in black chalk to which I always felt greatly attached while from home. It is true the features which it represented were indelibly engraved in my heart; but the portrait of her to whom I owe so much of the happiness that I have enjoyed was not the less dear to me. When I thought during the following night of the loss I had sustained in consequence of my own negligence, imagined the possible fate of the collection, and saw it in the hands of one of the numerous boatmen lounging along the shores, who might paste the drawings to the walls of his cabin, nail them to the steering-oars of his flatboat, or distribute them among his fellows, I felt little less vexed than I did some years before when the rats, as you know, devoured a much larger collection.

It was useless to fret myself, and so I began to devise a scheme for recovering the drawings. I wrote to Mr. Gamier and my venerable friend Charles Carré. Mr. Berthoud also wrote to a mercantile acquaintance. The letters were forwarded to

Natchez from the first landing-place at which we stopped, and in the course of time we reached the great eddy running by the levee, or artificial embankment, at New Orleans. But before I present you with the answers to the letters sent to our acquaintances at Natchez, allow me to offer a statement of our adventures upon the Mississippi.

After leaving the eddy at Natchez, we passed a long file of exquisitely beautiful bluffs. At the end of twenty hours we reached Bayou Sara, where we found two brigs at anchor, several steamers, and a number of flatboats, the place being of considerable mercantile importance. Here the "Columbus" left us to shift for ourselves, her commander being anxious to get to Baton Rouge by a certain hour, in order to secure a good cargo of cotton. We now proceeded along the great stream, sometimes floating and sometimes rowing. The shores gradually became lower and flatter, orange-trees began to make their appearance around the dwellings of the wealthy planters, and the verdure along the banks assumed a brighter tint. The thermometer stood at 68° in the shade at noon; Butterflies fluttered among the flowers, of which many were in full blow; and we expected to have seen Alligators half awake floating on the numberless logs that accompanied us in our slow progress. The eddies, were covered with Ducks of various kinds, more especially with the beautiful species that breeds by preference on the great sycamores that every now and then present themselves along our southern waters. Baton Rouge is a very handsome place, but at present I have no time to describe it. Levees now began to stretch along the river, and wherever there was a sharp point on the shore, negroes were there amusing themselves by raising shrimps, and now and then a catfish, with scooping-nets.

The river increased in breadth and depth, and the sawyers and planters, logs so called, diminished in number the nearer we drew towards the famed city. At every bend we found the plantations increased, and now the whole country on both sides became so level and destitute of trees along the water's edge that we could see over the points before us, and observe the great stream stretching along for miles. Within the levees the land is much lower than the surface of the river when the water is high; but at this time we could see over the levee from the deck of our boat only the upper windows of the planters' houses, or the tops of the trees about them, and the melancholy-looking cypresses covered with Spanish moss forming the background. Persons rode along the levees at full speed; Pelicans, Gulls, Vultures, and Carrion Crows sailed over the stream, and at times there came from the shore a breeze laden with the delicious perfume of the orange-trees, which were covered with blossoms and golden fruits.

Having passed Bayou Lafourche, our boat was brought to on account of the wind, which blew with violence. We landed, and presently made our way to the swamps, where we shot a number of those beautiful birds called Boat-tailed Grakles. The Mocking-birds on the fence stakes saluted us with so much courtesy and with such delightful strains that we could not think of injuring them; but we thought it no harm to shoot a whole covey of Partridges. In the swamps we met with warblers of various kinds, lively and beautiful, waiting in these their winter retreats for the moment when Boreas should retire to his icy home, and the gentle gales of the South should waft them toward their breeding-places in the North. Thousands of Swallows flew about us, the Cat-birds mewed in answer to their chatterings, the Cardinal Grosbeak elevated his glowing crest as he stood perched on the magnolia branch, the soft notes of the Doves echoed among the woods, nature smiled upon us, and we were happy.

On the fourth of January we stopped at Bonnet Carré, where I entered a house to ask some questions about birds. I was received by a venerable French gentleman, whom I found in charge of about a dozen children of both sexes, and who was delighted to hear that I was a student of nature. He was well acquainted with my old friend Charles Carré, and must, I thought, be a good man, for he said he never suffered any of his pupils to rob a bird of her eggs or young, although, said he with a smile, "they are welcome to peep at them and love them." The boys at once surrounded me, and from them I received satisfactory answers to most of my queries respecting birds.

The 6th of January was so cold that the thermometer fell to 30°, and we had seen ice on the running-boards of our keel-boat. This was quite unlooked for, and we felt uncomfortable; but before the middle of the day, all nature was again in full play. Several beautiful steamers passed us. The vegetation seemed not to have suffered from the frost; green peas, artichokes, and other vegetables were in prime condition. This reminds me that on one of my late journeys I ate green peas in December in the Floridas, and had them once a week at least in my course over the whole of the Union, until I found myself and my family feeding on the same vegetable more than a hundred miles to the north of the St. John's River in New Brunswick.

Early on the 7th, thousands of tall spars, called masts by the mariners, came in sight; and as we drew nearer, we saw the port filled with ships of many nations, each bearing the flag of its country. At length we reached the levee, and found ourselves once more at New Orleans. In a short time my companions dispersed, and I commenced a search for something that might tend to compensate me for the loss of my drawings.

On the 16th of March following, I had the gratification of receiving a letter from Mr. A. P. Bodley, of Natchez, informing me that my portfolio had been found and deposited at the office of the "Mississippi Republican," whence an order from me would liberate it. Through the kindness of Mr. Garnier, I received it on the 5th of April. So very generous had been the finder of it, that when I carefully examined the drawings in succession, I found them all present and uninjured, save one, which had probably been kept by way of commission.

The Original Painter

As I was lounging one fair and very warm morning on the levee at New Orleans, I chanced to observe a gentleman whose dress and other accompaniments greatly attracted my attention. I wheeled about, and followed him for a short space, when, judging by everything about him that he was a true original, I accosted him.

But here, kind reader, let me give you some idea of his exterior. His head was covered by a straw hat, the brim of which might cope with those worn by the fair sex in 1830; his neck was exposed to the weather; the broad frill of a shirt, then fashionable, flapped about his breast, whilst an extraordinary collar, carefully arranged, fell over the top of his coat. The latter was of a light green color, harmonizing well with a pair of flowing yellow nankeen trousers, and a pink waistcoat, from the bosom of which, amidst a large bunch of the splendid flowers of the magnolia, protruded part of a young Alligator, which seemed more anxious to glide through the muddy waters of some retired swamp than to spend its life swinging to and fro among folds of the finest lawn. The gentleman held in one hand a cage full of richly-plumed Nonpareils, whilst in the other he sported a silk umbrella, on which I could plainly read, "Stolen from I," these words being painted in large white characters. He walked as if conscious of his own importance—that is, with a good deal of pomposity, singing, "My love is but a lassie yet," and that with such thorough imitation of the Scotch emphasis that had not his physiognomy brought to my mind a denial of his being from "within a mile of Edinburgh," I should have put him down in my journal for a true Scot. But no: his tournure, nay, the very shape of his visage, pronounced him an American from the farthest parts of our eastern Atlantic shores.

All this raised my curiosity to such a height that I accosted him with, "Pray, sir, will you allow me to examine the birds you have in that cage?" The gentleman stopped, straightened his body, almost closed his left eye, then spread his legs apart, and, with a look altogether quizzical, answered, "Birds, sir; did you say birds?" I nodded, and he continued, "What the devil do you know about birds, sir?"

Reader, this answer brought a blush into my face. I felt as if caught in a trap; for I was struck by the force of the gentleman's question—which, by the way, was not much in discordance with a not unusual mode of granting an answer in the United States. Sure enough, thought I, little or perhaps nothing do I know of the nature of those beautiful denizens of the air; but the next moment vanity gave me a pinch, and urged me to conceive that I knew at least as much about birds as the august personage in my presence. "Sir," replied I, "I am a student of Nature, and admire her works, from the noblest figure of man to the crawling reptile which you have in your bosom."—"Ah!" replied he, "a-a-a naturalist, I presume!"—"Just so, my good sir," was my answer. The gentleman gave me the cage; and I observed, from the corner of one of my eyes, that his were cunningly inspecting my face. I examined the pretty Finches as long as I wished, returned the cage, made a low bow, and was about to proceed on my walk, when this odd sort of being asked me a question quite accordant with my desire of knowing more of him: "Will you come with me, sir? If you will, you shall see some more curious birds, some of which are from different parts of the world. I keep quite a collection." I assured him I should feel gratified, and accompanied him to his lodgings.

We entered a long room, where, to my surprise, the first objects that attracted my attention were a large easel with a full-length unfinished portrait upon it, a table with palettes and pencils, and a number of pictures of various sizes placed along the walls. Several cages containing birds were hung near the windows, and two young gentlemen were busily engaged in copying some finished portraits. I was delighted with all I saw. Each picture spoke for itself: the drawing, the coloring, the handling, the composition, and the keeping—all proved, that, whoever was the artist, he certainly was possessed of superior talents.

I did not know if my companion was the painter of the picture, but, as we say in America, I strongly guessed, and, without waiting any longer, paid him the compliments which I thought he fairly deserved. "Ay," said he, "the world is pleased with my work. I wish I were so too; but time and industry are required, as well as talents, to make a good artist. If you will examine the birds, I'll to my labor." So saying, the artist took up his palette, and was searching for a rest-stick; but not finding the one with which he usually supported his hand, he drew the rod of a gun, and was about to sit, when he suddenly threw down his implements on the table, and, taking the gun, walked to me and asked if "I had ever seen a percussion-lock." I had not, for that improvement was not yet in vogue. He not only explained the superiority of the lock in question, but undertook to prove that it was capable of acting effectually under water. The bell was rung, a flat basin of water was produced, the gun was charged with powder, and the lock fairly immersed. The report terrified the birds, causing them to beat against the gilded walls of their prisons. I remarked this to the artist. He replied, "The devil take the birds!—more of them in the market; why, sir, I wish to show you that I am a marksman as well as a painter." The easel was cleared of the large picture, rolled to the further end of the room, and placed against the wall. The gun was loaded in a trice, and the painter, counting ten steps from the easel, and taking aim at the supporting-pin on the left, fired. The bullet struck the head of the wooden pin fairly, and sent the splinters in all directions. "A bad shot, sir," said this extraordinary person. "The ball ought to have driven the pin farther into the hole, but it struck on one side; I'll try at the hole itself." After reloading his piece, the artist took aim again, and fired. The bullet this time had accomplished its object, for it had passed through the aperture and hit the wall behind. "Mr.—, ring the bell and close the windows," said the painter, and, turning to me, continued, "Sir, I will show you the *ne plus ultra* of shooting." I was quite amazed, and yet so delighted that I bowed my assent. A servant having appeared, a lighted candle was ordered. When it arrived, the artist placed it in a proper position, and retiring some yards, put out the light with a bullet, in the manner which I have elsewhere in this volume described. When light was restored, I observed the uneasiness of the poor little Alligator, as it strove to effect its escape from the artist's waistcoat. I mentioned this to him. "True, true," he replied. "I had quite forgot the reptile; he shall have a dram"; and unbuttoning his vest, unclasped a small chain, and placed the Alligator in the basin of water on the table.

Perfectly satisfied with the acquaintance which I had formed with this renowned artist, I wished to withdraw, fearing I might inconvenience him by my presence. But my time was not yet come. He bade me sit down, and paying no more attention to the young pupils in the room than if they had been a couple of cabbages, said, "If you have leisure and will stay awhile, I will show you how I paint, and will relate to you an incident of my life which will prove to you how sadly situated an artist is at times." In full expectation that more eccentricities were to be witnessed, or that the story would prove a valuable one, even to a naturalist, who is seldom a painter, I seated myself at his side, and observed with interest how adroitly he transferred the colors from his glistening palette to the canvas before him. I was about to compliment him on his facility of touch, when he spoke as follows:—

"This is, sir, or, I ought to say rather, this will be the portrait of one of our best navy officers—a man as brave as Caesar, and as good a sailor as ever walked the deck of a seventy-four. Do you paint, sir?" I replied, "Not yet."—"Not yet! what do you mean?"—"I mean what I say: I intend to paint as soon as I can draw better than I do at present."—"Good," said he; "you are quite right. To draw is the first object; but, sir, if you should ever paint, and paint portraits, you will often meet with difficulties. For instance, the brave Commodore of whom this is the portrait, although an excellent man at everything else, is the worst sitter I ever saw; and the incident I promised to relate to you, as one curious enough, is connected with his bad mode of sitting. Sir, I forgot to ask if you would take any refreshment—a glass of wine, or" I assured him I needed nothing more than his agreeable company, and he proceeded. "Well, sir, the first morning that the Commodore came to sit, he was in full uniform, and with his sword at his side. After a few moments of conversation, and when all was ready on my part, I bade him ascend this throne, place himself in the attitude which I contemplated, and assume an air becoming an officer of the navy. He mounted, placed himself as I had desired, but merely looked at me as if I had been a block of stone. I waited a few minutes, when, observing no change on his placid countenance, I ran the chalk over the canvas to form a rough outline. This done, I looked up to his face again, and opened a conversation which I thought would warm his warlike nature; but in vain. I waited and waited, talked and talked, until, my patience—sir, you must know I am not overburdened with phlegm—being almost run out, I rose, threw my palette and brushes on the floor, stamped, walking to and fro about the room, and vociferated such calumnies against our navy that I startled the good Commodore. He still looked at me with a placid countenance, and, as he has told me since, thought I had lost my senses. But I observed him all the while, and, fully as determined to carry my point as he would be to carry off an enemy's ship, I gave my oaths additional emphasis, addressed him as a representative of the navy, and, steering somewhat clear of personal insult, played off my batteries against the craft. The Commodore walked up to me, placed his hand on the hilt of his sword, and told me, in a resolute manner, that if I intended to insult the navy, he would instantly cut off my ears. His features exhibited all the spirit and animation of his noble nature, and as I had now succeeded in rousing the lion, I judged it time to retreat. So, changing my tone, I begged his pardon, and told him he now looked precisely as I wished to represent him. He laughed, and, returning to his seat, assumed a bold countenance. And now, sir, see the picture!"

At some future period I may present you with other instances of the odd ways in which this admired artist gave animation to his sitters. For the present, kind reader, we shall leave him finishing the Commodore, while we return to our proper studies.

A Long Calm at Sea

On the 17th of May, 1826, I left New Orleans on board the ship "Delos" commanded by Joseph Hatch, Esq., of Kennebunk, bound for Liverpool. The steamer "Hercules," which towed the ship, left us several miles outside of the Balize, about ten hours after our departure; but there was not a breath of wind, the waters were smoother than the prairies of the Opelousas, and notwithstanding our great display of canvas, we lay like a dead whale, floating at the mercy of the currents. The weather was uncommonly fair, and the heat excessive; and in this helpless state we continued for many days. About the end of a week we had lost sight of the Balize, although I was assured by the commander, that all this while the ship had rarely answered the helm. The sailors whistled for wind, and raised their hands in all directions, anxious as they were to feel some motion in the air; but all to no purpose; it was a dead calm, and we concluded that "Æolus" had agreed with "Neptune" to detain us, until our patience should be fairly tried, or our sport exhausted; for sport we certainly had, both on board and around the ship. I doubt if I can better contribute to your amusement at present than by giving you a short account of the occurrences that took place during this sleepy fit of the being on whom we depended for our progress toward merry England.

Vast numbers of beautiful Dolphins glided by the side of the vessel, glancing like burnished gold through the day, and gleaming like meteors by night. The captain and his mates were expert at alluring them with baited hooks, and not less so at piercing them with five-pronged instruments, which they called grains; and I was delighted with the sport, because it afforded me an opportunity of observing and noting some of the habits of this beautiful fish, as well as several other kinds.

On being hooked, the Dolphin flounces vigorously, shoots off with great impetuosity to the very end of the line, when, being suddenly checked, it often rises perpendicularly several feet out of the water, shakes itself violently in the air, gets disentangled, and thus escapes. But when well secured, it is held in play for a while by the experienced fisher, soon becomes exhausted, and is hauled on board. Some persons prefer pulling them in at once, but they seldom succeed, as the force with which the fish shakes itself on being raised out of the water is generally sufficient to enable it to extricate itself. Dolphins move in shoals, varying from four or five to twenty or more, hunting in packs in the waters, as Wolves pursue their prey on land. The object of their pursuit is generally the Flying-fish, now and then the Bonita; and when nothing better can be had, they will follow the little Rudder-fish, and seize it immediately under the stern of the ship. The Flying-fishes after having escaped for a while by dint of their great velocity, on being again approached by the Dolphin, emerge from the waters, and spreading their broad wing-like fins, sail through the air and disperse in all directions, like a covey of timid Partridges before the rapacious Falcon. Some pursue a direct course, others diverge on either side; but in a short time they all drop into their natural element. While they are travelling in the air, their keen and hungry pursuer, like a greyhound, follows in their wake, and performing a succession of leaps, many feet in extent, rapidly gains upon the quarry, which is often seized just as it falls into the sea.

Dolphins manifest a very remarkable sympathy with each other. The moment one of them is hooked or grained, those in company make up to it, and remain around until the unfortunate fish is pulled on board, when they generally move off together, seldom biting at anything thrown out to them. This, however, is the case only with the larger individuals, which keep apart from the young, in the same manner as is observed in several species of birds; for when the smaller Dolphins are in large shoals, they all remain under the bows of a ship, and bite in succession at any sort of line, as if determined to see what has become of their lost companions, in consequence of which they are often all caught.

You must not suppose that the Dolphin is without its enemies. Who, in this world, man or fish, has not enough of them? Often it conceives itself on the very eve of swallowing a fish, which, after all, is nothing but a piece of lead, with a few feathers fastened to it, to make it look like a Flying-fish, when it is seized and severed in two by the insidious Balacouda, which I have once seen to carry off by means of its sharp teeth, the better part of a Dolphin that was hooked, and already hoisted to the surface of the water.

The Dolphins caught in the Gulf of Mexico during this calm were suspected to be poisonous; and to ascertain whether this was really the case, our cook, who was an African negro, never boiled or fried one without placing beside it a dollar.

If the silver was not tarnished by the time the Dolphin was ready for the table, the fish was presented to the passengers, with an assurance that it was perfectly good. But as not a single individual of the hundred that we caught had the property of converting silver into copper, I suspect that our African sage was no magician.

One morning, that of the 22d of June, the weather sultry, I was surprised on getting out of my hammock, which was slung on deck, to find the water all around swarming with Dolphins, which were sporting in great glee. The sailors assured me that this was a certain "token of wind," and, as they watched the movements of the fishes, added, "ay, and of a fair breeze too." I caught several Dolphins in the course of an hour, after which scarcely any remained about the ship. Not a breath of air came to our relief all that day, no, nor even the next. The sailors were in despair, and I should probably have become despondent also, had not my spirits been excited by finding a very large Dolphin on my hook. When I had hauled it on board, I found it to be the largest I had ever caught. It was a magnificent creature. See how it quivers in the agonies of death! its tail flaps the hard deck, producing a sound like the rapid roll of a drum. How beautiful the changes of its colors! Now it is blue, now green, silvery, golden, and burnished copper! Now it presents a blaze of all the hues of the rainbow intermingled; but, alack! it is dead, and the play of its colors is no longer seen. It has settled into the deep calm that has paralyzed the energies of the blustering winds, and smoothed down the proud waves of the ocean.

The best bait for the Dolphin is a long strip of Shark's flesh. I think it generally prefers this to the semblance of the Flying-fish, which indeed it does not often seize unless when the ship is under way, and it is made to rise to the surface. There are times, however, when hunger and the absence of their usual food will induce the Dolphins to dash at any sort of bait; and I have seen some caught by means of a piece of white linen fastened to a hook. Their appetite is as keen as that of the Vulture, and whenever a good opportunity occurs, they gorge themselves to such a degree that they become an easy prey to their enemies the Balacouda and the Bottle-nosed Porpoise. One that had been grained while lazily swimming immediately under the stern of our ship, was found to have its stomach completely crammed with Flying-fish, all regularly disposed side by side, with their tails downwards—by which I mean to say that the Dolphin always *swallows its prey tail-foremost.* They looked in fact like so many salted Herrings packed in a box, and were to the number of twenty-two, each six or seven inches in length.

The usual length of the Dolphins caught in the Gulf of Mexico is about three feet, and I saw none that exceeded four feet two inches. The weight of one of the latter size was only eighteen pounds; for this fish is extremely narrow in proportion to its length, although rather deep in its form. When just caught, the upper fin, which reaches from the forehead to within a short distance of the tail, is of a fine dark blue. The upper part of the body in its whole length is azure, and the lower parts are of a golden hue, mottled irregularly with deep-blue spots. It seems that they at times enter very shallow water, as in the course of my last voyage along the Florida coast, some were caught in a seine, along with their kinsman the "Cavalier," of which I shall speak elsewhere.

The flesh of the Dolphin is rather firm, very white, and lies in flakes when cooked. The first caught are generally eaten with great pleasure, but when served many days in succession, they become insipid. It is not, as an article of food, equal to the Balacouda, which is perhaps as good as any fish caught in the waters of the Gulf of Mexico.

Still Becalmed

On the 4th of June, we were still in the same plight, although the currents of the Gulf had borne us to a great distance from the place where, as I have informed you, we had amused ourselves with catching Dolphins. These currents are certainly very singular, for they carried us hither and thither, at one time rendering us apprehensive of drifting on the coast of Florida, at another threatening to send us to Cuba. Sometimes a slight motion in the air revived our hopes, swelled our sails a little, and carried us through the smooth waters like a skater gliding on ice; but in a few hours it was again a dead calm.

One day several small birds, after alighting on the spars, betook themselves to the deck. One of them, a female Rice Bunting, drew our attention more particularly, for, a few moments after her arrival, there came down, as if in her wake, a beautiful Peregrine Falcon. The plunderer hovered about for a while, then stationed himself on the end of one of the yard-arms, and suddenly pouncing on the little gleaner of the meadows, clutched her and carried her off in exultation. But, reader, mark the date, and judge besides of my astonishment when I saw the Falcon feeding on the Finch while on wing, precisely with the same ease and composure as the Mississippi Kite might show while devouring high in air a Red-throated Lizard, swept from one of the magnificent trees of the Louisiana woods.

There was a favorite pet on board belonging to our captain, and which was nothing more nor less than the female companion of a cock—in other words, a common hen. Some liked her because she now and then dropped a fresh egg—a rare article at sea, even on board the "Delos"; others, because she exhibited a pleasing simplicity of character; others again, because, when they had pushed her overboard, it gave them pleasure to see the poor thing in terror strike with her feet, and strive to reach her floating home, which she would never have accomplished, however, had it not been for the humane interference of our captain, Mr. Joseph Hatch, of Kennebunk. Kind, good-hearted man! when, several weeks after, the same pet hen accidentally flew overboard, as we were scudding along at a furious rate, I thought I saw a tear stand in his eye, as she floated panting in our wake. But as yet we are becalmed, and heartily displeased at old "Æolus" for overlooking us.

One afternoon we caught two Sharks. In one of them, a female, about seven feet long, we found ten young ones, all alive, and quite capable of swimming, as we proved by experiment; for, on casting one of them into the sea, it immediately made off, as if it had been accustomed to shift for itself. Of another, that had been cut in two, the head half swam off out of our sight. The rest were cut in pieces, as was the old shark, as bait for the Dolphins, which I have already said are fond of such food.

Our captain, who was much intent on amusing me, informed me that the Rudder-fishes were plentiful astern, and immediately set to dressing hooks for the purpose of catching them. There was now some air above us, the cotton sheets aloft bulged out, the ship moved through the water, and the captain and I repaired to the cabin window. I was furnished with a fine hook, a thread line, and some small bits of bacon, as was the captain, and we dropped our bait among the myriads of delicate little fishes below. Up they came, one after another, so fast in succession that, according to my journal, we caught three hundred and seventy in about two hours. What a mess! and how delicious when roasted! If ever I am again becalmed in the Gulf of Mexico, I shall not forget the Rudder-fish. The little things scarcely measured three inches in length; they were thin and deep in form, and afforded excellent eating. It was curious to see them keep to the lee of the rudder in a compact body; and so voracious were they that they actually leaped out of the water at the sight of the bait, as "sunnies" are occasionally wont to do in our rivers. But the very instant that the ship became still, they dispersed around her sides, and would no longer bite. I made a figure of one of them, as indeed I tried to do of every other species that occurred during this deathlike calm. Not one of these fishes did I ever see when crossing the Atlantic, although many kinds at times come close to the stern of any vessel in the great sea, and are called by the same name.

Another time we caught a fine Porpoise, which measured about two yards in length. This took place at night, when the light of the moon afforded me a clear view of the spot. The fish, contrary to custom, was grained, instead of being harpooned; but in such a way and so effectually, through the forehead, that it was thus held fast, and allowed to flounce and beat about the bows of the ship, until the person who had struck it gave the line holding the grains to the captain, slid down upon the bobstays with a rope, and after a while managed to secure it by the tail. Some of the crew then hoisted it on board. When it arrived on deck, it gave a deep groan, flapped with great force, and soon expired. On opening it next morning, eight hours after death, we found its intestines still warm. They were arranged in the same manner as those of a pig; the paunch contained several cuttle-fishes partially digested. The lower jaw extended beyond the upper about three-fourths of an inch, and both were furnished with a single row of conical teeth, about half an inch long, and just so far separated as to admit those of one jaw between the corresponding ones of the other. The animal might weigh about four hundred pounds; its eyes were extremely small, its flesh was considered delicate by some on board; but in my opinion, if it be good, that of a large Alligator is equally so; and on neither do I intend to feast for some time. The captain told me that he had seen these Porpoises leap at times perpendicularly out of the water to the height of several feet, and that small boats have now and then been sunk by their falling into them when engaged with their sports.

During all this time flocks of Pigeons were crossing the Gulf, between Cuba and the Floridas; many a Rose-breasted Gull played around by day; Noddies alighted on the rigging by night; and now and then the Frigate bird was observed ranging high over head in the azure of the cloudless sky.

The directions of the currents were tried, and our captain, who had an extraordinary genius for mechanics, was frequently employed in turning powder-horns and other articles. So calm and sultry was the weather that we had a large awning spread, under which we took our meals and spent the night. At length we got so wearied of it that the very sailors, I thought, seemed disposed to leap overboard and swim to land. But at length, on the thirty-seventh day after our departure, a smart breeze overtook us. Presently there was an extraordinary bustle on board; about twelve the Tortugas light-house bore north of us, and in a few hours more we gained the Atlantic. Æolus had indeed awakened from his long sleep; and on the nineteenth day after leaving the Capes of Florida, I was landed at Liverpool.

LAST VOYAGE TO LOUISIANA

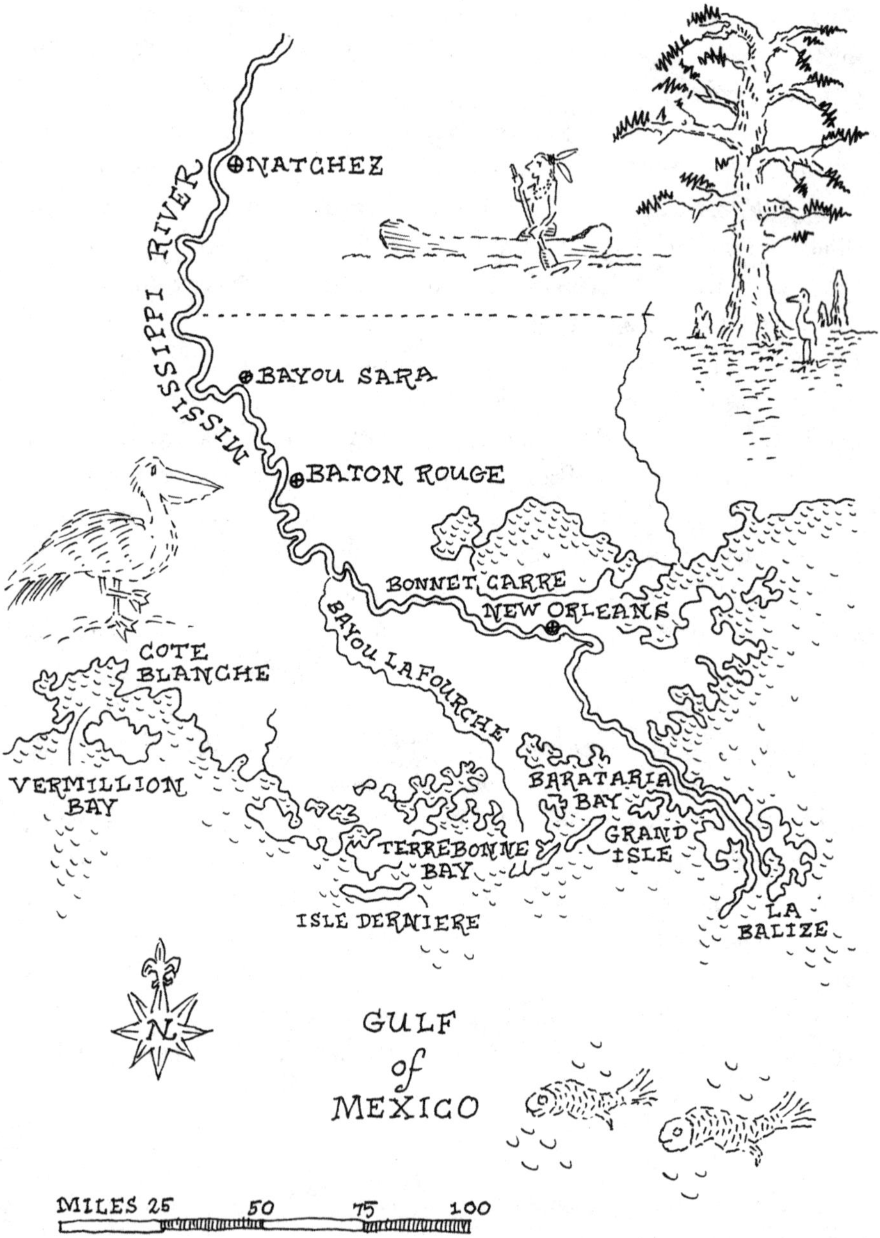

LAST VOYAGE

TO

LOUISIANA

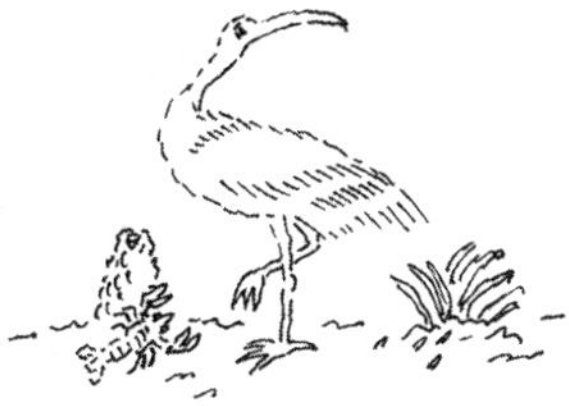

I hope to . . . have my carcass buried under the dark shade of some of our finest Louisiana Magnolias!

—JOHN JAMES AUDUBON, in a letter to Dr Thomas M. Brewer, July 5, 1837

When Audubon left on his last voyage to Louisiana in February 1837, he had been away for almost seven years. During that time the solitude, obscurity, and adversity he had fought to bring the *Birds of America* to life in print were fast fading into the past. His work was now roundly admired, praised, and purchased. There were still obstacles to overcome, and self-decreed deadlines to meet, but he and his family were beginning to regain a more than modest measure of prosperity, materially and socially. His name, if not a household word, was now known in artistic and scientific circles throughout England and the United States. In Washington, on November 9, 1836, a few months before the final Louisiana expedition, he and his son John, on their way south, were invited to dine at the White House with President Andrew Jackson. The main dish was wild turkey. In many respects, these were the glory days, but Audubon was not one to bask. If anything, the dawning of success made him more tenacious in his efforts to complete both the *Birds* and the *Biographies* to perfection.

Completeness depended on finding and drawing the missing birds, thus the need to scour new regions when the season was right. For three years after leaving the Felicianas with Lucy in January 1831, Audubon had thrown himself headlong into the search, continually on the move, spurring himself like a horse in the race to supply the engravers in London and keep the *Birds* on schedule. In October 1831 he had gone south to Charleston, where he met John Bachman, a Lutheran minister and amateur naturalist who was to become a lifelong friend and collaborator. Bachman, his wife Harriet, their nine children, and Harriet's sister, Maria Martin, lived in a spacious house where Audubon was welcomed immediately as a member of the family, a status that moved beyond the symbolic a few years later when both Audubon sons married daughters of Bachman. Maria Martin proved herself to be a gifted artist under Audubon's tutelage. Audubon made two trips along the eastern coast of Florida in 1832; the second, down to the Keys (and the Dry Tortugas) was especially productive in new species. Later that same year he headed north with Lucy, John, and Victor to ramble through Maine and New Brunswick. Both sons were now closely involved in the Audubon enterprise.

A month after the return from New Brunswick, Victor sailed to England to supervise the engravings, now entirely in the hands of Robert Havell Jr. In the summer of 1833, Audubon chartered a brand new schooner, the *Ripley*, for a trip further north, up past Nova Scotia and on to Newfoundland and Labrador. Along with John and four robust young men to help, he combed the coasts of Labrador for nearly three months, shooting and drawing, mainly seabirds and cliff dwellers. After spending the winter in Charleston with the Bachmans, Audubon, Lucy, and John left for England in May 1834. Audubon had planned to explore west Florida and the Gulf Coast, but Victor was overwhelmed with the burden of shouldering the affairs in England all by himself. He needed his father, and Audubon lost no time changing his plans. In England, Audubon settled with the whole family in London and bounded forward with the work at hand. He flattered Havell, touched up and filled in the latest trove of drawings, called door to door on sluggish subscribers, traveled back and forth between London and Edinburgh, and pushed on with the writing of the *Ornithological Biography*. The second volume of the *Biography* was published in December. The next year (1835), the third volume was published. As for the crucial *Birds of America*, the concluding plates of volume three had been printed by the end of 1835. The engravings for the last volume were already well underway when Audubon and his son sailed for the States in August 1836, leaving Lucy and Victor for the time being in London. The barrier islands of Louisiana once more loomed large in Audubon's eyes as he lifted himself higher in the saddle for the final turn to the finish line.

Arriving in New York early in September, Audubon stopped in Philadelphia on his way down to Charleston to look into a bundle of bird skins that had been collected by two young naturalists, the botanist Thomas Nuttall and the ornithologist John Kirk Townsend, on a recent expedition to the Pacific. Audubon managed to buy ninety-three of the skins, thus expanding his range of birds beyond the Rocky Mountains into the Far West. He had dreamed of going there himself all the way back to his descent down the Mississippi in 1820, but circumstances had worked against him time and time again. Barely believing his good luck, he and John finally headed to Charleston, stopping in Washington for the wild turkey dinner with President Jackson and the authorization to use a government revenue cutter for his foray into the Gulf of Mexico. The cutter that was eventually made available in New Orleans turned out to be the *Campbell*, captained by the same Napoleon Coste who had served as first officer on the ship that had taken Audubon to the Keys on his second voyage to Florida in 1832. In Charleston, once more cozily installed in the Bachman house, Audubon spent the winter drawing from the western skins while Maria Martin filled out the compositions with plants, butterflies, and other insects. These were then sent to Victor in London for the final plates.

While Audubon was making history with his painting, history was being made all around him. This would be a truism not worth dwelling on were it not that he had the gift of proximity with some of the major actors and events of his time. This was the case with both the beginning and the end of the last voyage to Louisiana. He had been warned against traveling to Florida because of the recent Seminole uprising. Audubon himself admired Osceola and thought Andrew Jackson's policies towards the southern Indians were tragic, condemning a proud people to poverty and dispossession. He certainly did not share the conventional attitude of John Bachman that the Seminoles were bloodthirsty criminals. In his mind, their bonds with the natural world were closer to what he had aspired to in his own life in the woods. In an episode Audubon had written two years earlier for the second volume of the *Ornithological Biography, St. John's River in Florida*, he recalled in high romantic prose the meeting of a solitary Seminole with fresh game to sell during his first voyage to Florida in 1832:

> Alas! thou fallen one, descendant of an ancient line of freeborn hunters, would that I could restore to thee thy birthright, thy natural independence, the generous feelings that were once fostered in thy brave bosom. But the irrevocable deed is done, and I can merely admire the perfect symmetry of his frame, as he dexterously throws on our deck the Trout and Turkeys which he has captured. He receives a recompense, and without smile or bow, or acknowledgment of any kind, off he starts with the speed of an arrow from his own bow.

When Audubon, his son John, and his good friend Edward Harris (who had come down from New Jersey to join the expedition) left Charleston in February

1837 on their way to the Gulf to meet up with the government cutter, they crossed through Alabama and saw a band of Creeks in chains in one of the forced removals of the southern tribes that culminated in the *Trail of Tears* of the Cherokees a year later. In his letter of February 24 to John Bachman (see below), Audubon writes down his impressions of the scene, and in so doing gives one of the most touching eyewitness accounts we possess of this dark hole of American history.

After stopping at Mobile, and then Pensacola, Audubon and his party made their way to New Orleans, where they remained three weeks waiting for Captain Coste and the *Campbell*. Audubon continued to hope he would be allowed to add western Florida to his plans, but this was not to be. In New Orleans he now enjoyed public acclaim. The state newspapers announced his visit. He dined in grand style with former governor André Bienvenue Roman (soon to be reelected), and he wrote proudly to Bachman that he had been acknowledged in the press as a "Native Citizen of Louisiana." Here was proof that this was a homecoming, and not only in his eyes. Audubon and the others left New Orleans on the *Campbell* for the Gulf the last week of March, and were soon joined by a smaller schooner, the *Crusador*, acting as a tender boat to provide easy access to the low-lying bays and marshes, and to offer extra protection. Texas had won its independence from Mexico a year earlier. Rubbing shoulders with history again, though not unduly fazed by the honor, Audubon would meet the new president of the Texas Republic, Sam Houston, presiding in the town named in his honor, on the last leg of the trip. There was some concern that the *Campbell* might be accosted by the Mexican flotilla, still a threat along the contested Texas coast. Audubon seems to have worried more about seasickness than warships. Born an exile, he had little respect for (and less fear of) manmade frontiers. He had been crossing dividing lines since he had left his father's plantation in the Caribbean as a small boy and sailed to France.

As the *Campbell* and the *Crusador* reached the Gulf on April 1 and began plying the coasts of Louisiana, the bird hunting took hold in earnest. Spring migration was at its peak, bringing in the long-distance northbound birds to mingle with the local standbys, and the waders of the swamps: snowy egrets, night herons, wood storks, and white ibis. The marine skies, an infinity of blue, filled with red-winged blackbirds, grackles, kingbirds, snipe, pelicans (brown and white), sandpipers, terns, and other seabirds and shorebirds, well over two hundred species. Though Audubon did not discover a single new bird to add to his collection, the trip rejuvenated him. In the company of his son and his good friend, he realized how much he had missed the wide-open wilderness, mosquitoes and mud be damned, while he had been working bent over a desk in the confining rooms of London and Edinburgh. The colonial place-names of the Louisiana wetlands were reminders that he was back in his element, a Frenchman at home in the lower South: Grande Terre, Grand Isle, Barataria Bay, Isle Caillou, Terrebonne Bay, Isle Dernière, Cote Blanche, Bayou Salle, Vermilion Bay.

Audubon's journal onboard the *Campbell* was lost, like so many others, or destroyed, but in his biography of the blue heron in volume four of the *Ornithological Biography* (1838) he reached into his notes for one of the few published accounts of what he had recorded on the spot:

> While on Cayo Island, in the Gulf of Mexico, on the 10th of April 1837, I observed large flocks of the Blue and Green Herons, *Ardea coerulea* and *A. virescens,* arriving from the westward about the middle of the day. They flew at a considerable height, and came down like so many hawks, to alight on the low bushes growing around the sequestered ponds; and this without any other noise than the rustling of their wings as they glided through the air towards the spot on which they at once alighted. There they remained until sunset, when they all flew off, so that none were seen there next day. This shews that although these species migrate both by day and night, they are quite diurnal during the period of their residence in any section of the country which they may have chosen for a season. It is more than probable that it has been from want of personal knowledge of the habits of these birds, that authors have asserted that all Herons are nocturnally inclined. This certainly is by no means the case, although they find it advantageous to travel by night during their migrations, which is a remarkable circumstance as opposed to their ordinary habits. In the instance above mentioned, I found the birds remarkably gentle, which was probably owing to fatigue.

Audubon wrote a brief uncharacteristically subdued version of his last Louisiana voyage in the introduction to the same volume four of the *Ornithological Biography.* The boyish enthusiasm he felt exploring the islands and inlets for the first time is far better captured in his lively letters to Lucy, John Bachman, and William MacGillivray, the Scottish naturalist who was helping him with the *Biography.*

After almost two months at sea, Audubon sailed back to the Mississippi on the *Campbell.* After stopping at the Balize, the *Campbell* was tugged up the river by a steamer, and docked in New Orleans on May 27. Along with the birdskins he always brought back from his expeditions, he carried along fourteen live turtles he managed to take, still living, to England. Other than the three weeks scouting the shorelines of Galveston Bay and Galveston Island, and the impromptu sally westward up Buffalo Bayou to stop in Houston, the entire trip had been spent along the Louisiana coast. When Audubon left New Orleans the last day of May, he would never see Louisiana again. In a sense, he did not need to. Louisiana was so richly woven into his mind, his work, and his life with Lucy, that it would never be lost. By the time he set foot on the steamboat taking him and John to Mobile, away from Louisiana for good, his life's work, at least the work that counted the most, was coming to an end. The final engraving of the fourth and last volume of *Birds of America,* packed with Louisiana species, was finished on June 16, 1838. The fifth and final volume of *Ornithological Biography* was published the next year, bringing the full text to over three thousand printed pages, with more references to Louisiana than to any other place in the United States. By 1839 Audubon had set his mind on another vast undertaking, the *Quadrupeds of America,* which had been agreed to by John Bachman as a joint effort. Audubon and his family moved to New York in September 1839 and soon settled in for good.

By this time, Audubon had decided to find a permanent home for Lucy in the North, as close to New York as possible. This would allow him to travel easily, and at the same time provide security and peace of mind for Lucy and the family. In the first letter below to John Bachman, Audubon had complained about the unhealthy summers in the "*Southern States,*" the implication being that he could no longer risk Lucy's well-being in the Deep South, especially since he had no intention of giving up his expeditions now that the planning for the *Quadrupeds* was well beyond the drawing board. On a subscription search for *Birds of America* in Richmond in April 1840, Audubon wrote to his son Victor that the smell of dogwood in full bloom reminded him of Louisiana, and all "the days passed never to return." But in the next line he added that he feared "the South will not answer our purpose as well as the Eastern districts." There may have been other reasons than hot and humid summers that argued against the South. Though Audubon had never been overly interested in politics, he was well aware of the rising slave debate, and had discussed the issue with his Quaker antislavery friends in Liverpool. Ever since the Nat Turner Rebellion in 1831, he had followed the changing mood of the country carefully, and no doubt wanted to make sure Lucy was well out of the inevitable line of fire. In 1841 he purchased land on the Hudson River and constructed a house called Minnie's Land, using the old Scottish term for mother and signing the house over to Lucy.

The move to Minnie's Land in early 1842, despite the frantic activity of working on the *Quadrupeds* and the American editions of the bird books, marked the beginning of a sad terminal downward cline. Audubon's two daughters-in-law, the Bachman sisters, died within a year of each other from tuberculosis, Maria Rebecca in September 1840, and Mary Eliza in May 1841. Overwhelmed with grief, Audubon never really recovered. For a time, the *Quadrupeds* supplied him with a refuge, and he resumed drawing with a vengeance, twelve to fourteen hours a day. In 1843, ostensibly to look for new species, but psychologically to escape, he organized a final expedition up the Missouri River to the Yellowstone. With a chosen crew of friends and helpers, Edward Harris, Isaac Sprague, John G. Bell, and Lewis Squires, he left Lucy and Minnie's Land in March and caught a ride in St. Louis on a steamboat belonging to the American Fur Company, surrounded on deck by a rowdy group of trappers leaving for the far-flung forts. After eight grueling (and to his relief, time-consuming) months on the western plains, he returned home exhausted in November.

Audubon's journal of the trip, the *Missouri River Journal,* is a fascinating document, containing some striking descriptions of life in the western wilderness. But Audubon more often gives the impression of fighting off his recent tragedy by losing himself in hunting, riding, swimming, dancing, writing down the day's events at night, anything to stop from thinking back, or too far ahead. The French have a fine phrase for Audubon's attitude, a *fuite en avant,* a "fleeing forward" into the limbo between the present and the future. Despite the vitalism of some of his writing, and his attempts to convince himself otherwise, the end result of the *Missouri River Journal* is a self-portrait of impending doom. Everything he once cherished in his explorations has changed for the worse. The once proud Indians, with only a few exceptions, are now degraded parasites. Whole tribes have been decimated by smallpox and other diseases of civilization, the survivors left behind to fish rotting buffalo out of the river, and to stumble by the frontier outposts in broken processions begging for food. He can no longer count on his robust health, and thinks of himself, now fifty-eight, as an old man with one tooth. He is tormented by nightmares. He tries vainly to keep up with his younger companions, risking his life in daredevil competitions.

Ironically, this was a voyage he had dreamed about for years, into the Far West, deep into what was left of old Upper Louisiana. Compared to the early years by himself in the woods and swamps of southern Louisiana, this time his heart had not been in the search. He came back to Bachman and the *Quadrupeds* with some superb drawings of common western animals, but otherwise empty-handed, without having found a single new mammal species to show for his efforts. For the next four years he worked tirelessly on his drawings, and was able to admire his work in the first volume, *The Viviparous Quadrupeds of North America* (by John James Audubon and the Rev. John Bachman), published in 1847. During the last years of his life, his sons John and Victor, and Bachman, gradually took over the entire project, which was completed in 1854, three years after Audubon's death. He had taken on the *Quadrupeds* with a show of tremendous enthusiasm, as he did everything else, but his interest in mammals had never inspired him like his lifelong love of birds, going back to his egg and nest collections as a boy on the Loire. Nothing would ever match the dizziness of discovery he had felt roaming the Felicianas in the 1820s, and had recovered, for two months at least, island-hopping with the seabirds along the Louisiana coast in the late spring of 1837.

SOURCES

The letters to John Bachman and Lucy are all taken from Howard Corning's second volume of *Letters of John James Audubon* (1930). The quotation above from the letter to Victor is taken from the same volume. The letter to William MacGillivray is taken directly from its publication in the *Edinburgh Journal of Natural History*, vol. 1 (for December 1838). The letter to Thomas Brewer is in the collection of John James Audubon Letters and Drawings 1805–1892 (MS Am 21) in the Houghton Library of Harvard University. I am grateful to the Library for kindly sending me a copy of the original letter. Brewer was a young physician Audubon had met in Boston the previous year. He later abandoned his medical career and became a noted ornithologist, specializing in oology, thus Audubon's promise of a shipment of eggs from the Louisiana voyage. Brewer remained a lifelong friend and correspondent.

Letters to John Bachman and One to Lucy

Written During (and Just After) Audubon's Trip to the Gulf of Mexico
in the Winter and Spring of 1837

TO REVEREND JOHN BACHMAN
Charleston, South Carolina

Mobile Feby 24th 1837—

MY DEAREST FRIENDS—

We left Charleston on the 17*th* Instant (Edd Harris, John & I) and arrived here last night. Our Journey was performed first by the Railroad to Augusta, a pretty Village in Georgia—the weather was extremely cold; nay the Ice on the morning of the 18th was one half Inch deep.—at Augusta we took the Mail, and luckily for us had no others than ourselves in the coach—The roads, in consequence of several previous days & nights of rain were as bad as can be; but we proceeded apace and had no accidents—having crossed Georgia, we entered the State of Alabama after crossing a bridge at Columbus—here the Swamps were shockingly bad, and we feared that our goods & chattels would have been wetted, but thanks to our Yankee drivers (*the very best in the World*) all was kept dry as cocks—the next morning we breakfasted at the Village of [——] where 100 Creek Warriors were confined in Irons, preparatory to leaving for ever the Land of their births!—Some miles onward we overtook about two thousands of these once free owners of the Forest, marching towards this place under an escort of Rangers, and militia mounted Men, destined for distant lands, unknown to them, and where alas, their future and latter days must be spent in the deepest of Sorrows, afliction and perhaps even phisical want—This view [was] produced on my mind an aflicting series of reflections more powerfuly felt than easy of description—the numerous groups of Warriors, of half clad females and of naked babes, trudging through the mire under the residue of their ever scanty stock of Camp furniture, and household utensiles—The evident regret expressed in the[ir] masked countenances of some and the tears of others—the howlings of their numerous dogs; and the cool demeanour of the chiefs,—all formed such a Picture as I hope I never will witness in reality—had Victor being with us, ample indeed would have been his means to paint Indians in sorrow—

We reached Montgomery at Night—remained there until 10 of the next day, and on board of a Steamer, made down the River Alabama—a Stream which though much smaller than the Mississippi resembles it very much.—like it, it is muddy, winding, and lined on its shores, by heavy cane brakes, and bluffs of various elevations & formations—

Our Intentions to visit the Families of our Friends the Lees were abandoned, and I wrote to them in stead—We heard from different persons that they were all well Doing and in good health—We saw many Southern Birds, but felt no difference in the climate—Indeed ever *here*, the weather is cool, and the country exhibits no appearance of spring—

Our first enquiry at this place was for Judge Hitchcock, but he is absent—a M*r* Martineau answered in his stead, and introduced us to the Collector of Mobile, who in turn presented us to Cap*n* Foster, who Commands a Cutter; a Jolly old Gentleman, who gave us to understand that he should like a Tour with us, *provided* I would obtain Commodore Dallas' permission to do so.—he gave us some pleasing information of desired birds &c, and we concluded to go to Pensacola Tomorrow, by Steamer to pay our regards to the Commodore.—our Intention is to spend but one day there; to return here, and await the receipt of answers to letters sent to M*r* Grimshaw and Cap*n* Coste, who we are told is on the New Orleans station, and now at that place—thus far you perceive we are unabled to form our plans; but expect to be able to do so very soon, when I shall not fail to give you all desirable information—Mobile is a small, compact, thriving place, of goodly appearance—there are about 10 Steamers that ply up the Mobile & Alabama Rivers—some to New Orleans *daily*—and also some to Pensacola *daily*—the country around is flat and swampy, and the accounts of the healthiness of this place, so varied that no one can depend on what is said on the subject; at the exception, that the population is about 13,000 people during winter, and that in July & August, it is reduced to about 5,000!—to me this, and what I have seen, is sufficient—My Mind therfor is made up never to seek refuge (much less health) in either the lower parts of Alabama or any *Southern States*—

Need I say to you all, how dearly glad we would be, to be enabled suddenly to accompany you to church on Sunday next? I believe not! [——] My spirits are not above par I assure you and this day, I have suffered much from the [——] of Drinking *Alabama Water*—Tomorrow I [——] to be cured, by a dance over the Waters of the Mexican Gulph—and then all will be right again.

John Bachman my friend, the salamanders are still asleep—Hares we have seen none of—Parokeets, by the hundreds, and also Wild Pigeons, and Hutchins' Geese—at Augusta Docr Wray was very kind to us—here we have seen Frank Lee, who looks well, and is Happy—I wish I was as fortunate! Mr Logan leaves out of Town and we have not had time to call upon him, but will do so on our return—our expenses to this place have amounted to *nearly* 200$—My sweet heart will be so good as to write to My Lucy, and give her the *interest* if any of this letter—I send my love to you all, and thousands of Kisses on the Wing—and Now God bless you all and believe me ever your truly attached & sincere & thankful friend

J. J. AUDUBON

MY MARIA, I will write you a long letter on my return from Pensacola and now say with Papa, God bless you all—Most affectionately

J. W. AUDUBON

TO REVEREND JOHN BACHMAN,
Charleston, S. Carolina

New Orleans 3d March 1837.—

MY DEAR BACHMAN.—

Having given you an account of our journey as far as Mobile in my last, I now will proceed with one of what has hapened since then.—

The next day we left Mobile for Pensacola by Steamer and reached the latter the same evening late—went to a most rascally house called "Collin's Hotel" May your star never shoot you there!—however morning came, it was Sunday, and poured of Rain—notwithstanding which, we delivered a few letters of introduction, and were taken on Board the U. S Frigate the Constallation, in Commodore's Dallas' Barge! The Commodore received friend Harris and I *quite well* (John remained at Mobile) we found him a very aimable person, and after he had read Mr Woodbury' letter he assured us that he would do all in his power to serve us & Science, and that in all probability he would put us on board of the Cutter, now commanded by Captain Robert Day! and that he would write to us at Mobile in 5 or 6 days.—he presented us to all his principal officers, shewed us every parts of his superb ship, gave us some first rate wine, and a bottle of Copenhagen *Snuff* that would make your nostrils expand with pleasure, and draw tears from your eyes—the stuff is so very potent.—We took our leave and returned to *Collins* where we sat contemplating the weather to our hearts content.—Pensacola is a small place at present; principally inhabited by Creole Spaniards of the lowest Class, and some few aimiable & talented famillies of Scotch, and Americans.—The place is said to be *perfectly healthy* The country is deeply sandy, and nothing but pine Barrens exist for about 80 miles back.—The Bay is grand and of good depth—The Bar at the entrance about 22 feet admits of Vessels of great Burthen—this is guarded by two powerful fortifications.—The Naval Depot or Navy Yard is 9 miles before the Village, we had not time to visit it.—Fish is abundant, and there I saw I think the finest oysters ever observed by me in any portion of the southern country.—Deer, Wild Turkeys and smaller Game is said to be very plentiful.—We saw thousands of Sallamander' Burrows; and here let me assure you that these Animals have an entrance to their burrows resembling that of European Rabbits, but smaller of course.—It is yet too early to procure them.—Rail Roads are in progress, and some projected to communicate with Blakeley (opposite Mobile distant 65 miles) and Montgomery in Alabama distant 175 or thereabouts Major Ingram of the Topographical Department politely shewed us all the plans on *paper*—much of the Iron Carrs, Engines &c are already on the spot—a New Town is laid out for sale, and an *Immense* Hotel is now being erected there.—but after all the back country is so poor, and the want of some navigable stream so great in my opinion, that I have great doubts of the ultimate boasted of advancement of either the new or the old town.—The former belongs to a company of New York speculators, and the rage for new Cities in this section is so great, that the lots already have brought great sums.—I have forgotten to tell you that Commodore Dallas shewed us the last received dispatches from Gen Jessup in which he gives great hopes of the Florida War being at an end! May this prove true—Harris and I Walked the whole of

Monday, and heard of a Bird breeding in that section, called the *Gris* which from the imperfect descriptions we have had, I conclude is My Brown Ibis, but of this we will tell you more another time.—We returned to Mobile on Tuesday—Called on M*r* Logan, but only saw his Wife.—Frank Lee, and John had passed their time pleasantly &—I found 2 letters from my Dear Wife dates 8*th* and 19*th* of Dec*r* and one from M*r* Grimshaw informing me that Coste was on the New Orleans station but absent and assisting a wrecked vessel on one of the Keys—that he would be *here* in less than one month—he is not under the control of C*n* Dallas, and we expect to sail with him for the Sabine and intermediate places—we hope to go with Cap*n* Day and Cap*n* Coste as far as Cape Sable and visit all the Keys [

] the *Western* coast of Floridas before we sail for the Westward.—Harris has gone back to Pensacola in the Revenue Cutter the Jackson, Cap*n* Foster, to remain and see the Commodore until something profitable is offered or said, and John & myself will await for news from Harris at this place, where I hope to procure a few subscribers.—We are snug and comfortable at M*r* Grimshaw (James Grimshaw Es*qr*) to whose care please to Write to us.—It pours of rain, and cannot go to Market this morning.—The first fair day we will do so and seak for Squirrels and Rabbits for You—Yesterday I received another letter from London dated 31 Dec*r*—This latter had rather perplexed me on account of the rapidity with which Victor and Havell have proceeded with the Work—but Victor will do his best, and I hope that the Drawings made at Charleston will arrive in time to be disposed of accordingly with my arrangements.—We have found our clothes here, from England, but the things from Charleston have not yet arrived. Why our English letters have been sent here, instead of to you, 1cannot guess—I beg of our Dear Sweet heart to write to My Wife and to copy whatever parts of my letters to you or to her, and send these promptly.—and now My Dear Friends God bless You all.—remember us all to every friend—I forgot to pay Friend Kunhardt my last *small* balance pray do so, and excuse me near him for the neglect—I find New Orleans so large, and so much improved in every point of View that I can scarcely recognize one street from another—We have lost however by Death the greater portion of our numerous former acquaintances here.—Ten thousands of Kisses for all the Dear Girls and Mamma too—and "Rabbitt" and "Sweet Meats"—and old friend M*rs* Davies, and the master in Phrenology!

Yours ever truly,

JOHN J. AUDUBON

To REVd JOHN BACHMAN
Charleston, South Carolina

New Orleans March 22d 1837—

MY DEAR BACHMAN—

We will leave this city, tomorrow Morning before day, on board of the Cutter, the Campbell, Commandant *Napoleon L. Coste,* and proceed directly though slowly on our researches—the Vessel is small, yet very roomy—i.e that she is less incumberded than any of her size I have yet seen.—Coste is full of the expedition, and so is the first Lieut*t* the other Lieut is I conceive a "Man a War' Man"—full of his duty on the weather side of the quarter Deck during his daily or nightly Watch!—16 Men—3 Boats, and provisions for about 2 Months, one *great Gun,* and Many Pikes, Cutlasses (*Cut-Lasses*) Pistols and Muskets, as can be crammed in so small a Craft.—We Intend Visiting the whole of Galveston Bay, and Islets thereabouts, spite of the Mexican's *Flotilla*—which after all I think is all "My Eye"—We hope for great things but God knows how many new species will be added to Our Fauna.—We are well provided with ammunitions &c but our *apples* have Turned into sour Cider, and I have some doubts whether our butter will not run through our biscuits ere we gobble the latter. John is packing up squirrels and a Wild Cat for you—I hope you never have seen the Like?—I have and think that you will be pleased with them—John or My Nephew Young Berthoud send you a bill of Lading—I am glad, and proud Too; that I have at last been Acknowledged by the public prints as a Native Citizen of Louisianna—and had it been supposed when we first arrived here that our stay would have been half as long as it had been—I really think that my Country men would have honoured us with a Public *Dinner!* Try to find out the Paragraphs in the New Orleans Courier (French & English) of the begaining of this month.—and above all my Dear Bachman pray attend to the following—

There is coming here another Cutter to take the place of the Campbell, and she may be here before our return, when the Campbell will scarcely be called for on this Coast.—*Coste* is very anxious to have, *in her* (The Campbell) the *Key West Station* (not the Savannah) and tells me that he would be delighted to take us round to Charleston, after having shewed us the different breeding places which he has discovered on the Western Florida Coast.—Now I should like this myself of all things, and the thought has crossed My Mind, that the *Florida War being at an end!!!* the Secretary of the treasury M*r* Levi Woodbury might grant us the privilege of that Vessel for the purpose mentioned above, If some one was at his Elbow, with strength and power enough to urge him to such an act of generosity towards a poor student of Nature, who in all probability is not likely ever afterwards to trouble his Go*vt* again—and Now that M*r* Poinset is the Secretary of War, and you very intimate with him, that were you to write to that Gentleman *at once*, and ask of him to speak to M*r* Woodbury, and ask the request I long for—Woodbury I *think* would hardly refuse *him* (M*r* Poinset)—all this must however be done at once, and not be put off even for a day if possible, as you Know our time is growing very short, and we are anxious to Make the best we can of it—Ask of M*r* Poinset that in case this last *petition* is granted by Woodbury to urge the latter to forward me a letter to the effect Care of James Grimshaw Esqr New Orleans.—We have a good number of trifling affairs to transact this day, and I now bid you all adieu for a While, and May God bless you all is the fervent wish of yours truly and most faithfully attached friend

JOHN J AUDUBON

Please have an extract of this sent to Lucy—

To M*rs* LUCY AUDUBON
London

New Orleans 23d March 1837—

MY DEAREST BELOVED!—

I have only a little time to say to thee, that we will leave this tomorrow Morning on board the Revenue Cutter, the Campbell, Cap*n* Napoleon Coste, for a Coasting Voyage of about Two Months.—That we *Three* are quite well! That I have received our Dear Victor's letters of the 6*th* and 9th of Jan.y. and that I am happy now and comfortable in the thought that he will Carry the publication according with the lists I have sent him.—I have heard of the arrival at Liverpool of the Mohawk on the 22*d* Jan.y and hope that the 9 Drawings sent by her were in good time for the Numbers intended for these.—I hope the *Superb* has also reached safely, and if So Victor will have enough until I return to You.—I have felt great uneasiness about your Precious healths since I have read the a/cs of the Influenza in London—May God preserve you both, and May he grant us the happiness of Meeting again well and Happy. It is not probable that I Shall hear from you now for Two Months but I will be patient and take good Care of John, Harris and Self.—*The Florida war is actually ended!* depend upon this intelligence.—We therefore hope to return by that *Coast* to Charleston after we have been to Galveston Bay &c on the Western Coast of *the Gulph.*—Not a New Subscriber at New Orleans as yet.—*Mr Fortsall* has not even asked me to his house once—I dined at Governor Roman in a large Company—he is a fine Man, and has written a few Kind things in the Papers here.—My "*Natal City*"!—Remember us to *Every body*, even to Bessy—Oh how glad I shall feel when I Land at No 4 Wimpole Street and Kiss again My best beloved and Dearest Friend & Wife and our Dear Victor.—but now adieu once more—for ever Yours faithfully

JOHN J AUDUBON

To REVd JOHN BACHMAN

Charleston, South Carolina

Mar 29—1837

MY DEAR BACHMAN.—

I have the great satisfaction of receiving your letter of the 15*th* Inst. yesterday, and send you my thanks for it, and a thousand Kisses for all the sweet hearts you and I and Johny have about you!

We are on board of the Cutter the Campbell.—about 2 miles below New Orleans, anchored on the Western side of the great River, and to day is the 29*th* of March of the year of our Lord 1837!—We are only waiting for 5 Sailors, and as our Cap*n* Coste, and the first Lieut*t* are gone to New Orleans in search of the hands we want, we hope to see the remainder of our Crew compleated this day—Cut Stick to night &c &c—Johny is in the swamps outlining Cypress Trees with the Camera for Victor—Harris is gone to town for Letters if any there are and here I am *scribling* this to you.—I thought you would want me pretty soon, but I can echo you at that, and I would willingly give up one year of my Life (*hereafter*) to have you at my side Just Now.—I would first ask what do you say to those *Squirrels?* and again to this Wild Cat? For I hope you have those which we have shipped to you before your eyes just at this moment.—About one other hundred questions I would ask of you to answer, but we are a pretty good distance asunder now, and must wait for the pleasure of meeting again in *this World!* A World which though Wicked enough in all conscience; *is perhaps* as good as Worlds unknown.

We took Harris on an Alligator Hunt on a *fine* Bayou. We Killed about 20 of these beautiful creatures, and brought only 7 on board.—Harris Killed several. he never had seen any before.—he Likes their flesh too, but no so Johny excepting the latter our *Mess* made a grand Dinner of the "Tail end" of one, and after all, alligators flesh is far from being bad.—God preserve us from ever "Riding" a live one—We had a fine frolic of this, but after all they are not to be fooled with.—Sand Hill Cranes are yet here—blue wings & Green wings.—Few, very few *Herons* have come to this latitude and longitude—but we have procured here the *Common American* Gull, L. Torhinas.

We have heard through Mr Grimshaw' English letters that all was well at London at a late date, but have had no letters ourselves.—Bats are plentiful and you shall I hope see some of them.—The Deer here is different from *yours*, and this also I trust you shall see.—in a Word you will be thought of as of a "Worthy Friend" during the whole of our expedition—My former letters will have given you many details which I cannot now repeat. One of your letters (I mean one of our beloved Maria Martin's) has not reached us—God bless her; and D. the Post!—

I have a most Kind Friend here collecting birds in Rum for me during our absence.—I trust that those at Charleston will not forget their promises on that score, and we already thank the latters for Certain "Brown Creepers" that are now snoozing in rum!—Harris Stands the Packets admirably—he is in facto one of the finest Men of Gods Creation—I wish he was my Brother!—

The Steamer *Fancy* was totally destroyed a few evenings ago on her way to Louisville Ky.—She belonged to my youngest brother in Law, Cargo and all. I fear his loss is considerable—The whole of the living on board saved themselves *only* with what they had on their backs—William Bakewell saved a Babe from the Waters, but the nurse was drowned. The Child belonged to one of the Passengers.—had this accident taken place at Night, it is more than probable that one half of the People must have been drowned or burnt to Death.

Have you written to My Lucy?—Do not forget this I pray you.—My Beloved Daughter must write also.—and so must our Dear aimiable friend & Sweet heart.—Do not forget to Write to Mr Poinset for me to ask of Mr Levi Woodbury to let me have the Campbell, Cap*n* N. L. Coste, to go to Charleston on our return, which 1am anxious to make around the Floridas—Write soon, and send Woodbury's answer or Orders (I hope) to this place care of James Grimshaw Esqr.—The failures at New Orleans have dampened the spirits of every one who speculates on Cotton, Land or Dollars.—

God bless you all.—Remember us to all Harris send you his best wishes & Love to the Ladies—and now My Dear Bachman believe me as ever Your friend

JOHN J AUDUBON

There is no Grog on board of the Campbell!!—What do say to that? –Snuff is yet partially afloat—but will be dropped astern very soon!—And once More God bless You all.—

To REVd JOHN BACHMAN
Charleston, South Carolina

Island of Barataria, Grande Terre, April 6th 1837
U. S Revenue Cutter Campbell—

MY DEAR BACHMAN—

I wrote a few lines to you from this place by a Schooner bound to New Orleans, but as winds & *Mail carriers* are not always to be depended on, I will try to have this ready for a Gentleman going from here to New Orleans by the Bayous 105 Miles, and who promises to have it put in the Post Office.—

We were detained a few miles below New Orleans for the want of sailors, until raising the wages to 40 Dolls we procured a few crew and some stout fellows, after which we sailed down the great stream to its Southwest pass or entrance.—The next morning we sailed (very foolishly) to the Northeast Pass, and sent an officer in a boat in search of a Capn Taylor. also of the Revenue Service, to whom the Collector at New Orleans had sent orders to Join us, and to assist us as a Pilot.—That day was lost.—The next morning we went shooting and Killed 4 *Marsh* Terns and some other birds—We had put Mrs Coste on Shore at a Fisher house to await the return of her husband, and the next morning early we *sailed* on our Expedition. The weather was fair and the sea smooth until we approached the Barr at *this* place, we however crossed it guided by Mr Tailor whom we towed, on board of his *Crusader,* a small Schooner of about 8 tons, acting as a tender on the Campbell. We anchored safely under the lee of Barataria Island, and have been here ever since—Shooting & fishing at a proper rate.—Johny & I shot 4 White Pelicans—Harris and the two latters a great number of different Tringas, Terns, Gulls &c—and so we have passed our time, at potting species, their habits, and skinning and placing specimens in Rum.—One cask is already filled.—We are all well.—we intend proceeding to Cayo Island 52 Miles West by the first fair wind, and then expect to do well, as it is said to be a great breeding ground.—Not a bat on our Island, only Racoons, otters, Wild Cats, and a few Rabbitts—we have not seen any thing more than tracts.—Not a New Bird as yet.—have Killed 5 Tringa hymantopus. Marsh Terns abundant. Cayenne and Common D*o*. Larus atricilla D*o*—White & Brown Pelicans—and a good variety of Ducks and the florida Cormorant.—few land Birds. Salt Water Marsh Hens and boat tailed Grackle breeding.—but enough as I have noted every incident worth notice, which you will read from the Journal.—We have now in contemplation to leave the Schooner as soon as we have reached Texas and seen Galveston (Galveston Bay) and return over Land to New Orleans, and there make ready either to go back to you at once, or proceed around the west Coast of Florida—We will however be guided by Circumstances and do all for the best—have this copied and forwarded to My Dear Wife, and if you please have us "reported" in the papers. We have been very Kindly treated by a Planter here who is a Partner of M*r* Forestall of New Orleans, who gave me a few lines of Introduction.—We have had fine vegetables, Milk & Corn Bread and *fresh Butter!*—This Island is about 10 miles long but scarcely a mile broad—it is low, and mostly marsh (hard however) with many ponds, Lagoons &c—it possesses one Sugar plantation and a few dilapidated Government buildings, began by *Jackson* but now abandoned and rotting.—This was "Lafitte's" (The Pirate) Strong hold.—The remains of his fortification, and the ground on which his houses stood are yet discernible.—Some say that much money is deposited there abouts—I wish it was all in the Charleston Bank placed to *our* Credit!—The Island is flat, and in 1830 was overflowed by the waves of the Gulph impelled by a Hurricane to the depth of 4 feet above the highest ground, and Castle &c was sent adrift towards the Main distant some 12 or 15 Miles.—The soil is good enough to [grow] produce Cotton or sugar.—and the place healthy and pleasant; and yet I should not like to be imprisoned at large upon it the remainder of my Life—It abounds with *Snakes* not however injurious excepting a very small *ground* Rattle species.—We have placed several in rum for Doc*r* Holbrook, and *Crabs* for yourself!—No Insects of note except *Musquitoes* and sand flies of which we could spare enough God Knows.—I forgot on former occasions to say to you that the *Bird of Washington* is found pretty abundant on the Lakes near New Orleans.—Governor Roman' Cousin assured Harris & I of this fact.—We are promised some in Rum.—Mr Zaringue (the cousin) has seen it dive after fish frequently and was the man who spoke of it thus. "Connaissez vous le Grand Aigle Brun Pecheur"?—he says that it breeds *on Trees.* he Knows the White headed Well.—and Dear Friend God bless you all.—remember [to] us to all about you. Harris Joins us in best regards &c and I remain as ever Your Friend most truly & sincerely attached

JOHN J AUDUBON

To REVd JOHN BACHMAN
Charleston, South Carolina

April 18th 1837—

MY DEAR BACHMAN

Here we are all safe & well in *Bayou Sallé Bay,* about 18 West of Mouth of the *Teche,* and in Attakapas.—We are now on board of our Tender Cutter the Crusader, ransacking the shores of this Coast, Most of which is flat & Marshy, and not so abundantly supplied with Birds as either of the (?) might wish it to be. however we have filled 3 casks with Valuables, but not with any new Bird!—We still intend moving Westward as far as Galveston Bay, and may return across the *Land.* I Hope this will reach you all Well, and that you may have received good accounts from My Dear Wife & Victor from England, the result of which I shall expect to find awaiting our Return at New Orleans—but when I cannot pricisely say.—We are all in good spirits and Work very hard I promise you—Kiss all the Dear Ones under Your Roof Write to Lucy and to Victor, and believe Me as ever Your friend, and God bless You all—

JOHN J AUDUBON

Letter to William MacGillivray

On Board the Crusader, Cote Blanche, 18th April 1837.

MY DEAR FRIEND,—Being just now snugly anchored in a bay, the description of which may prove agreeable to you, I sit down to give you an account of what I have been doing since I last wrote to you.

After visiting "Rabbit Island," on which, as I have already told you, not a single Rabbit or Hare is to be seen, we made our way between it and Friskey Point, by a narrow and somewhat difficult channel leading to the bay in which I now write. The shores around us are entirely formed of a bank, from twenty to thirty feet high, and composed of concrete shells of various kinds, among which the Common Oyster, however, predominates. This bank, which at present looks as if bleached by the sunshine and rains of centuries, is so white that it might well form a guiding line to the vessels which navigate this bay even in the darkest nights. The bay, however, is so shallow, that it is rarely entered by vessels larger than schooners of about seventy tons burthen, which visit its shores to take in the sugars and cottons grown in the neighbouring country.

The "Crusader" is a somewhat curious craft, small, snug withal, and considerably roguish-looking. She has not fewer than four "grunters" on her fore deck, her sails are of pure white cotton, and although she bears the lively flag of our country at the peak, her being painted purely black gives her the aspect, not merely of a smuggler, but of a pirate. But here she is, at the entrance of the canal of a sugar plantation, and close to another craft, much the worse for wear, and, for aught I know to the contrary, belonging to the captain alone, who, I would almost venture to assert, belongs to no country at all.

It is now four weeks since a razor came in contact with my chin. All my companions are equally hircine; or, if you please, hirsute. As to our clothing, were you to see us at this moment, you would be ready to exclaim, "What vagabonds these fellows are!" Coats and trousers plastered with mire, shirts no longer white, guns exhibiting the appearance of being in constant use, and all sorts of accoutrements that pertain to determined hunters, complete our *tout ensemble*. But, as I have said, here we are, and on shore must go. "Man the gig," quoth our captain. In a trice the gig is manned. One after another, for there are five or six of us, we swing ourselves into the after-sheets. The word is given, the oars are plied, and now we once more are on terra firma.

The crossing of large bays, cumbered with shallow bars and banks of oyster-shells, is always to me extremely disagreeable, and more especially when all these bars and banks do not contain a single living specimen of that most delectable shell-fish. Nay, I am assured by our pilot, who is no youngster, that ever since he first visited this extensive waste, not an oyster has been procured in these parts. But now in single file, like culprits or hungry travelers, we proceed along the margin of the canal. Ah, my dear friend, would that you were here just now to see the Snipes innumerable, the Blackbirds, the Gallinules, and the Curlews that surround us;—that you could listen, as I now do, to the delightful notes of the Mocking-bird, pouring forth his soul in melody as the glorious orb of day is fast descending toward the western horizon;—that you could watch the light gambols of the Night Hawk, or gaze on the Great Herons, which, after spreading their broad wings, croak aloud as if doubtful regarding the purpose of our visit to these shores! Ah! how well do I know you would enjoy all this; but alas! we are more than four thousand miles apart.

Hark! what's that? Nothing but a parcel of men coming to greet us. Here they are, seven or eight Negroes. Who lives here, my good fellows? Major Gordy, massa. Well, now shew us the way to the house. Yes, gentlemen, come along. So we follow our swarthy guides.

The plantations here are of great value, both on account of their proximity to the Gulf of Mexico, and the excellence of the soil, which, as in other parts of Louisiana, is composed of a fat, black mould. The Indian corn was at least six feet high, and looked most beautiful. As we approached the mansion of Major Gordy, I observed that it had a pleasant aspect, and was furnished with a fine garden, and a yard well stocked with cattle, together with a good number of horses and mules, just let loose from labour. A mill for grinding corn and making sugar particularly drew our notice to it, as the Crusader happened just then to be destitute of both articles; and as I saw some women milking the cows, my heart fairly leaped with joy, and the hope that ere long we might procure a full bowl of the delightful and salubrious beverage. The short twilight of our southern latitudes had now almost involved every object in that dim obscurity so congenial to most living creatures after the toils of the day, as allowing them to enjoy that placid quiet which is required to restore their faded energies.

Near the entrance of the mansion stood an elderly man, of tall stature and firm aspect, leaning on what I would call a desperate long gun. As I approached this Cote Blanche planter, I thought that something not so very friendly as I could have wished was expressed in his countenance. As he rested his heavy frame on his monstrous rifle, he neither moved his head, nor held out his hand to me, until I presented mine to him, saying, "My good sir, how do you do?" His answer was a rather suspicious look at me and my companions; but notwithstanding, and probably because he was on his ground, he asked us what was our wish, and then desired us to walk in.

Cote Blanche Bay, you must be informed, has for a number of years been infested by a set of rascally piratical vagabonds, who have committed extensive depredations, in consequence of which, a few years ago, a United States' revenue cutter was sent to protect the coast. I have no doubt that the major took us, to a man, as members of the gang who had more than once visited, not his house, but his plantation, on which they had played many wanton and atrocious freaks.

We now, however, had entered the house. Candles were lighted, and we at once came face to face, as it were. It curiously happened that our captain was without his uniform, and fully as rough-looking otherwise as any of us. I was, however, much pleased to see that the major himself was not much superior to us in respect of apparel; nor had his razor been employed for many days. I happened to have about me some unequivocal credentials, from the head departments of the United States, which, on my observing that some degree of suspicion still remained, I placed in his hands. He read them, spoke kindly to us, promised to forward our letters to the nearest post town, and invited us to consider his dwelling as our own. From that moment until we returned to our vessel, we were all as comfortable and merry as men can be when distant from their own dear homes.

Next morning we received from Major Gordy a barrel of sugar, another of corn meal, some pails of milk, and quantity of newly made butter, together with potatoes and other needful articles—and all this without our being allowed so much as to offer him the least recompense. The day after, we returned to breakfast by invitation, and found in the house several strangers, armed with rifles and double-barrelled guns. After we had been introduced to all around, we seated ourselves, and made a vigorous attack upon our host's eggs and bacon, coffee, tea, and milk. As this important business was proceeding, I was delighted to hear the following anecdotes, which I hope you, my dear friend, will relish as much as I did.

"Gentlemen," said our host, straightening himself in his arm-chair, "I am considerably suspicious as regards the strangers who happen to anchor within the range of my dominions. Indeed, gentlemen, I must acknowledge that even after you returned on board last night, I sent off some of my men in various directions, to let my neighbours know that a strange craft had anchored near the landing-place; and here, gentlemen, are those neighbours of mine; but as it happens that the name of the gentleman who calls himself "Naturalist" is well known to some of them, I now feel quite satisfied as respects the purposes you have in view. But let me tell you what happened to me some years ago.

"Such a shark-looking craft as the one you call the Crusader happened to drop its anchor abreast of my landing-place, about dusk one evening, and as I guessed that the fellows on board were not better than they should be, I watched their motions for a while from my back piazza. But nothing happened that night. Next morning, however, I heard the firing of guns down the meadows where my cattle and hogs were in the habit of feeding. So I took my rifle, walked towards the spot, and soon found, sure enough, that the rascals had killed a fine ox and several hogs, which they were dragging to the shore. Indeed, gentlemen, I saw the yawl crammed with the spoils of my plantation. Well, I took as good an aim as I could at the nearest man, and cracked away, but without hitting. At the report of my gun the fellows all took to their heels, and on getting on board hoisted sail and went off. I have never heard of them since. Well, gentlemen, about the same hour next morning, a black-looking barge, hardly as large as your Crusader, came to, off the very same spot, and although I watched it and every one on board nearly the whole night, and it was a beautiful moon-shiny one, not a soul of them came on shore until morning. Then, however, I saw some bustle on board. Several men got off in a very small affair, which was fastened a-stern of the large boat. I saw them land, and deliberately walk towards the meadows. No sooner had they reached the wettest part, and that is where my hogs generally root for food, than crack, crack, crack, went off their guns in all directions. You may well suppose how vexed I was at all this, and conceive how soon I mustered my men with clubs, and armed myself with my rifle. On reaching the ground, think, gentlemen, what were my thoughts, when I saw the fellows all advancing toward me and my people, as if they were the honestest men in the world. I was so mad when they came close up, that I had a great mind to shoot the one in front, for he looked for all the world as if he cared not a pin for any one. However, I did not shoot him, but asked him why he was shooting my hogs? 'Hogs! goodman, you are quite mistaken; we are shooting snipes until we come in contact with the rascally pirates who infest the coast, and lay waste your plantations. My name, my good Sir, is Captain ——, of the United States' Navy; and these are some of my men. Will you come on board, and breakfast with us on your own snipes?'" No wonder that the Major, having been subject to the visits of these marauders, should have taken us in the dusk, armed as we were, and withal not having precisely the aspect of sober citizens, for persons not quite so good as we should be. But I must now conclude, and in my next you shall hear something of the result of my expedition into the marshes.

Letter to Thomas Brewer

New York July 5th 1837—

My Dear Young Friend.—

I wrote to you from Charleston, immediately on my arrival there; and return from my late and last expedition.—I have had no answer from you, and I find that I cannot go to Boston ere I leave for Europe, I write again, and ask of you to answer to this at your very earliest convenience.—

I wish you to have the Cask or Casks containing Ornithological Specimens in Rum forwarded to me here at once for Rail Car & Steamer Car of Nicholas Berthoud Esq. so that I may send my cargo before my departure direct to London. I go to Liverpool, with my Youngest Son & Daughter in Law, the late Miss Bachman.—

As soon as my Boxes, Barrels, etc are on shore and hauled from the Ship which has brought them from New Orleans, I will send you a good mess of Eggs!—From your silence, I feel that you did not receive my last from Charleston. Please have the Kindness to send me a *Compleat* list of the Contents of the birds in Rum, and if your debtors call on Dr. Shattuck to whom I also write this day.—my Work will be finished in all Feb.y. next, and the Spring following, I hope to return to our beloved Country, to die in peace and have my carcass buried under the dark shade of some of our finest Louisiana Magnolias!

Present my Kindest regard to your Dear Parents, Brothers & Sisters for me.—to all enquiring Friends, and believe me ever Yours truly attached
John J. Audubon

To
Thomas M. Brewer M.D.
Boston.—

P.S. Should your have a Copy of your paper on the Habits of the Cow Bunting please send me as soon as you can by private opportunity The Office of Mr. Berthoud at 2 Hanover Street close to Wall Street.

LOUISIANA BIRDS

(Ornithological Biography)

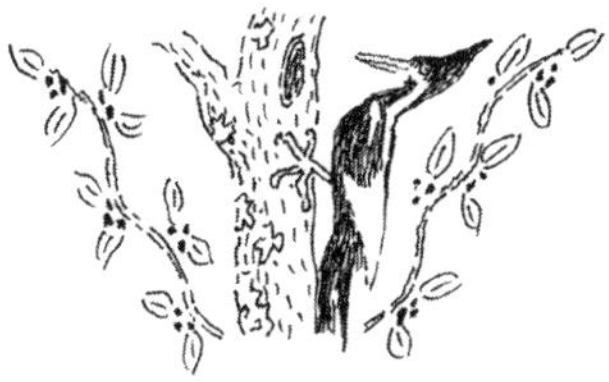

When Audubon decided in 1830 in Edinburgh to begin the process of publishing the *Ornithological Biography* as a series of companion volumes to the *Birds of America*, then being issued to subscribers in numbers of five engravings each, he was following in the footsteps of Alexander Wilson. Unlike Wilson, whose nine-volume *American Ornithology* (1804–1814) combined his paintings with short essays on the birds, Audubon was constrained by necessity, since he did not yet have enough biographies, to divide the engravings and the writings into two separate, simultaneous projects. He also wanted to avoid a British law that would have required him to deposit several copies of the *Birds* in national libraries had it contained textual matter. Later, when he brought out the seven-volume American edition of the *Birds* (1840–1844), he adopted Wilson's plan of printing plates and text together.

Wilson's essays were essential to Audubon. He had studied Wilson's work with care, going back to his first meeting with the naturalist in Louisville in 1810. Ornithologists had been describing living birds for centuries, including in colonial America the explorer Mark Catesby whose book the *Natural History of Carolina, Florida, and the Bahama Islands* (1731–1743) was a revelation in Europe. Wilson however was one of the first ornithologists to insist on the importance of direct observation of species in their natural habitat, and the first to insist on what he termed a "personal intimacy" with each bird. Audubon knew he had nothing to learn from Wilson concerning his painting; Audubon was clearly the superior artist. As a master of nature writing, however, Wilson cast a powerful spell on Audubon, along with William Bartram in his celebrated *Travels* (1791). After their deaths, both continued to reign supreme as experts on American birds. And both were associated with Philadelphia, where Audubon had been spurned in 1824; for Audubon they were private ghosts to conjure up and convince. More obviously, their books had given the new nation its first literary monuments of natural history. Audubon would be compared with them in every line he published.

Audubon definitely had the filiation with Wilson and Bartram in mind when he began his own bird biographies. It helped that all three were deeply involved in the revolutionary vision of nature that the Romantic movement, especially in England, had begun to foster. Bartram, in fact, had led the way. The major poets, Wordsworth and Coleridge and others, were seeking to revive man's fundamental bonds with the natural world by sounding its soul under optimal conditions: in solitude, often on foot, or on horseback, and in the spirit of total independence. This basic shift away from convention, hearsay, and the official halls of learning, towards the direct perceptions of the individual, open-minded and self-questioning, radically changed the way nature was seen and described. Though Romanticism is often misrepresented as an evasion, it was actually the opposite, a brazen search for greater realism, a willingness to explore all of man's experience, inner and outer, from dream to death. The effect of this on the naturalists of the new world, the Bartrams and Wilsons, and Audubon, was considerable, with the added dimension that the world they depicted was more primal than the domesticated landscapes of England. The result on their writing often takes the form of a duality of attitudes, unopposed but distinct. On one hand, they were awed by the daily mystery of existence, its mortal beauty and terror, all those deadly vital instincts of survival. On the other hand, they wanted to enter the living world as fully and as truthfully as possible, seeking to explore and to name every fact that crossed their path.

Anyone reading Bartram or Wilson for the first time cannot help but notice the way their language can soar into flights of astonishment at the sheer spectacle of life unfolding before their eyes, and at other times can stoop to a speck with the active curiosity of a child, a scientist, or an artist, testing it from every angle until it turns into a familiar presence, a gram of new knowledge, or a work of art. Bartram's gift for the exalted phrase can be found on almost any page of his journal. His description of a mockingbird rocketing upward is a good example, and one that would have pleased Audubon, for whom the mockingbird was a favorite and the subject of one of his finest biographies. Bartram begins with a general praise of birdsong, and then turns to the mockingbird, whose air (his song) is only exceeded by the speed of his wings: "The mock-bird (turdus polyglottus) who excells, distinguishes himself in a variety of action as well as air; from a turret he bounds aloft with the celerity of an arrow, as it were to recover or recall his very soul, expired in the last elevated strain." But Bartram can also adjust his style down to earth, pinning to the page a practical list of specific physical detail that could have found a place in a textbook. Here he is in Florida, writing about a crane the hunters had just killed and brought into camp for his inspection:

> This stately bird is about six feet in length from the toes to the extremity of the beak when extended, and the wings expand eight or nine feet; it is about five feet high when standing erect; the tail is remarkably short, but the flag or pendant feathers which fall down off the rump on each side, are very long and sharp pointed, of a delicate texture, and silky softness; the beak is very long, straight and sharp pointed; the crown of the head bare of feathers, of a reddish rose colour, thinly barbed with short, stiff, black hair; the legs and thighs are very long, and bare of feathers a great space above the knees: the plumage of this bird is generally of a pale ash colour, with shades or clouds of pale brown and sky blue, the brown prevails on the shoulders and back; the barrels of the prime quill-feathers are long and of a large diameter, leaving a large cavity when extracted from the wing: all the bones of this bird have a thin shell, and consequently a large cavity or medullary receptacle. When these birds move their wings in flight, when at a considerable distance or high above us, we plainly hear the quill-feathers, their shaft and webs upon another creak as the joints or working of a vessel in a tempestuous sea.

The last metaphor is vintage Bartram, a leap of the mind from one realm to another that Wilson, the free-spirited poet that he was, often displays in his descriptions. In the first volume of the *American Ornithology*, a copy of which he was carrying when he first met Audubon in Louisville, he uses his training as a weaver in his youth to unravel the secrets of an orchard oriole nest:

> It is formed exteriorly of a particular species of long, tough and flexible grass, knit or sewed thro and thro in a thousand directions, as if actually done with a needle. An old lady of my acquaintance to whom I was one day shewing this curious fabrication, after admiring its texture for sometime, asked me in a tone between joke and earnest, whether I did not think it possible to learn these birds to darn stockings? This nest is hemispherical, three inches deep by four

> in breadth; the concavity scarcely two inches deep by two in diameter. I had the curiosity to detach one of the fibres, or stalks, of dried grass from the nest, and found it to measure thirteen inches in length, and in that distance was thirty-four times hooked thro and returned, winding round and round the nest!

When Audubon turned to the young Scottish naturalist William MacGillivray in Edinburgh to help him with the bird biographies, he was partly motivated by the exceptional literary standards set by Bartram and Wilson. They were both poets, Bartram in prose, Wilson in prose and verse. Wilson in fact had been writing poetry since he was a young man in Scotland inspired by Burns. His texts in the *American Ornithology* sometimes included poetic tributes. It was hard to compete with a serious ornithologist who could place a fine poem on the kingbird, for example, at the heart of his scientific exposition. And of course Bartram and Wilson had the advantage of being native speakers, thus endowed with the verbal instincts and range of expression that are second nature to anyone using his mother tongue.

Audubon was probably misled in thinking he needed help with his writing. He had been speaking, reading, and writing English for twenty-five years when he began the *Ornithological Biography.* His letters and surviving journals contain errors of grammar and spelling, but his vocabulary in English is remarkably rich, capable of conveying almost anything he chooses. Someone who has learned a second language, especially an exile whose livelihood is at stake, can unearth layers of suggestion denied to many natives. At any rate, Audubon's powers of spirited communication, even flavored by his accent, were not lost on the sophisticated Wernerians in Edinburgh who continued to invite him to read his papers. McGillivray, with Lucy looking over his shoulder, overcorrected and over-conventionalized a number of passages in the *Biography;* there are too many effusive "dear readers," too much straining after effect in the introductory paragraphs. For the most part, however, Audubon's own unfettered voice prevails, flowing with enthusiasm, reluctant to hold back, with an engaging flair for finding the simple memorable phrase when needed. It is hard to forget him hearing the muffled sound of young puffins moving about unseen in their underground burrows as "voices from the grave." Given his skill as an artist, his vivid painting in words is not that surprising. What is more remarkable is his success in combining color and form with the rendering of movement. Here he is, in a Louisiana bayou, watching a flock of wood ibis take to the air after gorging themselves on frogs and fish:

> When digestion is partially accomplished, they all take to wing, rise in spiral circles to an immense height, and sail about for an hour or more, performing the most beautiful evolutions that can well be conceived. Their long necks and legs are stretched out to their full extent, the pure white of their plumage contrasts beautifully with the jetty black of the tips of their wings. Now in large circles they seem to ascend toward the upper regions of the atmosphere; now, they pitch towards the earth; and again, gently rising, they renew their gyrations. Hunger once more induces them to go in search of food, and, with extended front, the band sails rapidly towards another lake or bayou.

And here he is, near a Louisiana plantation, observing the attitude of a barred owl when approached:

> The position of the bird, which is generally erect, is immediately changed. It lowers its head and inclines its body, to watch the motions of the person beneath, throws forward the lateral feathers of its head, which thus has the appearance of being surrounded by a broad ruff, looks towards him as if half blind, and moves its head to and fro in so extraordinary a manner, as almost to induce a person to fancy that part dislocated from the body. It follows all the motions of the intruder with its eyes; and should it suspect any treacherous intentions, flies off to a short distance, alighting with its back to the person, and immediately turning about with a single jump, to recommence its scrutiny.

All in all, the writing of the *Ornithological Biography*, over nine years, was a stunning literary achievement, three thousand pages of natural history that have never been matched by a single individual working (for the most part) alone, without the support of an institute.

Audubon was determined to improve his style, often writing ten hours or more every day. In his mind, however, the quality that raised the *Biography* above all rivals was the mass of firsthand evidence he could bring to the task from his years of rigorous observations in the woods. Once again, Louisiana made all the difference. As Audubon well knew, Louisiana's geography had created a natural paradise for birds. The Mississippi Flyway brought endless flocks twice a year crossing back and forth by the millions over the Gulf on their annual migrations. The humid subtropical climate and maze upon maze of marshland, lakes, rivers, and bayous, brimming, bubbling, buzzing with life, provided nourishment all year round for the several hundred species that were permanent or semipermanent residents. The floodplain of southern Louisiana contained in Audubon's time a concentration of birdlife that was exceptional in America. This was the reason that brought Alexander Wilson there in 1810, two months after he had published the second volume of the *American Ornithology*, eager to renew his portfolio as quickly as possible for the volumes to follow. This was the trip that found him floating down the Ohio on his rowboat, *The Ornithologist*, where he docked in Louisville and met Audubon.

Had history taken a slightly different turn, William Bartram might have been traveling with Wilson, allowing us to imagine the three greatest American naturalists of their time, three generations on their feet all at once, standing together on the banks of the Ohio, talking shop. Bartram was Wilson's mentor in Philadelphia and one of his closest friends. Bartram had already entered into Louisiana briefly thirty years earlier at the end of his four-year exploration of the southern wilderness. When Wilson first organized his trip, he invited Bartram to join him on a more substantial expedition on foot into the South and West. Bartram had at first agreed, only to reconsider and decline, pleading his age. He was just shy of seventy-one. The two short months Wilson finally spent in the Louisiana Territory, making his way from Nashville to Natchez on the Natchez Trail, and then on down to New Orleans, was not long enough by far. As a matter of fact, his main success was the selling of his book. He found more new subscribers than new species.

Audubon was aware that both Bartram and Wilson had barely skimmed the surface of Louisiana when he set out down the Ohio (like Wilson) in 1820. This was his opportunity to bring to light the hidden birds they had been seeking. He needed only to seize it, and persevere, whatever the effort cost. Six years later he was on his way to England from New Orleans with Louisiana in his portfolio, and the beginnings of both *Birds of America* and the *Ornithological Biography* well on their way. Audubon's transforming plunge into the possibilities of Louisiana can be traced day by day in the *Louisiana Journal*. By the time he reached Oakley in June 1821, his entries on birds had lengthened. He reveled in every new bird he discovered, naming them one after the other, sometimes half a dozen a day. He took his time noting individual features as if he were preparing a lawyer's brief to make a case. These were the embryos of the bird biographies. Just as Louisiana provoked a seachange in the quality of his art, it trained him to be a confident ornithologist. Whenever he wrote about a Louisiana bird he considered Wilson had misnamed or misrepresented, and he did so often, he offered as proof his own careful and repeated investigations, not in a day or a week, but season after season.

The first bird biography he published was the brief account of the cliff swallow Samuel Mitchill had asked him to write for the New York Lyceum of Natural History in 1824. Audubon based the text on notes he had taken in his journal during the winter of 1820 in Louisiana. The same notes were used when he wrote about the cliff swallow in the *Ornithological Biography.* In fact, the text in the *Biography* is an almost verbatim copy of the Lyceum version. All through the *Biography* Audubon quotes directly from his journals, often, as he does in his biography of the whooping crane, drawing on the journals that were later lost or destroyed.

His first serious attempt at rewriting his notes into a full-scale biography was also an original Louisiana undertaking. During the ill-fated voyage to Philadelphia in 1824, Audubon had promised Charles Lucien Bonaparte a substantial report on the wild turkey for Bonaparte's project to expand Wilson's *American Ornithology* by adding the birds that Wilson had failed to include. The absence of the turkey was one of the glaring omissions in Wilson's work. Writing from Louisiana after he returned to Bayou Sara, Audubon duly sent the document to Bonaparte, who acknowledged the service in the wild turkey section that concluded the first volume of his supplement to the *American Ornithology*, published in 1825. Bonaparte expressed his gratitude at the outset of the section, referring to the "copious narrative" Audubon had sent him. Bonaparte then proceeded to pillage the document wholesale. Audubon never thought Bonaparte made it clear that his article would have been impossible without Audubon's contribution. Bonaparte had also hired an inferior artist (Alexander Rider) to remake an Audubon drawing of two grackles that Bonaparte had bought from Audubon himself for twenty-five dollars to be used in the same volume. Audubon was furious when he saw his name mentioned as one of the artists under the engraved plate of the misdrawn birds. Thereafter the grackles and the wild turkey were always linked in Audubon's mind as evidence of Bonaparte's devious intentions, the self-interest hidden behind all his sincere gesticulations of friendship. Fortunately, the original manuscript of the wild turkey paper survived, and was eventually transcribed and published in *The Auk* in July 1962.

The manuscript offers a fascinating glimpse into Audubon's ambitions and the knowledge he had acquired at the time. For one thing, he was quite capable of writing an extensive account in acceptable English (spelling and punctuation notwithstanding) without the help of a MacGillivray. For another, he was forging his conviction that despite the scorn he had suffered in Philadelphia concerning his lack of scientific training, his experience garnered on the ground placed him in a position well beyond the reach of his detractors. After a long, detailed description of the wild turkey's general behavior in the Louisiana woods, Audubon turns to the mating call of the female, and to his own integrity:

> When a female gives a call note, the Goblers in hearing all return the Sound, rolling note after note as fast as if about giving the Last with the first together; not Spreading their tail as when Courting near them or Practising on the Branches of the Trees on which they have roosted all Night, but much as the tame Turkeys do when a Singular Noise is offered to their ears and repeated often—if this Call of the female is Issuing from the Ground all the males fly imediately toward it whether or not in Sight of the Hen they then Spread & erect their Tail, throw their Heads back on the Shoulders and Walk Pompously; Shivering their Wings with a Puff from the Poulmons (lungs) at the same time—Stopping to Listen as well as to look where the female is and resume the Strutting again either Seen or unseen—but moving as fast as these Ceremonies will admit of—Whilst at this it often takes place that two Males see each other, in Such a Case Dreadfull Battles ensues in which *Many Males are Killed* (Recollect that John J. Audubon writes under an Influence Beyond the Power of Common Men to Conceive and that he Writes only What *he has seen* and not heard only) by Picking on the Head of the Weakest untill completely exausted.

Years later, when his fame was secure, and both *Birds of America* and the *Ornithological Biography* were almost finished, Audubon was still fuming about Bonaparte's inadequate gratitude. Visiting London in October 1837, Bonaparte paid a call on Audubon, who immediately wrote to Bachman, "Me thinks that he is over anxious to pump me, but I am no longer a green horn, and will not Write such accounts for him as I did when I sent him all the Habits of the Wild Turkey from Bayou Sara to New York, and for which he has received all the credit, and I scarcely any."

Thus when Audubon gathered himself fully to the writing of the *Ornithological Biography* in 1830, he could draw on at least ten years of testing his skills in descriptive prose. He had resurrected the Louisiana wild turkey text as the first of his papers for the Wernerians in November 1826. Others soon followed. By the time the first volume of the *Biography* was published in March 1831, he had already published half a dozen bird biographies in scientific journals.

All through the five volumes of the *Biography*, Audubon insisted on the authenticity of his work. In the article he published in *The Edinburgh Journal of Science* in 1828, *Account of the Method of Drawing Birds Employed by J.J. Audubon*, self-verification has become a badge of honor, a personal creed: "I have often listened to the different observations of men who accidently had made remarks on different species of birds, but seldom, except when with the rough hunters and squatters of the frontiers, have I discovered naked facts in such relations. This has dissuaded me from ever taking any account given me for granted, until corroborated either by my own ocular opportunities or accumulated repetitions."

With Wilson in the back of his mind, and the Philadelphia cabal led by Ord still smoldering, Audubon was not always subtle about setting himself apart from his illustrious predecessors. In the second paragraph of his text on the wood ibis in the third volume of the *Ornithological Biography*, he seems conscious of going too far, but he cannot stop:

> In treating of the bird now under your notice, Mr WILLIAM BARTRAM says, "This solitary bird does not associate in flocks, but is generally seen alone." This was published by WILSON, and every individual who has since written on the subject, has copied the assertion without probably having any other reason than that he believed the authors of it to state a fact. But the habits of this species are entirely at variance with the above quotation, to which I direct your attention not without a feeling of pain, being assured that MR BARTRAM could have made such a statement only because he had few opportunities of studying the bird in question in its proper haunts.

The few occasions where Audubon overreaches, and protests too much, fade into insignificance when measured against the wealth of information packed in the *Biography*, and not just concerning birds. Audubon's desire in the paintings to place each life-size bird in his natural habitat, perched on a branch of a local tree, or clinging to a flower bush, surrounded by native insects, mammals, and snakes, is reflected all through the biographies. Since Louisiana birds dominate the first volumes of *Birds of America*, and the biographies follow suit, the wildlife of Louisiana can be found in abounding plenty.

Audubon leads the reader deep into the woods and swamps, his purpose being to complement the paintings by showing us where and how he tracked and found his specimens. The paintings focus on the bird, or birds, caught in single frozen moments. They suffice, speaking for themselves. The biographies on the other hand are commentaries on cycles, locating the species in a world changing with the seasons, thick in the middle of feeding, fighting off predators, mating, reproducing, and raising the young. They are rare compositions of living space and place, guiding the reader to the origins of Audubon's inspiration. The Edinburgh Wernerians, his first major audience, were awed by the haunting strangeness of Louisiana's swamps, but Audubon found birds everywhere, on fence posts, trash heaps, farms, plantations, even in the big city of New Orleans. They were hunted, killed, eaten, overkilled, and left to rot on the ground. We might prefer the scenes of wilderness, like the Wernerians two hundred years ago, but Audubon's Louisiana in the biographies was a world of humans too, a society on the move. It might have been, in some respects, all too human. Audubon at times thought the same (including himself in the indictment), thus the almost total absence of men and women in the paintings. When they do appear (only twice) they are toy-like creatures in the background, diminished or caricatured (Audubon's comic-heroic self-portrait as a manikin in a fur hat straddling a log in the original watercolor of the golden eagle), wholly insignificant. The birds are well enough without us.

All through the *Biography* Audubon's first objective was to prove his credentials as a specialist, thus the proliferation of physical traits in exhaustive lists. He described patterns of flight; favorite feeding times and favorite foods; feather type and texture; gradations of color (male, female, hatchling, seasonal); the location, material, and construction of nests; egg shapes, colorings, and clusterings of spots; variations in song; communal behavior: anything that helped define the collective identity of a species. The accumulation of naked detail is remarkable, but what gives the biographies their special charm is Audubon's willingness to go beyond objective fact and consider subjective impressions of personality. He expressed his conviction in no uncertain terms in the first sentence of his text on the winter hawk in volume one of the *Biography:* "Every species of bird is possessed of a certain, not always definable, cast of countenance, peculiar to itself. Although it undergoes changes necessary for marking the passions of the individual, its joy, its anger, its terror or despondency, still it remains the same *specific look.*"

Two years later, in a letter to his son Victor, writing about a live golden eagle he has just acquired in Boston, Audubon narrows his theory down to the bird's eye:

> I have had the great gratification of purchasing this day the finest Golden Eagle alive which I have ever seen. It will be sent to you on paper by Mrs Gordon. I paid $14.75 a pretty good price but not $100 as the rascal in Phila wished to obtain from me for a far less handsome specimen—I will draw it tearing a young Fawn or a northern hare—I am glad to be now assured that the Ring Tailed Eagle is positively the young of this species—good Luck would have it that 2 of the tail feathers are yet undropt with the broad band of lightish Grey—shown in the Ring Tail & which 2 feathers a month hence would have been moulted.—Then the eye—aye—the Eye of Birds is like that of Man, each has its peculiar peculiarities. . . .

From the particular glitter in the eye, there was barely a small step towards the single individual bird, Audubon's pet sparrow hawk, Nero, to name only one, with his blue bill and grey eyes, "not unlike an owl," who developed his peculiar habits under Audubon's care, sang his own song, and was tragically killed by a hen on August 1, 1821. Nero is a good example of Alexander Wilson's philosophy of "personal intimacy" with birds, and perhaps too, once more, of Audubon's natural alignment with the Romantic thinkers and poets for whom the individual was the accomplishment of the race, and not the other way around, making it possible for a Wordsworth to write an epic poem, *The Prelude,* not to lionize the composite ideals of a national hero, but to trace the formation of a unique, unrepresentative mind, his own.

Another Romantic poet, John Keats, expressed the idea of an absolute identity, or personality (which he calls the soul), as something made, and struggled for, not given, in his beautiful definition of the world as a "vale of soul-making." Audubon did not know Keats, though he did know Keats's brother George (and his wife Georgiana), who in the winter of 1818 stayed as paying guests with the Audubons in Henderson, long enough to lose money on an investment Audubon had suggested. Both George and John never forgot, and never forgave, and Audubon himself was so troubled by the affair that he thought he saw Georgiana, or her ghost, on the streets of Liverpool in 1826. By one of those historical coincidences that beg belief, the chronicle letter in which John Keats announces and unfolds his theory of the "vale" was written to George from England, and begun while George was still in the Audubon house. Audubon would not have read the letter, but by the time he began the *Ornithological Biography*, ten years after the doomsday departure from Henderson, he knew something about struggle, suffering, and "soul-making," and could look back on Louisiana as the place where he had remade himself into a superior artist and a consummate ornithologist. Louisiana launched both the *Biography* and the *Birds of America* and pervades them both, from beginning to end. Louisiana was the body and soul of the one, and the soul and body of the other, just to vary the distinction.

There are scores of passages on Louisiana scattered all through the biographies, so many that it would be fairly easy to put together a single narrative stitching them together into one dazzling tapestry. Audubon's method was to identify every place where he had sighted a species, thus allowing him to range freely in the biographies from one region or latitude, as he was wont to say, to another, depending on the bird. A sighting in Louisiana may very well remind him of another, in Kentucky or Florida, or South Carolina. This was necessary to point out the vectors of migration, and the distances each species was prone to travel. There are a number of the biographies, however, in which Louisiana inspires the entire text, even when Audubon roams widely, as he does in the Great Blue Heron biography. In sticking to the principle of avoiding snippets, or any form of excerpt, the choice in *Audubon on Louisiana* has been to select nine complete biographies, all centered in the sphere of Audubon's Louisiana, and all narrated with a clear note of personal and communal triumph, namely because the birds described lived in Louisiana most of the year, at home, and were all painted by the self-proclaimed Louisiana *native* on the spot.

SOURCES

All selections from the *Ornithological Biography* are taken from the first edition, available online on the Biodiversity Heritage Library site. Just as a matter of curiosity, thanks to the blessings of the Internet, the particular first edition consulted was the personal copy of Charles Darwin held at the Natural History Museum Library, London. The plate numbers at the beginning of each biography correspond to the plates in the Double Elephant Folio edition of *The Birds of America*, thus the references in the texts to details of the paintings, including Audubon's mistaken assessment of the harlequin coral snake in the Chuck-Will's-Widow biography as "harmless." He should have added: "except when it bites." The quotations from William Bartram's *Travels* are taken from the Introduction and Part II, Chapter 7 of the first edition, available online on the Documenting the American South site at the University of North Carolina. The quotation from Alexander Wilson's *American Ornithology* is taken from the orchard oriole text in volume one of the first edition, available online on the Biodiversity Heritage Library site. The quotations from Charles Lucien Bonaparte's supplement to the *American Ornithology* are taken from the wild turkey chapter in volume one of the first edition, available online on the Digital Library for the Decorative Arts and Material Culture site at the University of Wisconsin–Madison. The quotation from Audubon's original manuscript of the wild turkey text used by Bonaparte is taken from the transcription by A.W. Schorger published in *The Auk* 79 (July 1962). The quotation from Audubon's *Account of the Method of Drawing Birds* is taken from the *Edinburgh Journal of Science 8* (Edinburgh, 1828): 48–54. The quotations from Audubon's letters are taken from the *Letters of John James Audubon 1826–1840*, edited by Horward Corning, published by the Club of Odd Volumes in Boston (1930) and reprinted by the Kraus Reprint Company of New York (1969). Keats's account of his "vale of soul-making" can be found in the chronicle letter dated February 14 to May 4, 1819, included in *The Selected Letters of John Keats*, edited by Grant F. Scott (based on the texts of Hyder E. Rollins), published by Harvard University Press (2002).

The Mocking Bird

Turdus Polyglottus, LINN.
PLATE XXI. Male and Female.

It is where the Great Magnolia shoots up its majestic trunk, crowned with evergreen leaves, and decorated with a thousand beautiful flowers, that perfume the air around; where the forests and fields are adorned with blossoms of every hue; where the golden Orange ornaments the gardens and groves; where Bignonias of various kinds interlace their climbing stems around the White-flowered Stuartia, and mounting still higher, cover the summits of the lofty trees around, accompanied with innumerable Vines, that here and there festoon the dense foliage of the magnificent woods, lending to the vernal breeze a slight portion of the perfume of their clustered flowers; where a genial warmth seldom forsakes the atmosphere; where berries and fruits of all descriptions are met with at every step;—in a word, kind reader, it is where Nature seems to have paused, as she passed over the Earth, and opening her stores, to have strewed with unsparing hand the diversified seeds from which have sprung all the beautiful and splendid forms which I should in vain attempt to describe, that the Mocking Bird should have fixed its abode, there only that its wondrous song should be heard.

But where is that favoured land?—It is in that great continent to whose distant shores Europe has sent forth her adventurous sons, to wrest for themselves a habitation from the wild inhabitants of the forest, and to convert the neglected soil into fields of exuberant fertility. It is, reader, in Louisiana that these bounties of nature are in the greatest perfection. It is there that you should listen to the love-song of the Mocking Bird, as I at this moment do. See how he flies round his mate, with motions as light as those of the butterfly! His tail is widely expanded, he mounts in the air to a small distance, describes a circle, and, again alighting, approaches his beloved one, his eyes gleaming with delight, for she has already promised to be his and his only. His beautiful wings are gently raised, he bows to his love, and again bouncing upwards, opens his bill, and pours forth his melody, full of exultation at the conquest which he has made.

They are not the soft sounds of the flute or of the hautboy that I hear, but the sweeter notes of Nature's own music. The mellowness of the song, the varied modulations and gradations, the extent of its compass, the great brilliancy of execution, are unrivalled. There is probably no bird in the world that possesses all the musical qualifications of this king of song, who has derived all from Nature's self. Yes, reader, all!

No sooner has he again alighted, and the conjugal contract has been sealed, than, as if his breast was about to be rent with delight, he again pours forth his notes with more softness and richness than before. He now soars higher, glancing around with a vigilant eye, to assure himself that none has witnessed his bliss. When these love-scenes, visible only to the ardent lover of nature, are over, he dances through the air, full of animation and delight, and, as if to convince his lovely mate that to enrich her hopes he has much more love in store, he that moment begins anew, and imitates all the notes which nature has imparted to the other songsters of the grove.

For a while, each long day and pleasant night are thus spent; but at a peculiar note of the female he ceases his song, and attends to her wishes. A nest is to be prepared, and the choice of a place in which to lay it is to become a matter of mutual consideration. The Orange, the Fig, the Pear-tree of the gardens are inspected; the thick briar patches are also visited. They appear all so well suited for the purpose in view, and so well does the bird know that man is not his most dangerous enemy, that instead of retiring from him, they at length fix their abode in his vicinity, perhaps in the nearest tree to his window. Dried twigs, leaves, grasses, cotton, flax, and other substances, are picked up, carried to a forked branch, and there arranged. The female has laid an egg, and the male redoubles his caresses. Five eggs are deposited in due time, when the male having little more to do than to sing his mate to repose, attunes his pipe anew. Every now and then he spies an insect on the ground, the taste of which he is sure will please his beloved one. He drops upon it, takes it in his bill, beats it against the earth, and flies to the nest to feed and receive the warm thanks of his devoted female.

When a fortnight has elapsed, the young brood demand all their care and attention. No cat, no vile snake, no dreaded hawk, is likely to visit their habitation. Indeed the inmates of the next house have by this time become quite attached to the lovely pair of Mocking Birds, and take pleasure in contributing to their safety. The dew-berries from the fields, and many kinds of fruit from the gardens, mixed with insects, supply the young as well as the parents with food. The brood is soon seen emerging from the nest, and in another fortnight, being now able to fly with vigour, and to provide for themselves, they leave the parent birds, as many other species do.

The above account does not contain all that I wish you to know of the habits of this remarkable songster; so, I shall shift the scene to the woods and wilds, where we shall examine it more particularly.

The Mocking Bird remains in Louisiana the whole year. I have observed with astonishment, that towards the end of October, when those which had gone to the Eastern States, some as far as Boston, have returned, they are instantly known by the "southrons," who attack them on all occasions. I have ascertained this by observing the greater shyness exhibited by the strangers for weeks after their arrival. This shyness, however, is shortly over, as well as the animosity displayed by the resident birds, and during the winter there exists a great appearance of sociality among the united tribes.

In the beginning of April, sometimes a fortnight earlier, the Mocking Birds pair, and construct their nests. In some instances they are so careless as to place the nest between the rails of a fence directly by the road. I have frequently found it in such places, or in the fields, as well as in briars, but always so easily discoverable that any person desirous of procuring one, might do so in a very short time. It is coarsely constructed on the outside, being there composed of dried sticks of briars, withered leaves of trees, and grasses, mixed with wool. Internally it is finished with fibrous roots disposed in a circular form, but carelessly arranged. The female lays from four to six eggs the first time, four or five the next, and when there is a third brood, which is sometimes the case, seldom more than three, of which I have rarely found more than two hatched. The eggs are of a short oval form, light green, blotched and spotted with umber. The young of the last brood not being able to support themselves until late in the season, when many of the berries and insects have become scarce, are stunted in growth;—a circumstance which has induced some persons to imagine the existence in the United States of two species of Mocking Bird, a larger and a smaller. This, however, in as far as my observation goes, is not correct. The first brood is frequently brought to the bird-market in New Orleans as early as the middle of April. A little farther up the country, they are out by the fifteenth of May. The second brood is hatched in July, and the third in the latter part of September. The nearer you approach to the sea-shores, the more plentiful do you find these birds. They are naturally fond of loose sands, and of districts scantily furnished with small trees, or patches of briars, and low bushes.

During incubation, the female pays such precise attention to the position in which she leaves her eggs, when she goes to a short distance for exercise and refreshment, to pick up gravel, or roll herself in the dust, that, on her return, should she find that any of them has been displaced, or touched by the hand of man, she utters a low mournful note, at the sound of which the male immediately joins her, and they are both seen to condole together. Some people imagine that, on such occasions, the female abandons the nest; but this idea is incorrect. On the contrary, she redoubles her assiduity and care, and scarcely leaves the nest for a moment; nor is it until she has been repeatedly forced from the dear spot, and has been much alarmed by frequent intrusions, that she finally and reluctantly leaves it. Nay, if the eggs are on the eve of being hatched, she will almost suffer a person to lay hold of her.

Different species of snakes ascend to their nests, and generally suck the eggs or swallow the young; but on all such occasions, not only the pair to which the nest belongs, but many other Mocking Birds from the vicinity, fly to the spot, attack the reptiles, and, in some cases, are so fortunate as either to force them to retreat, or deprive them of life. Cats that have abandoned the houses to prowl about the fields, in a half wild state, are also dangerous enemies, as they frequently approach the nest unnoticed, and at a pounce secure the mother, or at least destroy the eggs or young, and overturn the nest. Children seldom destroy the nests of these birds, and the planters generally protect them. So much does this feeling prevail throughout Louisiana, that they will not willingly permit a Mocking Bird to be shot at any time.

In winter, nearly all the Mocking Birds approach the farm-houses and plantations, living about the gardens or outhouses. They are then frequently seen on the roofs, and perched on the chimney-tops; yet they always appear full of animation. Whilst searching for food on the ground, their motions are light and elegant, and they frequently open their wings as butterflies do when basking in the sun, moving a step or two, and again throwing out their wings. When the weather is mild, the old males are heard singing with as much spirit as during the spring or summer, while the younger birds are busily engaged in practising, preparatory to the love season.

They seldom resort to the interior of the forest either during the day or by night, but usually roost among the foliage of evergreens, in the immediate vicinity of houses in Louisiana, although in the Eastern States they prefer low fir trees.

The flight of the Mocking Bird is performed by short jerks of the body and wings, at every one of which a strong twitching motion of the tail is perceived. This motion is still more apparent while the bird is walking, when it opens its tail like a fan and instantly closes it again. The common *cry* or *call* of this bird is a very mournful note, resembling that uttered on similar occasions by its first cousin the *Turdus rufus*, or, as it is commonly called, the *French Mocking Bird.* When travelling, this flight is only a little prolonged, as the bird goes from tree to tree, or at most across a field, scarcely, if ever, rising higher than the top of the forest. During this migration, it generally resorts to the highest parts of the woods near water courses, utters its usual mournful note, and roosts in these places. It travels mostly by day.

Few hawks attack the Mocking Birds, as on their approach, however sudden it may be, they are always ready not only to defend themselves vigorously and with undaunted courage, but to meet the aggressor half way, and force him to abandon his intention. The only hawk that occasionally surprises it is the *Falco Stanleii*, which flies low with great swiftness, and carries the bird off without any apparent stoppage. Should it happen that the ruffian misses his prey, the Mocking Bird in turn becomes the assailant, and pursues the Hawk with great courage, calling in the mean time all the birds of its species to its assistance; and although it cannot overtake the marauder, the alarm created by their cries, which are propagated in succession among all the birds in the vicinity, like the watchwords of sentinels on duty, prevents him from succeeding in his attempts.

The musical powers of this bird have often been taken notice of by European naturalists, and persons who find pleasure in listening to the song of different birds whilst in confinement or at large. Some of these persons have described the notes of the Nightingale as occasionally fully equal to those of our bird. I have frequently heard both species in confinement, and in the wild state, and without prejudice, have no hesitation in pronouncing the notes of the European Philomel equal to those of a *soubrette* of taste, which, could she study under a Mozart, might perhaps in time become very interesting in her way. But to compare her essays to the finished talent of the Mocking Bird, is, in my opinion, quite absurd.

The Mocking Bird is easily reared by hand from the nest, from which it ought to be removed when eight or ten days old. It becomes so very familiar and affectionate, that it will often follow its owner about the house. I have known one raised from the nest kept by a gentleman at Natchez, that frequently flew out of the house, poured forth its melodies, and returned at sight of its keeper. But notwithstanding all the care and management bestowed upon the improvement of the vocal powers of this bird in confinement, I never heard one in that state produce any thing at all approaching in melody to its own natural song.

The male bird is easily distinguished in the nest, as soon as the brood is a little fledged, it being larger than the female, and shewing more pure white. It does not shrink so deep in the nest as the female does, at the sight of the hand which is about to lift it. Good singing birds of this species often bring a high price. They are long-lived, and very agreeable companions. Their imitative powers are amazing, and they mimic with ease all their brethren of the forests or of the waters, as well as many quadrupeds. I have heard it asserted that they possess the power of imitating the human voice, but have never met with an instance of the display of this alleged faculty.

The Red-Tailed Hawk

Falco Borealis, Gmel.
PLATE LI. Male and Female.

The Red-tailed Hawk is a constant resident in the United States, in every part of which it is found. It performs partial migrations, during severe winters, from the Northern Districts towards the Southern. In the latter, however, it is at all times more abundant, and I shall endeavor to present you with a full account of its habits, as observed there.

Its flight is firm, protracted, and at times performed at a great height. It sails across the whole of a large plantation, on a level with the tops of the forest-trees which surround it, without a single flap of its wings, and is then seen moving its head sidewise to inspect the objects below. This flight is generally accompanied by a prolonged mournful cry, which may be heard at a considerable distance, and consists of a single sound resembling the monosyllable *Kae,* uttered in such a manner as to continue for three or four minutes, without any apparent inflection or difference of intensity. It would seem as if uttered for the purpose of giving notice to the living objects below that he is passing, and of thus inducing them to bestir themselves and retreat to a hiding-place, before they attain which he may have an opportunity of pouncing upon some of them. When he spies an animal, while he is thus sailing over a field, I have observed him give a slight check to his flight, as if to mark a certain spot with accuracy, and immediately afterwards alight on the nearest tree. He would then instantly face about, look intensely on the object that had attracted his attention, soon after descend towards it with wings almost close to his body, and dart upon it with such accuracy and rapidity as seldom to fail in securing it.

When passing over a meadow, a cotton-field, or one planted with sugar-canes, he performs his flight close over the grass or plants, uttering no cry, but marking the prey in the manner above described, and on perceiving it, ascending in a beautiful curved line to the top of the nearest tree, after which he watches and dives as in the former case. Should he not observe any object worthy of his attention, while passing over a meadow or a field, he alights, shakes his feathers, particularly those of the tail, and after spending a few minutes in pluming himself, leaves the perch, uttering his usual cry, and ascending in the air, performs large and repeated circular flights, carefully inspecting the field, to assure himself that there is in reality nothing in it that may be of use to him. He then proceeds to another plantation. At other times, as if not assured that his observations have been duly made, he rises in circles over the same field to an immense height, where he looks like a white dot in the heavens. Yet from this height he must be able to distinguish the objects on the ground, even when these do not exceed our little partridge or a young hare in size, and although their colour may be almost the same as that of surrounding bodies; for of a sudden his circlings are checked, his wings drawn close to his body, his tail contracted to its smallest breadth, and he is seen to plunge headlong towards the earth, with a rapidity which produces a loud rustling sound nearly equal to that of an Eagle on a similar occasion.

Should he not succeed in discovering the desired object in the fields, he enters the forest and perches on some detached tree, tall enough to enable him to see to a great distance around. His posture is now erect, he remains still and silent, moving only his head, as on all other occasions, to enable his keen eye to note the occurrences which may take place in his vicinity. The lively Squirrel is seen gaily leaping from one branch to another, or busily employed in searching for the fallen nuts on the ground. It has found one. Its bushy tail is beautifully curved along its back, the end of it falling off with a semicircular bend; its nimble feet are seen turning the nut quickly round, and its teeth are already engaged in perforating the hard shell; when, quick as thought, the Red-tailed Hawk, which has been watching it in all its motions, falls upon it, seizes it near the head, transfixes and strangles it, devours it on the spot, or ascends exultingly to a branch with the yet palpitating victim in his talons, and there feasts at leisure.

As soon as the little King-bird has raised its brood, and when its courage is no longer put in requisition for the defence of its young or its mate, the Red-tailed Hawk visits the farm-houses, to pay his regards to the poultry. This is done without much precaution, for, while sailing over the yard where the chickens, the ducklings, and the young turkeys are, the Hawk plunges upon any one of them, and sweeps it off to the nearest wood. When impelled by continued hunger, he now and then manages to elude the vigilance of the Martins, Swallows and King-birds, and watching for a good opportunity, falls upon and seizes an old fowl, the dying screams of which are heard by the farmer at the plough, who swears vengeance against the robber. He remembers that he has observed the Hawk's nest in the woods, and full of anger at the recollection of the depredations which the plunderer has already committed, and at the anticipation of its many visits during the winter, leaves his work and his horses, strides to his house, and with an axe and a rifle in his hands proceeds towards the tree, where the hopes of the Red-tailed Hawk are snugly nestled among the tall branches. The farmer arrives, eyes the gigantic tree, thinks for a moment of the labour which will be required for felling it, but resolves that he shall not be overreached by a Hawk. He throws aside his hat, rolls up his sleeves, and applies himself to the work. His brawny arms give such an impulse to the axe, that at every stroke large chips are seen to fall off on all sides. The poor mother-bird, well aware of the result, sails sorrowfully over and around. She would fain beg for mercy towards her young. She alights on the edge of the nest, and would urge her offspring to take flight. But the farmer has watched her motions. The axe is left sticking in the core of the tree, his rifle is raised to his shoulder in an instant, and the next moment the whizzing ball has pierced the heart of the Red-tailed Hawk, which falls unheeded to the earth. The farmer renews his work, and now changes sides. A whole hour has been spent in the application of ceaseless blows. He begins to look upwards, to judge which way the giant of the forest will fall, and having ascertained this, he redoubles his blows. The huge oak begins to tremble. Were it permitted to speak, it might ask why it should suffer for the deeds of another; but it is now seen slowly to incline, and soon after with an awful rustling produced by all its broad arms, its branches, twigs and leaves, passing like lightning through the air, the noble tree falls to the earth, and almost causes it to shake. The work of revenge is now accomplished: the farmer seizes the younglings, and carries them home, to be tormented by his children, until death terminates their brief career.

Notwithstanding the very common occurrence of such acts of retribution between man and the Hawk, it would be difficult to visit a plantation in the State of Louisiana, without observing at least a pair of this species hovering about, more especially during the winter months. Early in February, they begin to build their nest, which is usually placed within the forest, and on the tallest and largest tree in the neighbourhood. The male and female are busily engaged in carrying up dried sticks, and other materials, for eight or ten days, during which time their cry is seldom heard. The nest is large, and is fixed in the centre of a triply forked branch. It is of a flattish form, constructed of sticks, and finished with slender twigs and coarse grasses or Spanish moss. The female lays four or five eggs, of a dull white colour, splatched with brown and black, with a very hard, smooth shell. The male assists the female in incubating, but it is seldom that the one brings food to the other while thus employed.

I have seen one or two of these nests built in a large tree which had been left standing in the middle of a field; but occurrences of this kind are rare, on account of the great enmity shewn to this species by the farmers. The young are abundantly supplied with food of various kinds, particularly grey squirrels, which the parents procure while hunting in pairs, when nothing can save the squirrel from their attacks excepting its retreat into the hole of a tree; for should the animal be observed ascending the trunk or branch of a tree by either of the Hawks, this one immediately plunges toward it, while the other watches it from the air. The little animal, if placed against the trunk, when it sees the Hawk coming towards it, makes swiftly for the opposite side of the trunk, but is there immediately dived at by the other Hawk, and now the murderous pair chase it so closely, that unless it immediately finds a hole into which to retreat, it is caught in a few minutes, killed, carried to the nest, torn in pieces, and distributed among the young Hawks. Small hares, or, as we usually call them, *rabbits*, are also frequently caught, and the depredations of the Red-tailed Hawks at this period are astonishing, for they seem to kill every thing, fit for food, that comes in their way. They are great destroyers of tame Pigeons, and woe to the Cock or Hen that strays far from home, for so powerful is this Hawk, that it is able not only to kill them, but to carry them off in its claws to a considerable distance.

The continued attachment that exists between Eagles once paired, is not exhibited by these birds, which, after rearing their young, become as shy towards each other as if they had never met. This is carried to such a singular length, that they are seen to chase and rob each other of their prey, on all occasions. I have seen a couple thus engaged, when one of them had just seized a young rabbit or a squirrel, and was on the eve of rising in the air with it, for the purpose of carrying it off to a place of greater security. The one would attack the other with merciless fury, and either force it to abandon the prize, or fight with the same courage as its antagonist, to prevent the latter from becoming the sole possessor. They are sometimes observed flying either one after the other with great rapidity, emitting their continued cry of *kae*, or performing beautiful evolutions through the air, until one or other of them becomes fatigued, and giving way, makes for the earth, where the battle continues until one is overpowered and obliged to make off. It was after witnessing such an encounter between two of these powerful marauders, fighting hard for a young hare, that I made the drawing now before you, kind reader, in which you perceive the male to have greatly the advantage over the female, although she still holds the hare firmly in one of her talons, even while she is driven towards the earth, with her breast upwards.

I have observed that this species will even condescend to pounce on wood-rats and meadow-mice; but I never saw one of these birds seize even those without first alighting on a tree before committing the act.

During the winter months, the Red-tailed Hawk remains perched for hours together, when the sun is shining and the weather calm. Its breast is opposed to the sun, and it then is seen at a great distance, the pure white of that portion of its plumage glittering as if possessed of a silky gloss. They return to their roosting-places so late in the evening, that I have frequently heard their cry after sun-set, mingling with the jovial notes of Chuck-will's-widow, and the ludicrous laugh of the Barred Owl. In the State of Louisiana, the Red-tailed Hawk roosts amongst the tallest branches of the *Magnolia grandiflora*, a tree which there often attains a height of a hundred feet, and a diameter of from three to four feet at the base. It is also fond of roosting on the tall Cypress-trees of our swamps, where it spends the night in security, amidst the mosses attached to the branches.

The Red-tailed Hawk is extremely wary, and difficult to be approached by any one bearing a gun, the use of which it seems to understand perfectly; for no sooner does it perceive a man thus armed than it spreads its wings, utters a loud shriek, and sails off in an opposite direction. On the other hand, a person on horseback, or walking unarmed, may pass immediately under the branch on which it is perched, when it merely watches his motions as he proceeds. It seldom alights on fences, or the low branches of trees, but prefers the highest and most prominent parts of the tallest trees. It alights on the borders of clear streams to drink. I have observed it in such situations, immersing its bill up to the eyes, and swallowing as much as was necessary to quench its thirst at a single draught.

I have seen this species pounce on soft-shelled tortoises, and amusing enough it was to see the latter scramble towards the water, enter it, and save themselves from the claws of the Hawk by immediately diving. I am not aware that this Hawk is ever successful in these attacks, as I have not on any occasion found any portion of the skin, head, or feet of tortoises in the stomachs of the many Hawks of this species which I have killed and examined. Several times, however, I have found portions of bull-frogs in their stomach.

All our Falcons are pestered with parasitic flying ticks. Those found amongst the plumage of the Red-tailed Hawk, like all others, move swiftly sidewise between the feathers, issue from the skin, and shift from one portion of the body to another on wing, and do not abandon the bird for a day or two after the latter is dead. These ticks are large, and of an auburn colour.

The body of the Red-tailed Hawk is large, compact, and muscular. These birds protrude their talons beyond their head in seizing their prey, as well as while fighting in the air, in the manner shown in the Plate. I have caught several birds of this species by baiting a steel-trap with a live chicken.

The animal represented as held in one of the feet of the female, is usually called a *rabbit* in all parts of the United States, but is evidently a true hare. It never burrows, but has a *form* to rest in, and to which it returns in the manner of the common hare of Europe. I may hereafter present you, kind reader, with a full account of this American species, which occurs in great abundance in the United States.

I have only here to add, that amongst the American farmers the common name of our present bird is the *Hen-hawk,* while it receives that of *Grand mangeur de poules* from the Creoles of Louisiana.

Chuck-Will's-Widow

Caprimulgus carolinensis, Briss.
PLATE LIL. Male and Female.

Our Goatsuckers, although possessed of great power of wing, are particularly attached to certain districts and localities. The species now under consideration is seldom observed beyond the limits of the Choctaw Nation in the State of Mississippi, or the Carolinas, on the shores of the Atlantic, and may with propriety be looked upon as the southern species of the United States. Louisiana, Florida, the lower portions of Alabama and Georgia, are the parts in which it most abounds; and there it makes its appearance early in spring, coming over from Mexico, and probably still warmer climates.

About the middle of March, the forests of Louisiana are heard to echo with the well-known notes of this interesting bird. No sooner has the sun disappeared, and the nocturnal insects emerge from their burrows, than the sounds, "*chuck-will's-widow,*" repeated with great clearness and power six or seven times in as many seconds, strike the ear of every individual, bringing to the mind a pleasure mingled with a certain degree of melancholy, which I have often found very soothing. The sounds of the Goatsucker, at all events, forebode a peaceful and calm night, and I have more than once thought, are conducive to lull the listener to repose.

The deep ravines, shady swamps, and extensive pine ridges, are all equally resorted to by these birds; for in all such places they find ample means of providing for their safety during the day, and of procuring food under night. Their notes are seldom heard in cloudy weather, and never when it rains. Their roosting places are principally the hollows of decayed trees, whether standing or prostrate, from which latter they are seldom raised during the day, excepting while incubation is in progress. In these hollows I have found them, lodged in the company of several species of bats, the birds asleep on the mouldering particles of the wood, the bats clinging to the sides of the cavities. When surprised in such situations, instead of trying to effect their escape by flying out, they retire backwards to the farthest corners, ruffle all the feathers of their body, open their mouth to its full extent, and utter a hissing kind of murmur, not unlike that of some snakes. When seized and brought to the light of day, they open and close their eyes in rapid succession, as if it were painful for them to encounter so bright a light. They snap their little bill in the manner of Fly-catchers, and shuffle along as if extremely desirous of making their escape. On giving them liberty to fly, I have found them able to proceed until out of my sight. They passed between the trees with apparently as much ease and dexterity as if it had been twilight. I once cut two of the quill-feathers of a wing of one of these birds, and allowed it to escape. A few days afterwards I found it in the same log, which induces me to believe that they, like many other birds, resort to the same spot, to roost or spend the day.

The flight of the Chuck-will's-widow is as light as that of its relative, the well-known *Whip-poor-will*, if not more so, and is more graceful as well as more elevated. It somewhat resembles the flight of the Hen-harrier, being performed by easy flappings of the wings, interspersed with sailings and curving sweeps, extremely pleasing to the bystander. At the approach of night, this bird begins to sing clearly and loudly, and continues its notes for about a quarter of an hour. At this time it is perched on a fence-stake, or on the decayed branch of a tree in the interior of the woods, seldom on the ground. The sounds or notes which it emits seem to cause it some trouble, as it raises and lowers its head in quick succession at each of them. This over, the bird launches into the air, and is seen sweeping over the cotton fields or the sugar plantations, cutting all sorts of figures, mounting, descending, or sailing, with so much ease and grace, that one might be induced to call it the *Fairy of the night*. If it passes close to one, a murmuring noise is heard, at times resembling that spoken of when the bird is caught by day. It suddenly checks its course, inclines to the right or left, secures a beetle or a moth, continues its flight over the field, passes and repasses hundreds of times over the same ground, and now and then alights on a fence-stake, or the tallest plant in the place, from which it emits its notes for a few moments with increased vivacity. Now, it is seen following a road or a path on the wing, and alighting here and there to pick up the beetle emerging from its retreat in the ground; again, it rises high in air, and gives chase to the insects that are flying there, perhaps on their passage from one wood to another. At other times, I have seen it poise itself on its wings opposite the trunk of a tree, and seize with its bill the insects crawling on the bark, in this manner inspecting the whole tree, with motions as light as those by which the Humming Bird flutters from one flower to another. In this manner Chuck-will's-widow spends the greater part of the night.

The greatest harmony appears to subsist between the birds of this species, for dozens may be observed flying together over a field, and chasing insects in all directions, without manifesting any enmity or envy. A few days after the arrival of the male birds, the females make their appearance, and the love season at once commences. The male pays his addresses to the female with a degree of pomposity only equalled by the Tame Pigeon. The female, perched lengthwise on a branch, appears coy and silent, whilst the male flies around her, alights in front of her, and with drooping wings and expanded tail advances quickly, singing with great impetuosity. They are soon seen to leave the branch together and gambol through the air. A few days after this, the female, having made choice of a place in one of the most retired parts of some thicket, deposits two eggs, which I think, although I cannot be certain, are all that she lays for the season. This bird forms no nest. A little space is carelessly scratched amongst the dead leaves, and in it the eggs, which are elliptical, dull olive, and speckled with brown, are dropped. These are not found without great difficulty, unless when by accident a person passes within a few feet of the bird whilst sitting, and it chances to fly off. Should you touch or handle these dear fruits of happy love, and, returning to the place, search for them again, you would search in vain; for the bird perceives at once that they have been meddled with, and both parents remove them to some other part of the woods, where chance only could enable you to find them again. In the same manner, they also remove the young when very small.

This singular occurrence has as much occupied my thoughts as the equally singular manner in which the *Cow Bunting* deposits her eggs, which she does, like the *Common Cuckoo* of Europe, one by one, in the nests of other birds, of different species from her own. I have spent much time in trying to ascertain in what manner the Chuck-will's-widow removes her eggs or young, particularly as I found, by the assistance of an excellent dog, that neither the eggs nor the young were to be met with within at least a hundred yards from the spot where they at first lay. The Negroes, some of whom pay a good deal of attention to the habits of birds and quadrupeds, assured me that these birds push the eggs or young with their bill along the ground. Some farmers, without troubling themselves much about the matter, imagined the transportation to be performed under the wings of the old bird. The account of the Negroes appearing to me more likely to be true than that of the farmers, I made up my mind to institute a strict investigation of the matter. The following is the result.

When the Chuck-will's-widow, either male or female (for each sits alternately) has discovered that the eggs have been touched, it ruffles its feathers and appears extremely dejected for a minute or two, after which it emits a low murmuring cry, scarcely audible to me, as I lay concealed at a distance of not more than eighteen or twenty yards. At this time I have seen the other parent reach the spot, flying so low over the ground that I thought its little feet must have touched it, as it skimmed along, and after a few low notes and some gesticulations, all indicative of great distress, take an egg in its large mouth, the other bird doing the same, when they would fly off together, skimming closely over the ground, until they disappeared among the branches and trees. But to what distance they remove their eggs, I have never been able to ascertain; nor have I ever had an opportunity of witnessing the removal of the young. Should a person, coming upon the nest when the bird is sitting, refrain from touching the eggs, the bird returns to them and sits as before. This fact I have also ascertained by observation.

I wish I could have discovered the peculiar use of the *pectinated claw* which this bird has on each foot; but, reader, this remains one of the many desiderata in ornithology, and I fear, with me at least, will continue so.

The Chuck-will's-widow manifests a strong antipathy towards all snakes, however harmless they may be. Although these birds cannot in any way injure the snakes, they alight near them on all occasions, and try to frighten them away, by opening their prodigious mouth, and emitting a strong hissing murmur. It was after witnessing one of these occurrences, which took place at early twilight, that the idea of representing these birds in such an occupation struck me. The beautiful little snake, gliding along the dead branch, between two Chuck-will's-widows, a male and a female, is commonly called the *Harlequin Snake*, and is, I believe, quite harmless.

The food of the bird now under consideration consists entirely of all sorts of insects, among which the larger species of moths and beetles are very conspicuous. The long bristly feathers at the base of the mandibles of these birds no doubt contribute greatly to prevent the insects from escaping, after any portion of them has entered the mouth of the bird.

These birds become silent as soon as the young are hatched, but are heard again before their departure towards the end of summer. At this season, however, their cry is much less frequently heard than in spring. They leave the United States all of a sudden, about the middle of the month of August.

The Ivory-Billed Woodpecker

Picus Principalis, Linn.
PLATE LXVI. Male and Female.

I have always imagined, that in the plumage of the beautiful Ivory-billed Woodpecker, there is something very closely allied to the style of colouring of the great VANDYKE. The broad extent of its dark glossy body and tail, the large and well-defined white markings of its wings, neck, and bill, relieved by the rich carmine of the pendent crest of the male, and the brilliant yellow of its eye, have never failed to remind me of some of the boldest and noblest productions of that inimitable artist's pencil. So strongly indeed have these thoughts become ingrafted in my mind, as I gradually obtained a more intimate acquaintance with the Ivory-billed Woodpecker, that whenever I have observed one of these birds flying from one tree to another, I have mentally exclaimed, "There goes a Vandyke!" This notion may seem strange, perhaps ludicrous, to you, good reader, but I relate it as a fact, and whether or not it may be found in accordance with your own ideas, after you have inspected the plate in which is represented this great chieftain of the Woodpecker tribe, is perhaps of little consequence.

The Ivory-billed Woodpecker confines its rambles to a comparatively very small portion of the United States, it never having been observed in the Middle States within the memory of any person now living there. In fact, in no portion of these districts does the nature of the woods appear suitable to its remarkable habits.

Descending the Ohio, we meet with this splendid bird for the first time near the confluence of that beautiful river and the Mississippi; after which, following the windings of the latter, either downwards toward the sea, or upwards in the direction of the Missouri, we frequently observe it. On the Atlantic coast, North Carolina may be taken as the limit of its distribution, although now and then an individual of the species may be accidentally seen in Maryland. To the westward of the Mississippi, it is found in all the dense forests bordering the streams which empty their waters into that majestic river, from the very declivities of the Rocky Mountains. The lower parts of the Carolinas, Georgia, Alabama, Louisiana, and Mississippi, are, however, the most favourite resorts of this bird, and in those States it constantly resides, breeds, and passes a life of peaceful enjoyment, finding a profusion of food in all the deep, dark, and gloomy swamps dispersed throughout them.

I wish, kind reader, it were in my power to present to your mind's eye the favourite resort of the Ivory-billed Woodpecker. Would that I could describe the extent of those deep morasses, overshadowed by millions of gigantic dark cypresses, spreading their sturdy moss-covered branches, as if to admonish intruding man to pause and reflect on the many difficulties which he must encounter, should he persist in venturing farther into their almost inaccessible recesses, extending for miles before him, where he should be interrupted by huge projecting branches, here and there the massy trunk of a fallen and decaying tree, and thousands of creeping and twining plants of numberless species! Would that I could represent to you the dangerous nature of the ground, its oozing, spongy, and miry disposition, although covered with a beautiful but treacherous carpeting, composed of the richest mosses, flags, and water-lilies, no sooner receiving the pressure of the foot than it yields and endangers the very life of the adventurer, whilst here and there, as he approaches an opening, that proves merely a lake of black muddy water, his ear is assailed by the dismal croaking of innumerable frogs, the hissing of serpents, or the bellowing of alligators! Would that I could give you an idea of the sultry pestiferous atmosphere that nearly suffocates the intruder during the meridian heat of our dogdays, in those gloomy and horrible swamps! But the attempt to picture these scenes would be vain. Nothing short of ocular demonstration can impress any adequate idea of them.

How often, kind reader, have I thought of the difference of the tasks imposed on different minds, when, travelling in countries far distant from those where birds of this species and others as difficult to be procured are now and then offered for sale in the form of dried skins, I have heard the amateur or closet-naturalist express his astonishment that half-a-crown was asked by the person who had perhaps followed the bird when alive over miles of such swamps, and after procuring it, had prepared its skin in the best manner, and carried it to a market thousands of miles distant from the spot where he had obtained it. I must say, that it has at least grieved me as much as when I have heard some idle fop complain of the poverty of the Gallery of the Louvre, where he had paid nothing, or when I have listened to the same infatuated idler lamenting the loss of his shilling, as he sauntered through the Exhibition Rooms of the Royal Academy of London, or any equally valuable repository of art. But, let us return to the biography of the famed Ivory-billed Woodpecker.

The flight of this bird is graceful in the extreme, although seldom prolonged to more than a few hundred yards at a time, unless when it has to cross a large river, which it does in deep undulations, opening its wings at first to their full extent, and nearly closing them to renew the propelling impulse. The transit from one tree to another, even should the distance be as much as a hundred yards, is performed by a single sweep, and the bird appears as if merely swinging itself from the top of the one tree to that of the other, forming an elegantly curved line. At this moment all the beauty of the plumage is exhibited, and strikes the beholder with pleasure. It never utters any sound whilst on wing, unless during the love-season; but at all other times, no sooner has this bird alighted than its remarkable voice is heard, at almost every leap which it makes, whilst ascending against the upper parts of the trunk of a tree, or its highest branches. Its notes are clear, loud, and yet rather plaintive. They are heard at a considerable distance, perhaps half a mile, and resemble the false high note of a clarionet. They are usually repeated three times in succession, and may be represented by the monosyllable *pait, pait, pait*. These are heard so frequently as to induce me to say that the bird spends few minutes of the day without uttering them, and this circumstance leads to its destruction, which is aimed at, not because (as is supposed by some) this species is a destroyer of trees, but more because it is a beautiful bird, and its rich scalp attached to the upper mandible forms an ornament for the war-dress of most of our Indians, or for the shot-pouch of our squatters and hunters, by all of whom the bird is shot merely for that purpose.

Travellers of all nations are also fond of possessing the upper part of the head and the bill of the male, and I have frequently remarked, that on a steamboat's reaching what we call a *wooding-place*, the *strangers* were very apt to pay a quarter of a dollar for two or three heads of this Woodpecker. I have seen entire belts of Indian chiefs closely ornamented with the tufts and bills of this species, and have observed that a great value is frequently put upon them.

The Ivory-billed Woodpecker nestles earlier in spring than any other species of its tribe. I have observed it boring a hole for that purpose in the beginning of March. The hole is, I believe, always made in the trunk of a live tree, generally an ash or a hagberry, and is at a great height. The birds pay great regard to the particular situation of the tree, and the inclination of its trunk; first, because they prefer retirement, and again, because they are anxious to secure the aperture against the access of water during beating rains. To prevent such a calamity, the hole is generally dug immediately under the junction of a large branch with the trunk. It is first bored horizontally for a few inches, then directly downwards, and not in a spiral manner, as some people have imagined. According to circumstances, this cavity is more or less deep, being sometimes not more than ten inches, whilst at other times it reaches nearly three feet downwards into the core of the tree. I have been led to think that these differences result from the more or less immediate necessity under which the female may be of depositing her eggs, and again have thought that the older the Woodpecker is, the deeper does it make its hole. The average diameter of the different nests which I have examined was about seven inches within, although the entrance, which is perfectly round, is only just large enough to admit the bird.

Both birds work most assiduously at this excavation, one waiting outside to encourage the other, whilst it is engaged in digging, and when the latter is fatigued, taking its place. I have approached trees whilst these Woodpeckers were thus busily employed in forming their nest, and by resting my head against the bark, could easily distinguish every blow given by the bird. I observed that in two instances, when the Woodpeckers saw me thus at the foot of the tree in which they were digging their nest, they abandoned it for ever. For the first brood there are generally six eggs. They are deposited on a few chips at the bottom of the hole, and are of a pure white colour. The young are seen creeping out of the hole about a fortnight before they venture to fly to any other tree. The second brood makes its appearance about the 15th of August.

In Kentucky and Indiana, the Ivory-bills seldom raise more than one brood in the season. The young are at first of the colour of the female, only that they want the crest, which, however, grows rapidly, and towards autumn, particularly in birds of the first breed, is nearly equal to that of the mother. The males have then a slight line of red on the head, and do not attain their richness of plumage until spring, or their full size until the second year. Indeed, even then, a difference is easily observed between them and individuals which are much older.

The food of this species consists principally of beetles, larvae, and large grubs. No sooner, however, are the grapes of our forests ripe than they are eaten by the Ivory-billed Woodpecker with great avidity. I have seen this bird hang by its claws to the vines, in the position so often assumed by a Titmouse, and, reaching downwards, help itself to a bunch of grapes with much apparent pleasure. Persimons are also sought for by them, as soon as the fruit becomes quite mellow, as are hagberries.

The Ivory-bill is never seen attacking the corn, or the fruit of the orchards, although it is sometimes observed working upon and chipping off the bark from the belted trees of the newly-cleared plantations. It seldom comes near the ground, but prefers at all times the tops of the tallest trees. Should it, however, discover the half-standing broken shaft of a large dead and rotten tree, it attacks it in such a manner as nearly to demolish it in the course of a few days. I have seen the remains of some of these ancient monarchs of our forests so excavated, and that so singularly, that the tottering fragments of the trunk appeared to be merely supported by the great pile of chips by which its base was surrounded. The strength of this Woodpecker is such, that I have seen it detach pieces of bark seven or eight inches in length at a single blow of its powerful bill, and by beginning at the top branch of a dead tree, tear off the bark, to an extent of twenty or thirty feet, in the course of a few hours, leaping downwards with its body in an upward position, tossing its head to the right and left, or leaning it against the bark to ascertain the precise spot where the grubs were concealed, and immediately after renewing its blows with fresh vigour, all the while sounding its loud notes, as if highly delighted.

This species generally moves in pairs, after the young have left their parents. The female is always the most clamorous and the least shy. Their mutual attachment is, I believe, continued through life. Excepting when digging a hole for the reception of their eggs, these birds seldom, if ever, attack living trees, for any other purpose than that of procuring food, in doing which they destroy the insects that would otherwise prove injurious to the trees.

I have frequently observed the male and female retire to rest for the night, into the same hole in which they had long before reared their young. This generally happens a short time after sunset.

When wounded and brought to the ground, the Ivory-bill immediately makes for the nearest tree, and ascends it with great rapidity and perseverance, until it reaches the top branches, when it squats and hides, generally with great effect. Whilst ascending, it moves spirally round the tree, utters its loud *pait, pait, pait,* at almost every hop, but becomes silent the moment it reaches a place where it conceives itself secure. They sometimes cling to the bark with their claws so firmly, as to remain cramped to the spot for several hours after death. When taken by the hand, which is rather a hazardous undertaking, they strike with great violence, and inflict very severe wounds with their bill as well as claws, which are extremely sharp and strong. On such occasions, this bird utters a mournful and very piteous cry.

The Yellow-Throated Warbler

Sylvia Pensilis, Lath.
PLATE LXXXV. Male.

This beautiful bird absents itself from the State of Louisiana only for two months in the year, December and January. When they return in the beginning of February, they throw themselves by thousands into all the cypress woods and canebrakes, where they are heard singing from the first of March until late in autumn, sometimes in November.

Their habits are very different from those of the Warblers, and are more in general accordance with those of the Certhiæ. They move up and down, sidewise and spirally, along the trunks, branches, and even twigs of the tallest and largest Cypresses, or such other trees as are found intermingled with them. They are extremely active, in fact, fully as much so as the little Brown Creeper itself. Like it, they suddenly leave the uppermost branches or higher parts of the trunks, and diving downwards alight on the roots, and renew their search after small insects and larvae. I never saw any of them pursue insects on wing.

The nest of this species is prettily constructed. Its outer parts are composed of grey lichens and soft mosses, the interior of silky substances and a few fibres of the Spanish moss. The female lays four pure white eggs, having two or three purple dots near the larger end. I think they raise two broods during their stay in Louisiana, but cannot speak of this as certain. The nest is placed on a horizontal branch of a Cypress, twenty, thirty, or even fifty feet above the ground, and is with difficulty discovered from below, as it resembles a knot or a tuft of moss.

The song of the Yellow-throated Warbler would please you, kind reader. Of this I have not a doubt, as it is soft and loud, and is continued for two or three minutes at a time, not unlike that of the Painted Finch, or Indigo Bird. As it is heard in all parts of our most dismal Cypress Swamps, it contributes to soothe the mind of a person whose occupation may lead him to such places. I never saw this species on the ground. The male and the female are nearly alike in plumage, but the young birds, which hunt for insects in company, in the manner of Creepers or Titmice, do not acquire the yellow on the throat, nor the full brilliancy of their plumage, until the first spring.

These birds confine themselves to the Southern States, seldom moving farther towards the Middle Districts than North Carolina. They do not even ascend the Mississippi farther than the Walnut Hills. They are abundant in the neighbourhood of the Red River, and probably do not go farther south than Mexico, during their short absence from the United States.

Happening to shoot several of these birds on a large Chinquapin tree, growing on the edge of a hill close to a swamp, I have put a male on one of its twigs, which is furnished with a few fruits quite ripe and ready to leave their husks. In the Southern States this tree is rare. It generally prefers elevated places, and rocky declivities, with an arid soil. The wood resembles that of the Chestnut, but the trees being generally small, little use is made of it as timber. The fruit is eaten by children. This tree is abundant along the greater part of the range of the Alleghanies and its branches.

The Mississippi Kite

Falco Plumbeus, Gmel.
PLATE CXVII. Male and Female.

When, after many a severe conflict, the southern breezes, in alliance with the sun, have, as if through a generous effort, driven back for a season to their desolate abode the chill blasts of the north; when warmth and plenty are insured for a while to our happy lands; when clouds of anxious Swallows, returning from the far south, are guiding millions of Warblers to their summer residence; when numberless insects, cramped in their hanging shells, are impatiently waiting for the full expansion of their wings; when the vernal flowers, so welcome to all, swell out their bursting leaflets, and the rich-leaved Magnolia opens its pure blossoms to the Humming Bird;—then look up, and you will see the Mississippi Kite, as he comes sailing over the scene. He glances towards the earth with his fiery eye; sweeps along, now with the gentle breeze, now against it; seizes here and there the high-flying giddy bug, and allays his hunger without fatigue to wing or talon. Suddenly he spies some creeping thing, that changes, like the chameleon, from vivid green to dull-brown, to escape his notice. It is the red-throated panting lizard that has made its way to the highest branch of a tree in quest of food. Casting upwards a sidelong look of fear, it remains motionless, so well does it know the prowess of the bird of prey: but its caution is vain; it has been perceived, its fate is sealed, and the next moment it is swept away.

The Mississippi Kite thus extends its migrations as high as the city of Memphis, on the noble stream whose name it bears, and along our eastern shores to the Carolinas, where it now and then breeds, feeding the while on lizards, small snakes, and beetles, and sometimes, as if for want of better employ, teaching the Carrion Crows and Buzzards to fly. At other times, congregating to the number of twenty or more, these birds are seen sweeping around some tree, catching the large locusts which abound in those countries at an early part of the season, and reminding one of the Chimney Swallows, which are so often seen performing similar evolutions, when endeavouring to snap off the little dried twigs of which their nests are composed.

Early in May, the thick-leaved Bay-Tree (*Magnolia grandiflora*), affords in its high tops a place of safety, in which the Hawk of the South may raise its young. These are out by the end of July, and are fed by the parent birds until well practised in the art of procuring subsistence. About the middle of August, they all wing their way southward.

The affection which the old birds display towards their young, and the methods which they occasionally employ to insure the safety of the latter, are so remarkable, that, before I proceed to describe their general habits, I shall relate a case in which I was concerned.

Early one morning, whilst I was admiring the beauties of nature, as the vegetable world lay embalmed in dew, I heard the cry of a bird that I mistook for that of a Pewee Flycatcher. It was prolonged, I thought, as if uttered in distress. After looking for the bird a long time in vain, an object which I had at first supposed to be something that had accidentally lodged in a branch, attracted my attention, as I thought I perceived it moving. It did move distinctly, and the cry that had ceased from the time when I reached the spot where I stood, was repeated, evidently coming from the object in view. I now took it for a young one of the Chuck-Will's-Widow, as it sat lengthwise on the branch. I shot at it, but perhaps did not hit it, as it only opened and closed its wings, as if surprised. At the report of the gun, the old bird came, holding food in her claws. She perceived me, but alighted, and fed her young with great kindness. I shot at both, and again missed, or at least did not succeed, which might have happened from my having only small shot in my gun. The mother flew in silence, sailed over head just long enough to afford me time to reload, returned, and to my great surprise gently lifted her young, and sailing with it to another tree, about thirty yards distant, deposited it there. My feelings at that moment I cannot express. I wished I had not discovered the poor bird; for who could have witnessed, without emotion, so striking an example of that affection which none but a mother can feel; so daring an act, performed in the midst of smoke, in the presence of a dreaded and dangerous enemy. I followed, however, and brought both to the ground at one shot, so keen is the desire of possession!

The young had the head of a fawn-colour, but I took little more notice of it, depositing the two birds under a log, whence I intended to remove them on my return, for the purpose of drawing and describing them. I then proceeded on my excursion to a lake a few miles distant. On coming back, what was my mortification when I found that some quadruped had devoured both! My punishment was merited.

The Mississippi Kite arrives in Lower Louisiana about the middle of April in small parties of five or six, and confines itself to the borders of deep woods, or to those near plantations, not far from the shores of the rivers, lakes, or bayous. It never moves into the interior of the country, and in this respect resembles the *Falco furcatus*. Plantations lately cleared, and yet covered with tall dying girted trees, placed near a creek or bayou, seemed to suit it best.

Its flight is graceful, vigorous, protracted, and often extended to a great height, the Forked-tailed Hawk being the only species that can compete with it. At times it floats in the air, as if motionless, or sails in broad regular circles, when, suddenly closing its wings, it slides along to some distance, and renews its curves. Now it sweeps in deep and long undulations, with the swiftness of an arrow, passing almost within touching distance of a branch on which it has observed a small lizard, or an insect it longs for, but from which it again ascends disappointed. Now it is seen to move in hurried zig-zags, as if pursued by a dangerous enemy, sometimes seeming to turn over and over like a Tumbling Pigeon. Again it is observed flying round the trunk of a tree to secure large insects, sweeping with astonishing velocity. While travelling, it moves in the desultory manner followed by Swallows; but at other times it is seen soaring at a great elevation among the large flocks of Carrion Crows and Turkey Buzzards, joined by the Forked-tailed Hawk, dashing at the former, and giving them chase, as if in play, until these cowardly scavengers sweep downwards, abandoning this to them disagreeable sport to the Hawks, who now continue to gambol undisturbed. When in pursuit of a large insect or a small reptile, it turns its body sidewise, throws out its legs, expands its talons, and generally seizes its prey in an instant. It feeds while on wing, apparently with as much ease and comfort, as when alighted on the branch of a tall tree. It never alights on the earth; at least I have never seen it do so, except when wounded, and then it appears extremely awkward. It never attacks birds or quadrupeds of any kind, with the view of destroying them for food, although it will chase a fox to a considerable distance, screaming loudly all the while, and soon forces a Crow to retreat to the woods.

The nest of this species is always placed in the upper branches of the tallest trees. I thought it gave the preference to those tall and splendid magnolias and white oaks, which adorn our Southern States. The nest resembles that of the dilapidated tenement of the Common American Crow, and is formed of sticks slightly put together, along with branches of Spanish moss (*Usnea*), pieces of vine bark, and dried leaves. The eggs are two or three, almost globular, of a light greenish tint, blotched thickly over with deep chocolate-brown and black. Only one brood is raised in the season, and I think the female sits more than half the time necessary for incubation. The young I also think obtain nearly the full plumage of the old bird before they depart from us, as I have examined these birds early in August, when the migration was already begun, without observing much difference in their general colour, except only in the want of firmness in the tint of the young ones.

Once, early in the month of May, I found a nest of this bird placed on a fine tall white oak near a creek, and observed that the female was sitting with unceasing assiduity. The male I saw bring her food frequently. Not being able to ascend the tree, I hired a Negro, who had been a sailor for some years, to climb it and bring down the eggs or young. This he did by first mounting another tree, the branches of which crossed the lower ones of the oak. No sooner had he reached the trunk of the tree on which the nest was placed, than the male was seen hovering about and over it in evident displeasure, screaming and sweeping towards the intruder the higher he advanced. When he attained the branch on which the nest was, the female left her charge, and the pair, infuriated at his daring, flew with such velocity, and passed so close to him, that I expected every moment to see him struck by them. The black tar, however, proceeded quietly, reached the nest, and took out the eggs, apprising me that there were three. I requested him to bring them down with care, and to throw off the nest, which he did. The poor birds, seeing their tenement cast down to the ground, continued sweeping around us so low and so long, that I could not resist the temptation thus offered of shooting them.

The Mississippi Kite is by no means a shy bird, and one may generally depend on getting near it when alighted; but to follow it while on wing were useless, its flight being usually so elevated, and its sweeps over a field or wood so rapid and varied, that you might spend many hours in vain in attempting to get up with it. Even when alighted, it perches so high, that I have sometimes shot at it, without producing any other effect than that of causing it to open its wings and close them again, as if utterly ignorant of the danger to which it had been exposed, while it seemed to look down upon me quite unconcerned. When wounded, it comes to the ground with great force, and seldom attempts to escape, choosing rather to defend itself, which it does to the last, by throwing itself on its back, erecting the feathers of its head, screaming loudly in the manner of the Pigeon Hawk, disgorging the contents of its stomach, stretching out its talons, and biting or clenching with great vigour. It is extremely muscular, the flesh tough and rigid.

These birds at times search for food so far from the spot where their nest has been placed, that I have on several occasions been obliged to follow their course over the woods, as if in search of a wild bee's hive, before I could discover it. There is scarcely any perceptible difference between the sexes as to size, and in colour they are precisely similar, only the female has less of the ferruginous colour on her primaries than the male. The stomach is thin, rugous, and of a deep orange colour.

The American Sparrow-Hawk

Falco Sparverius, Linn.
PLATE CXLII. Male and Female.

We have few more beautiful hawks in the United States than this active little species, and I am sure, none half so abundant. It is found in every district from Louisiana to Maine, as well as from the Atlantic shores to the western regions. Every one knows the Sparrow-Hawk, the very mention of its name never fails to bring to mind some anecdote connected with its habits, and, as it commits no depredations on poultry, few disturb it, so that the natural increase of the species experiences no check from man. During the winter months especially it may be seen in the Southern States about every old field, orchard, barn-yard, or kitchen-garden, but seldom indeed in the interior of the forest.

Beautifully erect, it stands on the highest fence-stake, the broken top of a tree, the summit of a grain stack, or the corner of the barn, patiently and silently waiting until it spy a mole, a field-mouse, a cricket, or a grasshopper, on which to pounce. If disappointed in its expectation, it leaves its stand and removes to another, flying low and swiftly until within a few yards of the spot on which it wishes to alight, when all of a sudden, and in the most graceful manner, it rises towards it and settles with incomparable firmness of manner, merely suffering its beautiful tail to vibrate gently for a while, its wings being closed with the swiftness of thought. Its keen eye perceives something beneath, when down it darts, secures the object in its talons, returns to its stand, and devours its prey piece by piece. This done, the little hunter rises in the air, describes a few circles, moves on directly, balances itself steadily by a tremulous motion of its wings, darts towards the earth, but, as if disappointed, checks it course, reascends and proceeds. Some unlucky finch crosses the field beneath it. The Hawk has marked it, and, anxious to secure its prize, sweeps after it, the chase is soon ended, for the poor affrighted and panting bird becomes the prey of the ruthless hunter, who, unconscious of wrong, carries it off to some elevated branch of a tall tree, plucks it neatly, tears the flesh asunder, and having eaten all that it can pick, allows the skeleton and wings to fall to the ground, where they may apprise the traveller that a murder has been committed.

Thus, reader, are the winter months spent by this little marauder. When spring returns to enliven the earth, each male bird seeks for its mate, whose coyness is not less innocent than that of the gentle dove. Pursued from place to place, the female at length yields to the importunity of her dear tormenter, when side by side they sail, screaming aloud their love notes, which if not musical, are doubtless at least delightful to the parties concerned. With tremulous wings they search for a place in which to deposit their eggs secure from danger, and now they have found it.

On that tall mouldering headless trunk, the hawks have alighted side by side. See how they caress each other! Mark! The female enters the deserted Woodpecker's hole, where she remains some time measuring its breadth and depth. Now she appears, exultingly calls her mate, and tells him there could not be a fitter place. Full of joy they gambol through the air, chase all intruders away, watch the Grakles and other birds to which the hole might be equally pleasing, and so pass the time, until the female has deposited her eggs, six, perhaps even seven in number, round, and beautifully spotted. The birds sit alternately, each feeding the other and watching with silent care. After a while the young appear, covered with white down. They grow apace, and now are ready to go abroad, when their parents entice them forth. Some launch into the air at once, others, not so strong, now and then fall to the ground; but all continue to be well provided with food, until they are able to shift for themselves. Together they search for grasshoppers, crickets, and such young birds as, less experienced than themselves, fall an easy prey. The family still resort to the same field, each bird making choice of a stand, the top of a tree, or that of the Great Mullein. At times they remove to the ground, then fly off in a body, separate, and again betake themselves to their stands. Their strength increases, their flight improves, and the field-mouse seldom gains her retreat before the little Falcon secures it for a meal. The trees, of late so richly green, now disclose the fading tints of autumn; the cricket becomes mute, the grasshopper withers on the fences, the mouse retreats to her winter quarters, dismal clouds obscure the eastern horizon, the sun assumes a sickly dimness, hoarfrosts cover the ground, and the long night encroaches on the domains of light. No longer are heard the feathered choristers of the woods, who throng towards more congenial climes, and in their rear rushes the Sparrow-Hawk.

Its flight is rather irregular, nor can it be called protracted. It flies over a field, but seldom farther at a time; even in barren lands, a few hundred yards are all the extent it chooses to go before it alights. During the love season alone it may be seen sailing for half an hour, which is, I believe, the longest time I ever saw one on the wing. When chasing a bird, it passes along with considerable celerity, but never attains the speed of the Sharp-shinned Hawk or of other species. When teazing an Eagle or a Turkey Buzzard, its strength seems to fail in a few minutes, and if itself chased by a stronger hawk, it soon retires into some thicket for protection. Its migrations are pursued by day, and with much apparent nonchalance.

The cry of this bird so much resembles that of the European Kestrel, to which it seems allied, that, were it rather stronger in intonation, it might be mistaken for it. At times it emits its notes while perched, but principally when on the wing, and more continually before and after the birth of its young, the weaker cries of which it imitates when they have left the nest and follow their parents.

The Sparrow Hawk does not much regard the height of the place in which it deposits its eggs, provided it be otherwise suitable, but I never saw it construct a nest for itself. It prefers the hole of a Woodpecker, but now and then is satisfied with an abandoned crow's nest. So prolific is it, that I do not recollect having ever found fewer than five eggs or young in the nest, and, as I have already said, the number sometimes amounts to seven. The eggs are nearly globular, of a deep buff-colour, blotched all over with dark brown and black. This Hawk sometimes raises two broods in the season, in the Southern States, where in fact it may be said to be a constant resident; but in the Middle and Eastern States, seldom if ever more than one. Nay, I have thought that in the South the eggs of a laying are more numerous than in the North, although of this I am not quite certain. So much attached are they to their stand, that they will return to it and sit there by preference for months in succession. My friend Bachman informed me that, through this circumstance, he has caught as many as seven in the same field, each from its favourite stump.

Although the greater number of these Hawks remove southward at the approach of winter, some remain even in the State of New York during the severest weather of that season. These keep in the immediate neighbourhood of barns, where now and then they secure a rat or a mouse for their support. Sometimes this species is severely handled by the larger Hawks. One of them who had caught a Sparrow, and was flying off with it, was suddenly observed by a Red-tailed Hawk, which in a few minutes made it drop its prey: this contented the pursuer and enabled the pursued to escape.

THEODORE LINCOLN, Esq. of Dennisville, Maine, informed me that the Sparrow-Hawk is in the habit of attacking the Republican Swallow, while sitting on its eggs, deliberately tearing the bottle-neck-like entrance of its curious nest, and seizing the occupant for its prey. This is as fit a place as any to inform you, that the father of that gentleman, who has resided at Dennisville upwards of forty years, found the swallow just mentioned abundant there, on his arrival in that then wild portion of the country.

In the Floridas the Sparrow-Hawk pairs as early as February, in the Middle States about April, and in the northern parts of Maine seldom before June. Few are seen in Nova Scotia, and none in Newfoundland, or on the western coast of Labrador. Although abundant in the interior of East Florida, I did not observe one on any of the keys which border the coast of that singular peninsula. During one of my journeys down the Mississippi, I frequently observed some of these birds standing on low dead branches over the water, from which they would pick up the beetles that had accidentally fallen into the stream.

No bird can be more easily raised and kept than this beautiful Hawk. I once found a young male that had dropped from the nest before it was able to fly. Its cries for food attracted my notice, and I discovered it lying near a log. It was large, and covered with soft white down, through which the young feathers protruded. Its little blue bill and yet grey eyes made it look not unlike an owl. I took it home, named it Nero, and provided it with small birds, at which it would scramble fiercely, although yet unable to tear their flesh, in which I assisted it. In a few weeks it grew very beautiful, and became so voracious, requiring a great number of birds daily, that I turned it out, to see how it would shift for itself. This proved a gratification to both of us: it soon hunted for grasshoppers and other insects, and on returning from my walks I now and then threw a dead bird high in the air, which it never failed to perceive from its stand, and towards which it launched with such quickness as sometimes to catch it before it fell to the ground. The little fellow attracted the notice of his brothers, brought up hard by, who, accompanied by their parents, at first gave it chase, and forced it to take refuge behind one of the window-shutters, where it usually passed the night, but soon became gentler towards it, as if forgiving its desertion. My bird was fastidious in the choice of food, would not touch a Woodpecker, however fresh, and as he grew older, refused to eat birds that were in the least tainted. To the last he continued kind to me, and never failed to return at night to his favourite roost behind the window-shutter. His courageous disposition often amused the family, as he would sail off from his stand, and fall on the back of a tame duck, which, setting up a loud quack, would waddle off in great alarm with the Hawk sticking to her. But, as has often happened to adventurers of similar spirit, his audacity cost him his life. A hen and her brood chanced to attract his notice, and he flew to secure one of the chickens, but met one whose parental affection inspired her with a courage greater than his own. The conflict, which was severe, ended the adventures of poor Nero.

I have often observed birds of this species in the Southern States, and more especially in the Floridas, which were so much smaller than those met with in the Middle and Northern Districts, that I felt almost inclined to consider them different; but after studying their habits and voice, I became assured that they were the same. Another species allied to the present, and alluded to by Wilson, has never made its appearance in our Southern States.

The Great Blue Heron

Ardea Herodias, Linn.
PLATE CCXI. Male.

The State of Louisiana has always been my favourite portion of the Union, although Kentucky and some other States have divided my affections; but as we are on the banks of the fair Ohio, let us pause a while, good Reader, and watch the Heron. In my estimation, few of our waders are more interesting than the birds of this family. Their contours and movements are always graceful, if not elegant. Look on the one that stands near the margin of the pure stream:—see his reflection dipping as it were into the smooth water, the bottom of which it might reach had it not to contend with the numerous boughs of those magnificent trees. How calm, how silent, how grand is the scene! The tread of the tall bird himself no one hears, so carefully does he place his foot on the moist ground, cautiously suspending it for a while at each step of his progress. Now his golden eye glances over the surrounding objects, in surveying which he takes advantage of the full stretch of his graceful neck. Satisfied that no danger is near, he lays his head on his shoulders, allows the feathers of his breast to droop, and patiently awaits the approach of his finned prey. You might imagine what you see to be the statue of a bird, so motionless is it. But now, he moves; he has taken a silent step, and with great care he advances; slowly does he raise his head from his shoulders, and now, what a sudden start! his formidable bill has transfixed a perch, which he beats to death on the ground. See with what difficulty he gulps it down his capacious throat! and now his broad wings open, and away he slowly flies to another station, or perhaps to avoid his unwelcome observers.

The "Blue Crane" (by which name this species is generally known in the United States) is met with in every part of the Union. Although more abundant in the low lands of our Atlantic coast, it is not uncommon in the countries west of the Alleghany Mountains. I have found it in every State in which I have travelled, as well as in all our "Territories." It is well known from Louisiana to Maine, but seldom occurs farther east than Prince Edward's Island in the Gulf of St Lawrence, and not a Heron of any kind did I see or hear of in Newfoundland or Labrador. Westward, I believe, it reaches to the very bases of the Rocky Mountains. It is a hardy bird, and bears the extremes of temperature surprisingly, being in its tribe what the Passenger Pigeon is in the family of Doves. During the coldest part of winter the Blue Heron is observed in the State of Massachusetts and in Maine, spending its time in search of prey about the warm springs and ponds which occur there in certain districts. They are not rare in the Middle States, but more plentiful to the west and south of Pennsylvania, which perhaps arises from the incessant war waged against them.

Extremely suspicious and shy, this bird is ever on the look-out. Its sight is as acute as that of any falcon, and it can hear at a considerable distance, so that it is enabled to mark with precision the different objects it sees, and to judge with accuracy of the sounds which it hears. Unless under very favourable circumstances, it is almost hopeless to attempt to approach it. You may now and then surprise one feeding under the bank of a deep creek or bayou, or obtain a shot as he passes unawares over you on wing; but to walk up towards one would be a fruitless adventure. I have seen many so wary, that, on seeing a man at any distance within half a mile, they would take to wing; and the report of a gun forces one off his grounds from a distance at which you would think he could not be alarmed. When in close woods, however, and perched on a tree, they can be approached with a good chance of success.

The Blue Heron feeds at all hours of the day, as well as in the dark and dawn, and even under night, when the weather is clear, his appetite alone determining his actions in this respect; but I am certain that when disturbed during dark nights it feels bewildered, and alights as soon as possible. When passing from one part of the country to another at a distance, the case is different, and on such occasions they fly under night at a considerable height above the trees, continuing their movements in a regular manner.

The commencement of the breeding season varies, according to the latitude, from the beginning of March to the middle of June. In the Floridas it takes place about the first of these periods, in the Middle Districts about the 15th of May, and in Maine a month later. It is at the approach of this period only that these birds associate in pairs, they being generally quite solitary at all other times; nay, excepting during the breeding season, each individual seems to secure for itself a certain district as a feeding ground, giving chase to every intruder of its own species. At such times they also repose singly, for the most part roosting on trees, although sometimes taking their station on the ground, in the midst of a wide marsh, so that they may be secure from the approach of man. This unsocial temper probably arises from the desire of securing a certain abundance of food, of which each individual in fact requires a large quantity.

The manners of this Heron are exceedingly interesting at the approach of the breeding season, when the males begin to look for partners. About sunrise you see a number arrive and alight either on the margin of a broad sandbar or on a savannah. They come from different quarters, one after another, for several hours; and when you see forty or fifty before you, it is difficult for you to imagine that half the number could have resided in the same district. Yet in the Floridas I have seen hundreds thus collected in the course of a morning. They are now in their full beauty, and no young birds seem to be among them. The males walk about with an air of great dignity, bidding defiance to their rivals, and the females croak to invite the males to pay their addresses to them. The females utter their coaxing notes all at once, and as each male evinces an equal desire to please the object of his affection, he has to encounter the enmity of many an adversary, who, with little attention to politeness, opens his powerful bill, throws out his wings, and rushes with fury on his foe. Each attack is carefully guarded against, blows are exchanged for blows; one would think that a single well-aimed thrust might suffice to inflict death, but the strokes are parried with as much art as an expert swordsman would employ; and, although I have watched these birds for half an hour at a time as they fought on the ground, I never saw one killed on such an occasion; but I have often seen one felled and trampled upon, even after incubation had commenced. These combats over, the males and females leave the place in pairs. They are now mated for the season, at least I am inclined to think so, as I never saw them assemble twice on the same ground, and they become comparatively peaceable after pairing.

It is by no means a constant practice with this species to breed in communities, whether large or small; for although I have seen many such associations, I have also found many pairs breeding apart. Nor do they at all times make choice of the trees placed in the interior of a swamp, for I have found heronries in the pine-barrens of the Floridas, more than ten miles from any marsh, pond, or river. I have also observed nests on the tops of the tallest trees, while others were only a few feet above the ground: some also I have seen on the ground itself, and many on cactuses. In the Carolinas, where Herons of all sorts are extremely abundant, perhaps as much so as in the lower parts of Louisiana or the Floridas, on account of the numerous reservoirs connected with the rice plantations, and the still more numerous ditches which intersect the rice-fields, all of which contain fish of various sorts, these birds find it easy to procure food in great abundance. There the Blue Herons breed in considerable numbers, and if the place they have chosen be over a swamp, few situations can be conceived more likely to ensure their safety, for one seldom ventures into those dismal retreats at the time when these birds breed, the effluvia being extremely injurious to health, besides the difficulties to be overcome in making one's way to them.

Imagine, if you can, an area of some hundred acres, overgrown with huge cypress trees, the trunks of which, rising to a height of perhaps fifty feet before they send off a branch, spring from the midst of the dark muddy waters. Their broad tops, placed close together with interlaced branches, seem intent on separating the heavens from the earth. Beneath their dark canopy scarcely a single sunbeam ever makes its way; the mire is covered with fallen logs, on which grow matted grasses and lichens, and the deeper parts with nympheæ and other aquatic plants. The Congo snake and water-moccasin glide before you as they seek to elude your sight, hundreds of turtles drop, as if shot, from the floating trunks of the fallen trees, from which also the sullen alligator plunges into the dismal pool. The air is pregnant with pestilence, but alive with musquitoes and other insects. The croaking of the frogs, joined with the hoarse cries of the Anhingas and the screams of the Herons, forms fit music for such a scene. Standing knee-deep in the mire, you discharge your gun at one of the numerous birds that are breeding high over head, when immediately such a deafening noise arises, that, if you have a companion with you, it were quite useless to speak to him. The frightened birds cross each other confusedly in their flight; the young attempting to secure themselves, some of them lose their hold, and fall into the water with a splash; a shower of leaflets whirls downwards from the tree-tops, and you are glad to make your retreat from such a place. Should you wish to shoot Herons, you may stand, fire, and pick up your game as long as you please; you may obtain several species, too, for not only does the Great Blue Heron breed there, but the White, and sometimes the Night Heron, as well as the Anhinga, and to such places they return year after year, unless they have been cruelly disturbed.

The nest of the Blue Heron, in whatever situation it may be placed, is large and flat, externally composed of dry sticks, and matted with weeds and mosses to a considerable thickness. When the trees are large and convenient, you may see several nests on the same tree. The full complement of eggs which these birds lay is three, and in no instance have I found more. Indeed, this is constantly the case with all the large species with which I am acquainted, from *Ardea cœrulea* to *Ardea occidentalis;* but the smaller species lay more as they diminish in size, the Louisiana Heron having frequently four, and the Green Heron five, and even sometimes six. Those of the Great Blue Heron are very small compared with the size of the bird, measuring only two and a half inches by one and seven-twelfths; they are of a dull bluish-white, without spots, rather rough, and of a regular oval form.

The male and the female sit alternately, receiving food from each other, their mutual affection being as great as it is towards their young, which they provide for so abundantly, that it is not uncommon to find the nest containing a quantity of fish and other food, some fresh, and some in various stages of putrefaction. As the young advance they are less frequently fed, although still as copiously supplied whenever opportunity offers; but now and then I have observed them, when the nests were low, standing on their haunches, with their legs spread widely before them, and calling for food in vain. The quantity which they require is now so great that all the exertions of the old birds appear at times to be insufficient to satisfy their voracious appetite; and they do not provide for themselves until fully able to fly, when their parents chase them off, and force them to shift as they can. They are generally in good condition when they leave the nest; but from want of experience they find it difficult to procure as much food as they have been accustomed to, and soon become poor. Young birds from the nest afford tolerable eating; but the flesh of the old birds is by no means to my taste, nor so good as some epicures would have us to believe, and I would at any time prefer that of a Crow or young Eagle.

The principal food of the Great Blue Heron is fish of all kinds; but it also devours frogs, lizards, snakes, and birds, as well as small quadrupeds, such as shrews, meadow-mice, and young rats, all of which I have found in its stomach. Aquatic insects are equally welcome to it, and it is an expert flycatcher, striking at moths, butterflies, and libellulæ, whether on the wing or when alighted. It destroys a great number of young Marsh-Hens, Rails, and other birds; but I never saw one catch a fiddler or a crab; and the only seeds that I have found in its stomach were those of the great water-lily of the Southern States. It always strikes its prey through the body, and as near the head as possible. When the animal is strong and active, it kills it by beating it against the ground or a rock, after which it swallows it entire. While on the St John's River in East Florida, I shot one of these birds, and on opening it on board, found in its stomach a fine perch quite fresh, but of which the head had been cut off. The fish, when cooked, I found excellent, as did Lieutenant PIERCY and my assistant Mr WARD, but Mr LEEHMAN would not so much as taste it. When on a visit to my friend JOHN BULOW, I was informed by him, that although he had several times imported gold fishes, from New York, with the view of breeding them in a pond, through which ran a fine streamlet, and which was surrounded by a wall, they all disappeared in a few days after they were let loose. Suspecting the Heron to be the depredator, I desired him to watch the place carefully with a gun; which was done, and the result was, that he shot a superb specimen of the present species, in which was found the last gold fish that remained.

In the wild state it never, I believe, eats dead fish of any sort, or indeed any other food than that killed by itself. Now and then it strikes at a fish so large and strong as to endanger its own life; and I once saw one on the Florida coast, that, after striking a fish, when standing in the water to the full length of its legs, was dragged along for several yards, now on the surface, and again beneath. When, after a severe struggle, the Heron disengaged itself, it appeared quite overcome, and stood still near the shore, his head turned from the sea, as if afraid to try another such experiment. The number of fishes, measuring five or six inches, which one of these birds devours in a day, is surprising: Some which I kept on board the Marion would swallow, in the space of half an hour, a bucketful of young mullets; and when fed on the flesh of green turtles, they would eat several pounds at a meal. I have no doubt that, in favourable circumstances, one of them could devour several hundreds of small fishes in a day. A Heron that was caught alive on one of the Florida keys, near Key West, looked so emaciated when it came on board, that I had it killed to discover the cause of its miserable condition. It was an adult female that had bred that spring; her belly was in a state of mortification, and on opening her, we found the head of a fish measuring several inches, which, in an undigested state, had lodged among the entrails of the poor bird. How long it had suffered could only be guessed, but this undoubtedly was the cause of the miserable state in which it was found.

I took a pair of young Herons of this species to Charleston. They were nearly able to fly when caught, and were standing erect a few yards from the nest, in which lay a putrid one that seemed to have been trampled to death by the rest. They offered little resistance, but grunted with a rough uncouth voice. I had them placed in a large coop, containing four individuals of the *Ardea occidentalis*, who immediately attacked the new-comers in the most violent manner, so that I was obliged to turn them loose on the deck. I had frequently observed the great antipathy evinced by the majestic white species towards the blue in the wild state, but was surprised to find it equally strong in young birds which had never seen one, and were at that period smaller than the others. All my endeavours to remove their dislike were unavailing, for when placed in a large yard, the White Herons attacked the Blue, and kept them completely under. The latter became much tamer, and were more attached to each other. Whenever a piece of turtle was thrown to them, it was dexterously caught in the air and gobbled up in an instant, and as they became more familiar, they ate bits of biscuit, cheese, and even rhinds of bacon.

When wounded, the Great Blue Heron immediately prepares for defence, and woe to the man or dog who incautiously comes within reach of its powerful bill, for that instant he is sure to receive a severe wound, and the risk is so much the greater that birds of this species commonly aim at the eye. If beaten with a pole or long stick, they throw themselves on their back, cry aloud, and strike with their bill and claws with great force. I have shot some on trees, which, although quite dead, clung by their claws for a considerable time before they fell. I have also seen the Blue Heron giving chase to a Fish Hawk, whilst the latter was pursuing its way through the air towards a place where it could feed on the fish which it bore in its talons. The Heron soon overtook the Hawk, and at the very first lounge made by it, the latter dropped its quarry, when the Heron sailed slowly towards the ground, where it no doubt found the fish. On one occasion of this kind, the Hawk dropped the fish in the water, when the Heron, as if vexed that it was lost to him, continued to harass the Hawk, and forced it into the woods.

The flight of the Great Blue Heron is even, powerful, and capable of being protracted to a great distance. On rising from the ground or on leaving its perch, it goes off in silence with extended neck and dangling legs, for eight or ten yards, after which it draws back its neck, extends its feet in a straight line behind, and with easy and measured flappings continues its course, at times flying low over the marshes, and again, as if suspecting danger, at a considerable height over the land or the forest. It removes from one pond or creek, or even from one marsh to another, in a direct manner, deviating only on apprehending danger. When about to alight, it now and then sails in a circular direction, and when near the spot it extends its legs, and keeps its wings stretched out until it has effected a footing. The same method is employed when it alights on a tree, where, however, it does not appear to be as much at its ease as on the ground. When suddenly surprised by an enemy, it utters several loud discordant notes, and mutes the moment it flies off.

This species takes three years in attaining maturity, and even after that period it still increases in size and weight. When just hatched they have a very uncouth appearance, the legs and neck being very long, as well as the bill. By the end of a week the head and neck are sparingly covered with long tufts of silky down, of a dark grey colour, and the body exhibits young feathers, the quills large with soft blue sheaths. The tibio-tarsal joints appear monstrous, and at this period the bones of the leg are so soft, that one may bend them to a considerable extent without breaking them. At the end of four weeks, the body and wings are well covered with feathers of a dark slate colour broadly margined with ferruginous, the latter colour shewing plainly on the thighs and the flexure of the wing; the bill has grown wonderfully, the legs would not now easily break, and the birds are able to stand erect on the nest or on the objects near it. They are now seldom fed oftener than once a-day, as if their parents were intent on teaching them that abstinence without which it would often be difficult for them to subsist in their after life. At the age of six or seven weeks they fly off, and at once go in search of food, each by itself.

In the following spring, at which time they have grown much, the elongated feathers of the breast and shoulders are seen, the males shew the commencement of the pendent crest, and the top of the head has become white. None breed at this age, in so far as I have been able to observe. The second spring, they have a handsome appearance, the upper parts have become light, the black and white marks are much purer, and some have the crest three or four inches in length. Some breed at this age. The third spring, the Great Blue Heron is as represented in the plate.

The males are somewhat larger than the females, but there is very little difference between the sexes in external appearance. This species moults in the Southern States about the beginning of May, or as soon as the young are hatched, and one month after the pendent crest is dropped, and much of the beauty of the bird is gone for the season. The weight of a full grown Heron of this kind, when it is in good condition, is about eight pounds; but this varies very much according to circumstances, and I have found some having all the appearance of old birds that did not exceed six pounds. The stomach consists of a long bag, thinly covered by a muscular coat, and is capable of containing several fishes at a time. The intestine is not thicker than the quill of a swan, and measures from eight and a half to nine feet in length.

The Wood Ibis

Tantalus Loculator, Linn.
PLATE CCXVI. Male.

This very remarkable bird, and all others of the same genus that are known to occur in the United States, are constant residents in some part of our Southern Districts, although they perform short migrations. A few of them now and then stray as far as the Middle States, but instances of this are rare; and I am not aware that any have been seen farther to the eastward than the southern portions of Maryland, excepting a few individuals of the Glossy and the White Ibises, which have been procured in Pennsylvania, New Jersey, and New York. The Carolinas, Georgia, the Floridas, Alabama, Lower Louisiana, including Opellousas, and Mississippi, are the districts to which they resort by preference, and in which they spend the whole year. With the exception of the Glossy Ibis, which may be looked upon as a bird of the Mexican territories, and which usually appears in the Union singly or in pairs, they all live socially in immense flocks, especially during the breeding season. The country which they inhabit is doubtless the best suited to their habits; the vast and numerous swamps, lagoons, bayous, and submersed savannahs that occur in the lower parts of our Southern States, all abounding with fishes and reptiles; and the temperature of these countries being congenial to their constitutions.

In treating of the bird now under your notice, Mr WILLIAM BARTRAM says, "This solitary bird does not associate in flocks, but is generally seen alone." This was published by WILSON, and every individual who has since written on the subject, has copied the assertion without probably having any other reason than that he believed the authors of it to state a fact. But the habits of this species are entirely at variance with the above quotation, to which I direct your attention not without a feeling of pain, being assured that Mr BARTRAM could have made such a statement only because he had few opportunites of studying the bird in question in its proper haunts.

The Wood Ibis is rarely met with single, even after the breeding season, and it is more easy for a person to see a hundred together at any period of the year, than to meet with one by itself. Nay, I have seen flocks composed of several thousands, and that there is a natural necessity for their flocking together I shall explain to you. This species feeds entirely on fish and aquatic reptiles, of which it destroys an enormous quantity, in fact more than it eats; for if they have been killing fish for half an hour and have gorged themselves, they suffer the rest to lie on the water untouched, when it becomes food for alligators, crows, and vultures, whenever these animals can lay hold of it. To procure its food, the Wood Ibis walks through shallow muddy lakes or bayous in numbers. As soon as they have discovered a place abounding in fish, they dance as it were all through it, until the water becomes thick with the mud stirred from the bottom by their feet. The fishes, on rising to the surface, are instantly struck by the beaks of the Ibises, which, on being deprived of life, they turn over and so remain. In the course of ten or fifteen minutes, hundreds of fishes, frogs, young alligators, and water-snakes cover the surface, and the birds greedily swallow them until they are completely gorged, after which they walk to the nearest margins, place themselves in long rows, with their breasts all turned towards the sun, in the manner of Pelicans and Vultures, and thus remain so for an hour or so. When digestion is partially accomplished, they all take to wing, rise in spiral circlings to an immense height, and sail about for an hour or more, performing the most beautiful evolutions that can well be conceived. Their long necks and legs are stretched out to their full extent, the pure white of their plumage contrast beautifully with the jetty black of the tips of their wings. Now in large circles they seem to ascend toward the upper regions of the atmosphere; now, they pitch towards the earth; and again, gently rising, they renew their gyrations. Hunger once more induces them to go in search of food, and, with extended front, the band sails rapidly towards another lake or bayou.

Mark the place, reader, and follow their course through cane-brake, cypress-swamp, and tangled wood. Seldom do they return to the same feeding place on the same day. You have reached the spot, and are standing on the margin of a dark-watered bayou, the sinuosities of which lead your eye into a labyrinth ending in complete darkness. The tall canes bow to each other from the shores; the majestic trees above them, all hung with funereal lichen, gently wave in the suffocating atmosphere; the bullfrog, alarmed, shrinks back into the water; the alligator raises his head above its surface, probably to see if the birds have arrived, and the wily cougar is stealthily advancing toward one of the Ibises, which he expects to carry off into the thicket. Through the dim light your eye catches a glimpse of the white-plumaged birds, moving rapidly like spectres to and fro. The loud clacking of their mandibles apprises you of the havoc they commit among the terrified inhabitants of the waters, while the knell-like sounds of their feet come with a feeling of dread. Move, gently or not, move at all, and you infallibly lose your opportunity of observing the actions of the birds. Some old male has long marked you; whether it has been with eye or with ear, no matter. The first stick your foot cracks, his hoarse voice sounds the alarm. Off they all go, battering down the bending canes with their powerful pinions, and breaking the smaller twigs of the trees, as they force a passage for themselves.

Talk to me of the stupidity of birds, of the dullness of the Wood Ibis! Say it is fearless, easily approached, and easily shot. I listen, but it is merely through courtesy; for I have so repeatedly watched its movements, in all kinds of circumstances, that I am quite convinced we have not in the United States a more shy, wary, and vigilant bird than the Wood Ibis. In the course of two years spent, I may say, among them, for I saw some whenever I pleased during that period, I never succeeded in surprising one, not even under night, when they were roosting on trees at a height of nearly a hundred feet, and sometimes rendered farther secure by being over extensive swamps.

My Journal informs me, that, one autumn while residing near Bayou Sara, being intent on procuring eight or ten of these birds, to skin for my learned and kind friend the Prince of Muignano, I took with me two servants, who were first-rate woodsmen, and capital hands at the rifle, and that notwithstanding our meeting with many hundreds of Wood Ibises, it took us three days to shoot fifteen, which were for the most part killed on wing with rifle-balls, at a distance of about a hundred yards. On that occasion we discovered that a flock roosted regularly over a large corn field covered with huge girted trees, the tops of which were almost all decayed. We stationed ourselves apart in the field, concealed among the tall ripened corn, and in silence awaited the arrival of the birds. After the sun had disappeared, the broad front of a great flock of Ibises was observed advancing towards us. They soon alighted in great numbers on the large branches of the dead trees; but whenever one of the branches gave way under their weight, all at once rose in the air, flew about several times, and alighted again. One of my companions, having a good opportunity, fired, and brought two down with a single bullet; but here the sport was ended. In five minutes after, not an Ibis was within a mile of the place, nor did any return to roost there for more than a month. When on the margin of a lake, or even in the centre of it—for all the lakes they frequent are exceedingly shallow—the first glimpse they have of a man induces them to exert all their vigilance; and should he after this advance a few steps, the birds fly off.

The name of "Wood Ibis" given to this bird, is not more applicable to it than to any other species; for every one with which I am acquainted resorts quite as much to the woods at particular periods. All our species may be found on wet savannahs, on islands surrounded even by the waters of the sea, the Florida Keys for example, or in the most secluded parts of the darkest woods, provided they are swampy, or are furnished with ponds. I have found the Wood, the Red, the White, the Brown, and the Glossy Ibises, around ponds in the centre of immense forests; and in such places, even in the desolate pine-barrens of the Floridas; sometimes several hundred miles from the sea coast, on the Red River, in the State of Louisiana, and above Natchez, in that of Mississippi, as well as within a few miles of the ocean. Yet, beyond certain limits, I never saw one of these birds.

One of the most curious circumstances connected with this species is, that although the birds are, when feeding, almost constantly within the reach of large alligators, of which they devour the young, these reptiles never attack them; whereas if a Duck or a Heron comes within the reach of their tails, it is immediately killed and swallowed. The Wood Ibis will wade up to its belly in the water, round the edges of "alligators' holes," without ever being injured; but should one of these birds be shot, an alligator immediately makes towards it and pulls it under water. The gar-fish is not so courteous, but gives chase to the Ibises whenever an opportunity occurs. The Snapping Turtle is also a great enemy to the young birds of this species.

The flight of the Wood Ibis is heavy at its rising from the ground. Its neck at that moment is deeply curved downward, its wings flap heavily but with great power, and its long legs are not stretched out behind until it has proceeded many yards. But as soon as it has attained a height of eight or ten feet, it ascends with great celerity, generally in a spiral direction, in silence if not alarmed, or, if frightened, with a rough croaking guttural note. When fairly on wing, they proceed in a direct flight, with alternate flappings and sailings of thirty or forty yards, the sailings more prolonged than the flappings. They alight on trees with more ease than Herons generally do, and either stand erect or crouch on the branches, in the manner of the Wild Turky, the Herons seldom using the latter attitude. When they are at rest, they place their bill against the breast, while the neck shrinks as it were between the shoulders. In this position you may see fifty on the same tree, although some individual of the party will be constantly on the look-out, and ready to sound the alarm.

In the spring months, when these birds collect in large flocks, before they return to their breeding places, I have seen thousands together, passing over the woods in a line more than a mile in extent, and moving with surprising speed at the height of only a few yards above the trees. When a breeding place has once been chosen, it is resorted to for years in succession; nor is it easy to make them abandon it after they have deposited their eggs, although, if much annoyed, they never return to it after that season.

Besides the great quantity of fishes that these Ibises destroy, they also devour frogs, young alligators, wood-rats, young rails and grakles, fiddlers and other crabs, as well as snakes and small turtles. They never eat the eggs of the alligator, as has been alleged, although they probably would do so, could they demolish the matted nests of that animal, a task beyond the power of *any* bird known to me. I never saw one eat any thing which either it or some of its fellows had not killed. Nor will it eat an animal that has been dead for some time, even although it may have been killed by itself. When eating, the clacking of their mandibles may be heard at the distance of several hundred yards.

When wounded, it is dangerous to approach them, for they bite severely. They may be said to be very tenacious of life. Although usually fat, they are very tough and oily, and therefore are not fit for food. The Negroes, however, eat them, having, previous to cooking them, torn off the skin, as they do with Pelicans and Cormorants. My own attempts, I may add, were not crowned with success. Many of the Negroes of Louisiana destroy these birds when young, for the sake of the oil which their flesh contains, and which they use in greasing machines.

The French Creoles of that State name them "Grands Flamans," while the Spaniards of East Florida know them by the name of "Gannets." When in the latter country, at St Augustine, I was induced to make an excursion, to visit a large pond or lake, where I was assured there were Gannets in abundance, which I might shoot off the trees, provided I was careul enough. On asking the appearance of the Gannets, I was told that they were large white birds, with wings black at the end, a long neck, and a large sharp bill. The description so far agreeing with that of the Common Gannet or Solan Goose, I proposed no questions respecting the legs or tail, but went off. Twenty-three miles, Reader, I trudged through the woods, and at last came in view of the pond; when, lo! its borders and the trees around it were covered with Wood Ibises. Now, as the good people who gave the information spoke according to their knowledge, and agreeably to their custom of calling the Ibises Gannets, had I not gone to the pond, I might have written this day that Gannets are found in the interior of the woods in the Floridas, that they alight on trees, &c. which, if *once* published, would in all probability have gone down to future times through the medium of compilers, and all perhaps without acknowledgment.

The Wood Ibis takes four years in attaining full maturity, although birds of the second year are now and then found breeding. This is rare, however, for the young birds live in flocks by themselves, until they have attained the age of about three years. They are at first of a dingy brown, each feather edged with paler; the head is covered to the mandibles with short downy feathers, which gradually fall off as the bird advances in age. In the third year, the head is quite bare, as well as a portion of the upper part of the neck. In the fourth year, the bird is as you see it in the plate. The male is much larger and heavier than the female, but there is no difference in colour between the sexes.

SOURCES AND OTHER MATTERS

In working on Audubon over the years, I have relied primarily on two seminal biographies, *Audubon, the Naturalist,* by Francis Hobart Herrick (1917, revised edition 1938) and *John James Audubon: A Biography,* by Alice Ford (1964, revised edition 1988). To a lesser extent, for different reasons, I have profited from Stanley Clisby Arthur's *Audubon: An Intimate Life of the American Woodsman* (1937). Arthur knew Louisiana better than Herrick and Ford, and he possessed an ornithologist's knowledge of Louisiana birds. His listing of the dates and places in Louisiana where Audubon painted his original watercolors continues to be a valuable source, rightly singled out in George H. Lowery's magisterial *Louisiana Birds* (1955).

Over the last twenty years, Audubon's star has risen again, and in addition to several new books on his bird paintings, there has been a steady flood of new biographies, at least twelve by my count. Six are American, two are British, and four are French. I have read, or consulted, all of them. A number of specialized studies have also appeared recently, notably Daniel Patterson's meticulous edition of the first European diary, *John James Audubon's Journal of 1826: The Voyage to The Birds of America* (2011). Similar care and scholarship can be found in Patterson's new edition of the Missouri River journal, or journals, *The Missouri River Journals of John James Audubon* (2016), and in Peter B. Logan's edition of Audubon's journal made during his trip to Labrador in the summer of 1833, *Audubon: America's Greatest Naturalist and His Voyage of Discovery to Labrador* (2016).

Unfortunately, most of the new biographies overlook, or slight, Audubon's formative years in Nantes. Readers wanting a well-researched account of Audubon's French background should consult Yvon Chatelin's *Audubon: Peintre, Naturaliste, Aventurier* (2001). Though the recent biographies are all more or less reliable, none of them adds much to Herrick and Ford. The one exception is Carolyn De-

Latte's *Lucy Audubon: A Biography* (updated edition 2008). I need to emphasize that reading DeLatte confirmed me in all my convictions concerning Audubon and Louisiana. DeLatte spends half her book on the period of nearly ten years during which Lucy made her home in Louisiana, and proves without a doubt that for Audubon the Louisiana years changed his life, work, and prospects for good.

As far as collections of Audubon's writings are concerned, there have been many of them, going back to the nineteenth century. They range from the general to the more thematic, such as Kathryn Hall Proby's anthology of Audubon's Florida writings, *Audubon in Florida* (1974) and Alice Ford's reweaving of selected texts into an informal Audubon autobiography, *Audubon, by Himself* (1969). The first major collection of Audubon's writings was a French anthology, *Scènes de la Nature dans les Etats-Unis*, compiled by Eugène Bazin in 1857. It is seldom mentioned, and seems to be unknown to the recent American biographers. Since Bazin, there have been a dozen or so general collections of Audubon, including an extensive selection of Audubon's writings supplied by Lucy Audubon for Robert Buchanan's *The Life and Adventures of John James Audubon, the Naturalist* (1868). The most important of the early collections remains *Audubon and His Journals*, edited and introduced by his granddaughter Maria R. Audubon, and first published in 1897. It is still in print, and has been a major source for many of the collections that followed.

A new generation of Audubon collections began in 1992 with my two-volume French edition, *Journaux et Récits*, published by Atalante and the City of Nantes. It included a thousand pages of translated text, a long introduction, and editorial commentary interspersed throughout to help the reader follow Audubon's career. In 1996, I published another collection of Audubon's writings in the Penguin Nature Classics series, *Selected Journals and Other Writings*. The hallmark of both *Journaux et Récits* and *Selected Journals* was the desire to demonstrate the wide range, and depth, of Audubon's writing. *Selected Journals* included five bird biographies and a choice of Audubon's correspondence, along with the selection of journals and episodes. *Selected Journals* also contained a long introduction (and editorial commentary) to guide the reader from section to section. Three years later, in 1999, the Library of America brought out a volume of Audubon's *Writings and Drawings*, edited by Christoph Irmscher. Irmscher quietly adopted the concept of *Selected Journals* and covered much the same ground in his selections, but he provided no introduction and very little editorial guidance. His notes, however, are substantial. Finally, there is the *Audubon Reader* edited by Richard Rhodes for Everyman's Library in 2006. Since Rhodes probably did not want to duplicate the selections in the Penguin and Library of America anthologies, his choices are limited, and not fully representative.

Two books I have had constantly at my side while working on *Audubon on Louisiana* are Roberta J. M. Olson's wise and brilliantly illustrated New York Historical Society edition of Audubon's original watercolors, *Audubon's Aviary* (2012, with an essay by Marjorie Shelley and a contribution by Alexander Mazzitelli), and Joseph Kaster's delightfully written and well-informed biographical history of American naturalists, *A Species of Eternity* (1977). No lover of Audubon should be without them.

In selecting the writings for *Audubon on Louisiana* I have done what I did for my three other books on Audubon, and have gone back, as far as possible, to the original manuscripts, transcriptions, and publications. Not wanting to pincushion the text with footnotes, I have multiplied cross-references wherever necessary. Every directly quoted passage can be easily traced to the original document by consulting the headnotes and the source notes at the beginning of each section. The epigraph for *Audubon on Louisiana*, Audubon's exclamation "Magnificent Louisiana!," is taken from his letter of November 18, 1829, to Dr. Richard Harlan included in Francis Hobart Herrick's biography (the entire letter is reproduced above in the headnote to "Louisiana Episodes").

INDEX

Note: JJA refers to John James Audubon. Spellings follow Audubon's usage.